Springer Undergraduate Mathematics Series

Springer

London
Berlin
Heidelberg
New York
Hong Kong
Milan
Paris
Tokyo

Advisory Board

P.J. Cameron *Queen Mary and Westfield College*
M.A.J. Chaplain *University of Dundee*
K. Erdmann *Oxford University*
L.C.G. Rogers *University of Cambridge*
E. Süli *Oxford University*
J.F. Toland *University of Bath*

Other books in this series

Nicholas F. Britton

Essential
Mathematical Biology

With 92 Figures

 Springer

Nicholas Ferris Britton, MA, DPhil
Centre for Mathematical Biology, Department of Mathematical Sciences, University
of Bath, Claverton Down, Bath BA2 7AY, UK

Cover illustration elements reproduced by kind permission of:
Aptech Systems, Inc., Publishers of the GAUSS Mathematical and Statistical System, 23804 S.E. Kent-Kangley Road, Maple Valley, WA 98038,
   USA. Tel: (206) 432 - 7855 Fax (206) 432 - 7832 email: info@aptech.com URL: www.aptech.com
American Statistical Association: Chance Vol 8 No 1, 1995 article by KS and KW Heiner 'Tree Rings of the Northern Shawangunks' page 32 fig 2
Springer-Verlag: Mathematica in Education and Research Vol 4 Issue 3 1995 article by Roman E Maeder, Beatrice Amrhein and Oliver Gloor
   'Illustrated Mathematics: Visualization of Mathematical Objects' page 9 fig 11, originally published as a CD ROM 'Illustrated Mathematics' by
   TELOS: ISBN 0-387-14222-3, German edition by Birkhauser: ISBN 3-7643-5100-4.
Mathematica in Education and Research Vol 4 Issue 3 1995 article by Richard J Gaylord and Kazume Nishidate 'Traffic Engineering with Cellular
   Automata' page 35 fig 2. Mathematica in Education and Research Vol 5 Issue 2 1996 article by Michael Trott 'The Implicitization of a Trefoil
   Knot' page 14.
Mathematica in Education and Research Vol 5 Issue 2 1996 article by Lee de Cola 'Coins, Trees, Bars and Bells: Simulation of the Binomial Pro-
   cess' page 19 fig 3. Mathematica in Education and Research Vol 5 Issue 2 1996 article by Richard Gaylord and Kazume Nishidate 'Contagious
   Spreading' page 33 fig 1. Mathematica in Education and Research Vol 5 Issue 2 1996 article by Joe Buhler and Stan Wagon 'Secrets of the
   Madelung Constant' page 50 fig 1.

British Library Cataloguing in Publication Data
Britton, N.F.
    Essential mathematical biology (Springer undergraduate
    mathematics series)
    1. Biomathematics.
    I. Title
    570.1'51
ISBN 185233536X

Library of Congress Cataloging-in-Publication Data
Britton, N.F.
    Essential mathematical biology / N.F. Britton
       p.   cm. -- (Springer undergraduate mathematics series)
    Includes index.
    ISBN 1-85233-536-X (alk. paper)
    1. Biomathematics. I. Title. II. Series.
QH323.5 .B745 2002
570'.1'51—dc21   2002036455

Springer Undergraduate Mathematics Series ISSN 1615-2085
ISBN 1-85233-536-X Springer-Verlag London Berlin Heidelberg
a member of BertelsmannSpringer Science+Business Media GmbH
http://www.springer.co.uk

Typesetting: Camera-ready by author
12/3830-543210 Printed on acid-free paper SPIN 10844058

Preface

This book aims to cover the topics normally included in a first course on Mathematical Biology for mathematics or joint honours undergraduates in the UK, the USA and other countries. Such a course may be given in the second year of an undergraduate degree programme, but more often appears in the third year.

Mathematical Biology is not as hierarchical as many areas of Mathematics, and therefore there is some flexibility over what is included. As a result the book contains more than enough for two one-semester courses, e.g. one based on Chapters 1 to 4, mainly using difference equations and ordinary differential equations, and one based on Chapters 5 to 8, mainly using partial differential equations. However there are some classic areas that are covered in almost every course, the most obvious being population biology, often including epidemiology, and mathematical ecology of one or two species. Population genetics is also a classic area of application, although it does not appear in every first course. In spatially non-uniform systems there is even more choice, but reaction-diffusion equations are almost always included, with applications at the molecular and population level. Some large areas have been excluded through lack of space, such as bio-fluid dynamics and most of mathematical physiology, each of which could fill an undergraduate textbook on its own.

To cover the whole book the student will need a background in linear algebra, vector calculus, difference equations, and ordinary and partial differential equations, although a one-semester course without vector calculus and partial differential equations could easily be constructed. Methods only are required, and the necessary results are collected together in appendices. Some additional material appears on an associated website, at

http://www.springer.co.uk/britton/

v

This site includes more exercises, more detailed answers to the exercises in the book, and links to other useful sites. Some parts of it require more mathematical background than the book itself, including stochastic processes and continuum mechanics.

I would like to thank Jim Murray for his support and advice since my undergraduate days. I am grateful to all the (academic and non-academic) staff of the Department of Mathematical Sciences at the University of Bath for making this a genial place to work, the past and present members of the Centre for Mathematical Biology at the University of Bath for stimulating discussions, my students for their feedback on the lecture notes from which this book was developed, and Springer Verlag. Finally I would like to thank Suzy and Rachel for their love and support.

Contents

List of Figures

1
Single Species Population Dynamics

- The first goal of this chapter is to understand how the rate of growth or decay of an isolated population is determined. This depends on modelling birth and death processes.

- Models of birth and death processes have a long history. Complicating factors are the dependence of birth and death rates on the age structure of the population, the effects of competition for resources, and delays in responding to environmental changes.

- We are interested in controlling as well as understanding population dynamics, and we shall discuss management strategies for fisheries.

- Spatial effects are increasingly being seen as important in fields such as conservation biology, and we shall introduce a simple method for investigating these.

1.1 Introduction

Attempts to understand population processes date back to the Middle Ages and earlier. Often, human populations were the focus of interest. Sir William Petty in about 1300 composed a table "shewing how the People might have doubled in the several ages of the World", starting with 8 people one year after the Flood, 2700 years before the birth of Christ, and doubling at first every ten years but then at successively longer intervals of time to arrive at 320 million (not a bad estimate) by the year 1300. Real data show that doubling times were about 1000 years over this period of time, but more importantly that they have become successively *shorter* to reach about 35 years in the later 20th century. The rate of population growth has therefore been faster than exponential, as we can see in Figure 1.1.

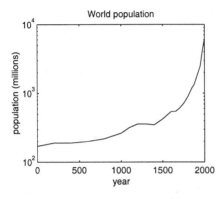

Figure 1.1   World human population growth over the last 2000 years. Exponential growth would give a straight line on this semilogarithmic plot.

Alternatively, we can think of this curve as exponential growth, but with an exponent that has (in general) increased over time. Population processes with parameters that change over time are called *non-stationary*; for simplicity, we shall restrict attention almost everywhere to *stationary* processes.

Human populations are exceptional in exhibiting exponential or faster growth over such a long time period. Most biological populations, except when colonising virgin territory, are regulated by competition for limiting resources or by other means. We shall look at population dynamics first in the absence of regulation, and then including regulatory effects.

1.2 Linear and Nonlinear First Order Discrete Time Models

Let us assume that the size N_n of a population at time n completely determines its size at time $n+1$. The use of discrete time is sometimes rather artificial, but it may be appropriate if the population is censused at intervals, so that data for births and deaths are only available for discrete time periods. Such a model is sometimes called a *metered* model, especially when used in the fisheries context. Consider a continuously breeding population, such as a human population, censused at intervals. Let the probability of any given individual dying between censuses (the *per capita mortality*) be d, and let the average number of births to any given individual in the same time period (the *per capita production* or *reproduction*) be b. Then the total number of deaths is dN_n, the total number of births bN_n, and so $N_{n+1} = (1 + b - d)N_n = \lambda N_n$, say. Alternatively, a discrete time model may be used for creatures that reproduce at a specific time of year, the breeding season, as we shall see in the next subsection.

The simplest first order model is therefore linear,

$$N_{n+1} = \lambda N_n, \tag{1.2.1}$$

known as the *Malthusian* equation in discrete time. The parameter λ is called the *(net) growth ratio*. If the process is stationary, λ constant, its solution with initial condition N_0 given is $N_n = N_0 \lambda^n$, which is geometrically growing with growth ratio λ if $\lambda > 1$. This is known as *Malthusian growth*. In 1798, Thomas Malthus published an Essay on the Principle of Population in which he stated that "population, when unchecked, increases in a geometric ratio". He went on to say that subsistence increases in an arithmetic ratio, and to discuss the dire consequences of this difference. The phrase "when unchecked" is crucial, and Malthus recognised that such checks were constantly in place. However, the unchecked growth equation (1.2.1) and the equivalent equation in continuous time have come to bear his name.

1.2.1 The Biology of Insect Population Dynamics

Insects often have well-defined annual non-overlapping generations. For example, adults may lay eggs in spring or summer, and then die. The eggs hatch out into larvae, which eat and grow and then overwinter in a pupal stage. The adults emerge from the pupae the following spring. We have to decide at what time of year to take a census. Let us count adults at the breeding season, and let the average number of eggs laid by each adult be R_0. The parameter R_0 is the *basic reproductive ratio*, defined to be the average number of offspring produced over a lifetime, all of these offspring in this case being produced at the same time. It is clear that this is a crucial parameter, and it will recur many times in this book. It is often used to represent not the total number of offspring produced, but the number that would survive to breed in the absence of the particular effect that is being studied. In an unchecked insect population with no premature mortality, Equation (1.2.1) therefore applies, with $\lambda = 1 + b - d$, $b = R_0$ and $d = 1$, so that the growth ratio λ is the basic reproductive ratio R_0.

In a real insect population, some of the R_0 offspring produced by each adult will not survive to be counted as adults in the next census. Let us assume that a fraction $S(N)$ do survive, this survival rate depending on N. The Malthusian equation is replaced by

$$N_{n+1} = R_0 S(N_n) N_n, \tag{1.2.2}$$

which may alternatively be written

$$N_{n+1} = N_n F(N_n) = f(N_n). \tag{1.2.3}$$

Here $F(N)$ is the *per capita production* and $f(N)$ the *production* of a population of size N. The model is called *density-dependent* if the per capita production F depends on N, which we assume to be the case. Density-dependent effects can also occur if the per capita fertility rather than the per capita survival depends on the density.

The model may be criticised for being deterministic, and for taking no account of predators, prey, or competing species, or of abiotic influences such as the weather. It assumes that population size is regulated by density-dependent factors. The question of whether density-dependent or density-independent factors are more important was a source of great controversy in the middle decades of the last century, although it is clear that if the population is regulated, density-dependent factors must have a role to play. However, the purpose of models of this form is to investigate whether simple assumptions on how the population is regulated are supported by the data.

1.2.2 A Model for Insect Population Dynamics with Competition

We shall test the hypothesis that insect populations are regulated by intraspecific (within-species) competition for some resource which is in short supply. Typical resources are food and space. We shall interpret R_0 as the number of adults each adult in one generation would produce in the subsequent generation in the absence of competition. For simplicity in the exposition we shall assume that competition affects survival rather than fertility, and discuss it in terms of the survival fraction S or the per capita production F. There are various idealised forms of intraspecific competition that can be considered, defined as follows.

– *No competition*: then $S(N) = 1$ for all N.

– *Contest competition*: here there is a finite number of units of resource (these could, for example, be a fixed number of safe refuges). Each individual which obtains one of these units of resource survives to breed, and produces

R_0 offspring in the subsequent generation; all others die without producing offspring. Thus $S(N) = 1$ for N less than some critical value N_c, and $S(N) = N_c/N$ for $N > N_c$.

– *Scramble competition*: here every individual is assumed to get an equal share of a limited resource. If this amount of resource is sufficient for survival to breeding, then all survive and produce R_0 offspring in the next generation; if not, all die. Thus $S(N) = 1$ for N less than some critical value N_c, and $S(N) = 0$ for $N > N_c$.

Contest and scramble competition may alternatively be seen in terms of fecundity. Below the threshold there is no competition and all individuals produce R_0 offspring, the maximum possible. Above the threshold, N_c individuals produce R_0 offspring and the rest produce none at all under contest competition, while no offspring whatever are produced under scramble competition.

The difficulty is that these ideals do not occur in real populations, and so real data are not always easily classified as contest or scramble competition. A threshold density is not usually seen. Moreover, zero survival is unrealistic, at least for large populations. Such populations may crash in a season, but will not generally fall to zero. A classification of intraspecific competition is therefore often made on the basis of the asymptotic behaviour of $S(N)$, or $F(N)$, as $N \to \infty$. Idealised contest competition exhibits *exact compensation* for large enough N, where increased mortality between times n and $n + 1$ compensates exactly for any increase in numbers at time n. This occurs if $F(N) \to c/N$, $f(N) \to c$, a constant, as $N \to \infty$. But other behaviour is also seen in the data. If $F(N) \to c/N^b$ as $N \to \infty$, this represents *under-compensation* where increased mortality less than compensates for any increase in numbers, if $0 < b < 1$, and *over-compensation* if $b > 1$. A classification in terms of under- or over-compensation is much more useful than a simple dichotomy between contest and scramble competition, but we shall refer to competition as contest if b is around 1 and scramble if b is large. All these types of competition are *compensatory*, with $F(N)$ decreasing with N. Negative values of b are also sometimes seen for some ranges of N (although not of course as $N \to \infty$), so that $F(N)$ is *increasing* with N; this is known as *depensation* and will be discussed later.

A model which exhibits all kinds of compensatory behaviour depending on the parameters is given by the Hassell equation

$$N_{n+1} = f(N_n) = \frac{R_0 N_n}{(1 + aN_n)^b}. \tag{1.2.4}$$

with R_0 and a positive and $b \geq 0$. For $b = 0$ there is no competition, for $0 < b < 1$ under-compensation, for $b = 1$ exact compensation, for $b > 1$ over-compensation, and as $b \to \infty$ scramble competition, as we can see in comparing Figure 1.2(b) and Figure 1.3(b) below.

The model may be criticised for taking a rather arbitrary form for f, but it is sufficient to investigate how the basic reproductive ratio R_0 and the kind of competition that occurs (represented by the parameter b) might affect the qualitative behaviour of the system. We wish to see if the findings of the model agree with biological data, and gain insight into real insect populations.

Analysis of the model is easier if we reduce the number of parameters by defining $x_n = aN_n$. The equation becomes

$$x_{n+1} = f(x_n; R_0, b) = \frac{R_0 x_n}{(1 + x_n)^b}. \tag{1.2.5}$$

Note that the parameter a has disappeared and therefore does not affect the dynamics of the system at all. We shall assume that $b \neq 0$; otherwise the problem reduces to a linear one. We sketch $x_{n+1} = x_n$ and $x_{n+1} = f(x_n; R_0, b)$. Then a cobweb map may be used to show how the population changes through time, as explained in Section A.1.1 of the appendix.

The kind of behaviour expected from such an equation is explained in Chapter A of the appendix, in particular in Section A.2.4. Most important is that for small values of R_0 and b there is a stable steady state, but as either R_0 or b increases the asymptotic behaviour of solutions of the equation become periodic and then chaotic, through a *cascade of period-doubling bifurcations*.

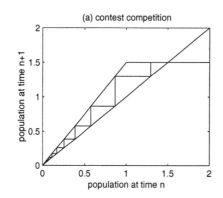
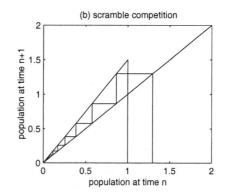

Figure 1.2   Cobweb maps for idealised contest and scramble competition Populations are scaled with the critical population N_c. The census is taken on the adult population at breeding. In (a), the population tends to a constant level, while in (b) it crashes to extinction.

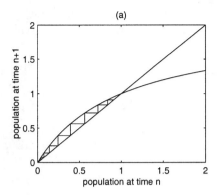 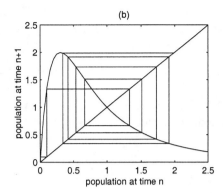

Figure 1.3   Cobweb maps for the Hassell equation (1.2.5) with (a) $b = 1$ and (b) $b > 1$. In (a) there is exact compensation, a smoothed version of the idealised contest competition, and the population tends monotonically to a steady state, while in (b) there is over-compensation, and it tends to an oscillation of period 2. Over-compensatory competition is generally destabilising.

Example 1.1

Find the steady states of Equation (1.2.5) for $R_0 < 1$ and $R_0 > 1$. Show that for $R_0 > 1$ and $0 < b \leq 1$ (exact or under-compensation), the non-trivial steady state is monotonically stable. Make a sketch of the (R_0, b)-plane, marking the regions of parameter space where the steady state is monotonically stable, oscillatorily stable and oscillatorily unstable.

For $R_0 < 1$ there is no non-trivial steady state solution and the trivial steady state is stable; for $R_0 > 1$ there is a single non-trivial steady state solution given by

$$x^* = R_0^{1/b} - 1.$$

There is a transcritical bifurcation   (see Section A.2 of the appendix) at $(x^*, R_0) = (0, 1)$, where the trivial and non-trivial solutions coincide and an interchange of stability takes place. Section A.1.2 of the appendix shows us that x^* is oscillatorily unstable, oscillatorily stable, monotonically stable or monotonically unstable according to whether $f'(x^*; R_0, b)$ is in the intervals $(-\infty, -1)$, $(-1, 0)$, $(0, 1)$ or $(0, \infty)$. But

$$f'(x^*; R_0, b) = \frac{R_0}{(1 + x^*)^b} - \frac{R_0 b x^*}{(1 + x^*)^{b+1}}$$

$$= 1 - \frac{b x^*}{1 + x^*} = 1 - b + \frac{b}{1 + x^*} = 1 - b + \frac{b}{R_0^{1/b}}.$$

For $b \leq 1$, $0 < f'(x^*; R_0, b) < 1$, and x^* is monotonically stable. In general, $f'(x^*; R_0, b) < 1$, so changes in the qualitative behaviour of solutions are only possible when $f'(x^*; R_0, b)$ passes through 0 or -1. The condition for $f'(x^*; R_0, b) = 0$, the borderline between monotonically and oscillatorily stable behaviour, is

$$R_0 = \left(\frac{b}{b-1}\right)^b,$$

and for $f'(x^*; R_0, b) = -1$, the borderline between oscillatorily stable and periodic behaviour, is

$$R_0 = \left(\frac{b}{b-2}\right)^b.$$

These curves and the corresponding behaviour near the steady state x^* are sketched below.

To compare this with biological data, values of R_0 and b have been estimated for various insect populations and the results plotted in parameter space.

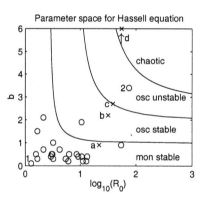

Figure 1.4   Parameter space for Equation (1.2.5) showing stability of the steady state. Just into the oscillatorily unstable region the solution is periodic, but as we move away from the stability boundary there is a cascade of period-doubling bifurcations to chaos (see Section A.2.4 of the appendix). Field populations are denoted by circles, laboratory populations by crosses.

The following points should be noted.

− Most points are well within the stable region.

− Laboratory populations (crosses) are less stable than field ones (circles). This is because of

  − the scramble for resources in confined and homogeneous conditions (high b), and

  − the low density-independent mortality (high R_0).

− Point 2 represents the Colorado beetle, which exhibits over-compensating density-dependent competition (high b) and high net rates of population increase; the prediction of the model is in accordance with observed fluctuations in Colorado beetle populations.

- Point 1 represents the grey larch bud moth, and is well within the stable region. However, real populations of this species have a periodic oscillation. Clearly, some important effect has been omitted in the modelling here, possibly a predator-prey or host-parasitoid relationship.

- Points a, b and c represent laboratory populations of two species of weevil, and the laboratory experiments show the behaviour expected from their position on the diagram.

- The point marked d should appear at the value of R_0 shown, but at $b = 10$, and is therefore in the chaotic region. , In 1954 Nicholson performed a series of experiments on blowflies, under various feeding conditions, that have become classic examples in ecological texts. This point represents one of the experiments where the blowfly *larvae* were forced to scramble for resources, which exhibited irregular behaviour. Later we shall consider another experimental condition where resources for the *adult* blowflies were limited.

By and large, there is good agreement of model predictions with data, although later studies have not always given such good results. One important conclusion is that high reproductive ratios R_0 and highly over-compensating density dependence (high b) are capable of provoking periodic behaviour or chaotic fluctuations in population density. This would have been very difficult to establish without recourse to a mathematical model of the situation. Indeed, before the possibility of chaos had been demonstrated mathematically, biologists had assumed that any irregular fluctuations in their data must have been due to stochastic effects. The question of whether there do exist chaotic systems in ecology is still an area of active research, not least because of the difficulty in detecting chaos in time-series data that are inherently noisy. However, several laboratory host-parasitoid systems do seem to exhibit chaos; good fits have been obtained between the data and chaotic mathematical models.

EXERCISES

1.1. Consider the nonlinear equation for population growth

$$x_{n+1} = \frac{\lambda x_n}{1 + x_n}$$

where $\lambda > 0$.

a) Does this equation exhibit over-, under- or exact compensation?

b) Does it have a non-trivial steady state?

c) Find the stability of the steady state(s) of the equation, and discuss any bifurcations that take place.

d) Draw cobweb maps for each case that you have identified.

e) Check your answers by using the substitution $y_n = 1/x_n$ to solve the equation exactly. (Hint: try a solution in the form $y_n = Ar^n + B$.)

1.2. A population is assumed to be governed by the equation $N_{n+1} = f(N_n)$, where the function f is as shown in Figure 1.5, and there is a single steady state N^*. A variant of the cobwebbing method uses the graphs of f and its reflection $\bar{f}$ to produce successive iterates N_n, as in part (a) of the figure.

a) What is the point marked S in part (a) of the figure?

b) A somewhat different situation is shown in part (b) of the figure. Interpret the points A, B and C and use cobwebbing to decide what happens to the population N_n.

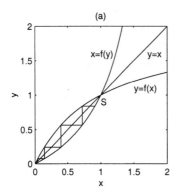 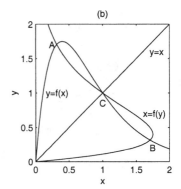

Figure 1.5   Cobwebbing variant.

1.3. a) Derive the condition for linearised stability of the steady solution $N_n = N^*$ of the equation $N_{n+1} = h(N_n)$.

b) Consider an organism such as the Pacific salmon that has a two year life cycle, so that after one year newborn individuals become immature young, and after two years mature adults. Let the numbers of young and adults in year n be y_n and a_n, and let

$$a_{n+1} = f(y_n), \quad y_{n+1} = g(a_n),$$

where f and g are both increasing functions, $f(0) = g(0) = 0$, $f(y) < y$ for all $y > 0$, and $g(a) \to g_0$, a constant, as $a \to \infty$. Explain these hypotheses in biological terms.

c) Show that the fixed points of this system are the intersections of the curves $y = g(a)$ and $y = f^{-1}(a)$. Derive the condition for stability in the form $g'(a^*) < (f^{-1})'(a^*)$ and deduce from geometrical considerations that these points, ordered in the obvious way, are generically alternately stable and unstable. If there are three fixed points $S_0 = (0,0)$, $S_1 = (y_1^*, a_1^*)$ and $S_2 = (y_2^*, a_2^*)$, in that order, show that S_1 is generically unstable and state *three* possibilities for the asymptotically stable behaviour of the system as $n \to \infty$.

1.4. It has been suggested that a means of controlling insect pests is to introduce and maintain a number of sterile insects in the population. One model for the resulting population dynamics is given by

$$N_{n+1} = f(N_n) = R_0 N_n \frac{N_n}{N_n + S} \frac{1}{1 + aN_n}$$

where $R_0 > 1$, $a > 0$ and S is the constant sterile insect population. The idea is that S is under our control, and we wish to choose it to accomplish certain ends. The problem is therefore one in *control theory*.

a) Explain the $N_n/(N_n + S)$ factor in the model.

b) Find the equation satisfied by the steady states N^* of the model, and sketch the relationship between N^* and S in the form of a graph of S as a function of N^*.

c) What is the least value S_c of S that will drive the insects to extinction?

d) Sketch cobweb maps for $S < S_c$ and $S > S_c$. What kind of bifurcation occurs as S passes through S_c?

1.3 Differential Equation Models

In this section we shall consider models for populations which change continuously with time. Let b be the *per capita production* or *reproduction rate*, i.e. the probability that any given individual in the population produces an offspring in the next small interval of time δt is given by $b\delta t + O(\delta t^2)$. Similarly, let d be the per capita *death rate* or *mortality rate*, i.e. the probability that any given individual dies in the next small interval of time δt is given by $d\delta t + O(\delta t^2)$. Then, if the population is so large that stochastic effects may be neglected,

$$N(t + \delta t) = N(t) + bN(t)\delta t - dN(t)\delta t + O(\delta t^2).$$

Subtracting $N(t)$ from each side of the equation, dividing through by δt and letting $\delta t \to 0$, we obtain the linear equation

$$\frac{dN}{dt} = (b - d)N = rN, \tag{1.3.6}$$

with initial conditions

$$N(0) = N_0,$$

which has solution

$$N(t) = N_0 e^{(b-d)t} = N_0 e^{rt}.$$

This is called *Malthusian growth*, and Equation (1.3.6) is called the *Malthusian equation* in continuous time. The parameter $r = b - d$ is called the *net per capita growth rate* or the *Malthusian parameter*. There are no density-dependent effects like intraspecific competition, and b, d and r are constant. The basic reproductive ratio, or expected number of offspring produced per individual, is $R_0 = b/d$, since an individual is producing at rate b for an expected time $1/d$. The solution reduces to that of the Malthusian equation in discrete time if we define $r = \log \lambda$, or $\lambda = e^r$, and meter the model at unit intervals of time $t = n$. The dividing line between stable and unstable behaviour is when $r = 0$, or $\lambda = 1$, or $R_0 = 1$. As intuition suggests, $R_0 > 1$ tells us that the population is growing, but λ (or r) tells us how fast it does so, through the relationship $r = d(R_0 - 1)$. In contrast to λ or R_0, r has dimensions of inverse time, and is therefore a true rate rather than a ratio.

If density-dependent effects are included, the per capita birth and death rates for the population depend on the size of the population, and the equation for population growth is given by

$$\frac{dN}{dt} = f(N),$$

to be solved subject to $N(0) = N_0$. If the population size is zero then there is no growth, so that $f(0) = 0$. We shall often write $f(N) = NF(N)$, as in the discrete case, to emphasise this, so that the equation becomes

$$\frac{dN}{dt} = f(N) = NF(N).$$

$f(N)$ is known as the *net growth rate* of a population of size N. $F(N)$ is known as the *net per capita growth rate*.

How do we incorporate the density-dependent effects? The steps in the *empirical approach* are as follows.

- Look at the data and decide what its essential characteristics are.

- Write down the simplest form for the function f (involving certain parameters) which will give a solution with such characteristics. (The principle of *Occam's razor* [1] is being invoked here.)

- Choose the parameters in the model to give a best fit to the data.

The main characteristic of the data in Figure 1.6(b), typical for microorganisms and many other populations, is an S-shaped growth curve to an *environmental carrying capacity* K. The existence of a carrying capacity suggests using a function $f(N)$ that is positive for $N \in (0, K)$ but satisfies $f(0) = f(K) = 0$. The simplest choice is

$$\frac{dN}{dt} = f(N) = rN(1 - \frac{N}{K}). \tag{1.3.7}$$

This is known as the *logistic equation*, and its solution is known as a *logistic curve* with parameters r and K.

Separating variables, Equation (1.3.7) gives

$$t = \frac{1}{r} \int_{N_0}^{N} \frac{K dN}{N(K - N)} = \frac{1}{r} \log \frac{(K - N_0)N}{(K - N)N_0},$$

so that

$$N(t) = \frac{K N_0 e^{rt}}{K - N_0 + N_0 e^{rt}}.$$

This solution is sketched below, and compared with data for the growth of the population of the USA and a colony of bacteria in the laboratory.

The logistic equation was first used in models for human populations by Verhulst in 1838, who followed a suggestion from his mentor Quetelet that the resistance to growth should be quadratic, and is also known as the *Verhulst equation*. Quetelet's motivation for the quadratic term was the analogy with motion in a resisting medium, where the resistance term may be modelled as quadratic in the velocity. The logistic was revived by Pearl and Reed in 1920. Pearl and others thought that fitting such a curve to a population time series would provide realistic short-term forecasts as well as estimates of the ultimate steady state population K. There were some successes; the outcome of the 1930 USA census was predicted with an error that was probably less than that of the census itself. However, an estimate of two thousand million for the upper limit of world population, made in 1924, was proved false by 1930, and a revised estimate of 2.6 thousand million, made in 1936, was proved false by 1955. Even

---

[1] A principle named after the 14th century philosopher William of Occam, present-day Ockham in Surrey, England. Expressed in more modern terms by Einstein, "Everything should be made as simple as possible, but not simpler".

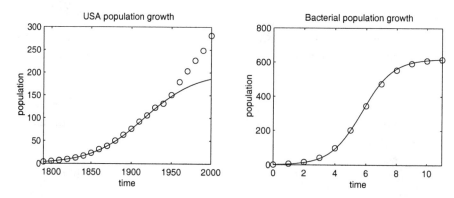

Figure 1.6   Data for growth of (a) USA population and (b) bacteria. The logistic curve fitted to (a) was the best fit in 1940; it can be seen that the prediction for the second half of the twentieth century was poor. The growth curve in (b) is fitted excellently with a logistic.

Pearl, its most enthusiastic proponent, realised that the logistic could not be valid indefinitely, and assumed that from time to time technical or medical advances would create a new and higher value of K which the curve would then tend towards. There are clearly difficulties in using it as a predictive tool for human populations. Non-human populations are generally better behaved, and very often tend towards a well-characterised upper limit, which can be predicted by using a logistic equation. However, there is a plethora of S-shaped curves that could be used to fit such population growth data. The logistic does better than many others in some situations, but the main advantage in using it is often its mathematical simplicity.

One important feature of the logistic model is that it is *compensatory*, as were the models in Section 1.2. In other words, the net per capita growth rate $F(N)$ decreases with N. A *depensatory* growth rate is one where $F(N)$ increases with N over some range. For example, in a study of the common guillemot on Skomer Island, South Wales, it was shown that breeding success was depensatory. As long as the density was not too high, per capita breeding success increased with density. It seems that denser populations have more success in fighting off predatory gulls. Depensation is also known as the *Allee effect* and is usually due to cooperation between the individuals concerned. In an extreme case, known as *critical depensation*, the growth rate is negative for small values of N, and there is a minimum viable population size.

EXERCISES

1.5. In this section we derived the logistic equation for limited population growth using an empirical approach. An alternative is the *ecosystems* or *resource-based approach* which goes back at least to both Lotka and Volterra, two of the founders of mathematical ecology. The ecosystems approach will be dealt with more fully in Chapter 2, but we outline the method here.

Let the per capita growth rate of a population depend on some *resource*. Let this resource exist in two states, either free (available for use by the members of the population) or bound (already in use). Let the density of free resource be R; then

$$\frac{dN}{dt} = NG(R).$$

Let the resource be *abiotic* (non-biological), and therefore not subject to birth and death. (The archetypal example in the ecosystems approach is a mineral resource, but another possibility is something like nest-sites.) Let the total amount of resource, free and bound, be a constant $C > 0$, and let the amount of bound resource depend on the population. Then

$$R = C - H(N).$$

It remains to model G and H. Clearly G is an increasing function satisfying $G(0) < 0$; the simplest model is $G(R) = \alpha R - \beta$, $\alpha > 0$, $\beta > 0$. H is also an increasing function but satisfies $H(0) = 0$; the simplest model is $H(N) = \gamma N$, $\gamma > 0$.

a) Show that we still obtain the logistic equation for the growth of the population.

b) Give expressions for the Malthusian parameter r and the carrying capacity K.

c) What happens if the total amount C of resource is insufficient?

d) Give an advantage of each approach to modelling limited growth.

1.6. A population N is growing according to a logistic differential equation, and $N(t_1) = n_1$, $N(t_1 + \tau) = n_2$, $N(t_1 + 2\tau) = n_3$. Show that the carrying capacity is given by

$$K = \frac{1/n_1 + 1/n_3 - 2/n_2}{1/(n_1 n_3) - 1/n_2^2}.$$

1.7. A model for population growth is given by

$$\frac{dN}{dt} = f(N) = rN\left(\frac{N}{U} - 1\right)\left(1 - \frac{N}{K}\right),$$

where r, K and U are positive parameters with $U < K$.

a) Sketch the function $f(N)$.

b) Discuss the behaviour of $N(t)$ as $t \to \infty$, and show that the model exhibits critical depensation.

1.8. For some organisms finding a suitable mate may cause difficulties at low population densities, and a more realistic equation for population growth than a linear one in the absence of intraspecific competition may be $\dot{N} = rN^2$, with $r > 0$, to be solved with initial conditions $N(0) = N_0$.

a) Show that this model exhibits depensation.

b) Solve this problem and show that the solution becomes infinite in finite time.

c) The model above is improved to

$$\frac{dN}{dt} = rN^2(1 - \frac{N}{K}).$$

Without solving this equation find the steady state solutions and say whether they are stable or unstable.

d) Derive the model of part (c) from that of part (b) by a resource-based method. Assume that the total amount of resource (free and bound) is sufficient to sustain a population. If the amount of free resource is non-negative initially, show, by a graphical argument or otherwise, that it always remains non-negative.

1.4 Evolutionary Aspects

So far we have considered the parameters in our models as immutable, but in fact they change as they are honed by natural selection. In this section we shall indicate very briefly some aspects of *evolutionary ecology*, which seeks to incorporate evolutionary effects into ecological models. We shall not consider the genetics of sexually reproducing organisms until Chapter 4, and so we shall assume that the organisms reproduce clonally, each individual producing offspring that are genetically identical to itself. Let us imagine that we have a genetically uniform population growing logistically, with parameters r_1 and K_1, and that a mutation occurs that produces an individual with parameters r_2 and K_2. It is intuitively clear that if both $r_2 > r_1$ and $K_2 > K_1$ the mutant will displace the resident, but we shall assume that there is a *trade-off*, that the organism can "choose" to put its resources into increasing r at the expense of K or vice versa. Hence either (a) $r_2 > r_1$ and $K_2 < K_1$ or (b) $r_2 < r_1$ and $K_2 > K_1$. The crucial question is whether or not the mutant will invade the steady state consisting of the resident only. A possible model for the situation (which we shall not justify in detail) is

$$\frac{dN_1}{dt} = r_1 N_1 \left(1 - \frac{N}{K_1}\right), \quad \frac{dN_2}{dt} = r_2 N_2 \left(1 - \frac{N}{K_2}\right),$$

where $N = N_1 + N_2$. The Jacobian matrix at the steady state $(K_1, 0)$ is given by

$$J^* = \begin{pmatrix} -r_1 & -r_1 \\ 0 & r_2(1 - K_1/K_2) \end{pmatrix},$$

which has eigenvalues $-r_1$ and $r_2(1 - K_1/K_2)$. The mutant invades if $K_2 > K_1$, in case (b) but not in case (a). The result implies that organisms should always be *K-selected* rather than *r-selected*, evolving to increase K at the expense of r. The fact that many organisms have high r and low K points to a deficiency in the model. Most importantly, it neglects all environmental unpredictability, whereas we shall see later that r-selected organisms do well in patchy and variable environments.

1.5 Harvesting and Fisheries

Consider a population of fish that grows without harvesting according to $N_{n+1} = f(N_n)$. This is based on the assumption that the fish have annual discrete, non-overlapping generations, so that the context is similar to that in Section 1.2.1 for insect populations. Continuously-growing populations are more easily analysed, and an example of these is given in Exercise 1.11. The stock-recruitment function f gives the number of young to be recruited into the adult class next year, if the number of adults this year is N. The assumption that the generations are non-overlapping could be relaxed by using an age-structured model, but we shall not do this here. Now let a harvest Y_n be taken in the $(n+1)$th interval. Then

$$N_{n+1} = f(N_n) - Y_n.$$

We assume that we can treat the population dynamics and the harvesting separately, so that first the N_n adults in one year lead to a stock recruitment of $N_{n+1} = f(N_n)$ adults the following year, and then the harvesting decreases it from $f(N_n)$ to $f(N_n) - Y_n$.

− If the amount taken is under our control, how should we choose it?

The branch of mathematics that deals with such questions is *control theory*. The question is quite different from most of those we have looked at so far, where our motive has been to analyse and understand the behaviour of a biological system; now we wish to control it as well. It is a problem in natural resources management, which apart from fisheries has applications in forestry, wildlife conservation and pest management (as in Exercise 1.4), among others.

The first step is deciding our *utility*, i.e. what it is that we would like to maximise. Let us assume that we wish to find a *maximum sustainable yield*, or MSY, i.e. we wish to come to a steady state (N^*, Y^*) where Y^* is as large as possible. Then

$$Y^* = f(N^*) - N^*, \tag{1.5.8}$$

The maximum of this function occurs where

$$f'(N^*) = 1,$$

so we try to adjust Y so that N approaches a value where $f'(N) = 1$. In many cases this is easier said than done; for example, in wild fish populations N is not observable, so we have to use an indirect approach to decide how to choose Y. Two quantities that we can observe in a fishery are the effort put into fishing and the yield obtained. We assume that the rate at which fish are caught is proportional to the effort E that is put into catching them and the number of fish available, so that while fishing is taking place,

$$\frac{dN}{dt} = -qEN,$$

where the constant q is the *catchability coefficient*. Separating the variables and integrating over the length τ of the fishing season,

$$-q\tau E = [\log(N)]_{f(N)}^{f(N)-Y}.$$

Now let us assume that we are at a steady state, so that $f(N) - Y = f(N^*) - Y^* = N^*$, and $E = E^*$. The equation becomes

$$q\tau E^* = \log(F(N^*)), \tag{1.5.9}$$

where F is defined by $f(N) = NF(N)$ as usual. The yield equation (1.5.8) becomes

$$Y^* = N^*F(N^*) - N^*. \tag{1.5.10}$$

Eliminating the non-observable N^* between Equations (1.5.9) and (1.5.10), we obtain the *yield-effort relationship*

$$\exp(q\tau E^*) = F\left(\frac{Y^*}{\exp(q\tau E^*) - 1}\right). \qquad (1.5.11)$$

Now let us be specific about the population dynamics, and take the Beverton–Holt stock-recruitment curve for fisheries,

$$f(N) = \frac{R_0 N}{1 + N/K}. \qquad (1.5.12)$$

The assumptions that this is based on are explored in Exercise 1.9.

In this case we can solve for yield in terms of effort,

$$Y^* = K\left(R_0 \exp(-q\tau E^*) - 1\right)\left(\exp(q\tau E^*) - 1\right). \qquad (1.5.13)$$

This is shown in Figure 1.7(a). The maximum sustained yield is obtained when $q\tau E^* = \frac{1}{2}\log R_0$, and is given by

$$Y^* = K(\sqrt{R_0} - 1)^2.$$

This leads to the following fisheries policy.

- Increase effort until the yield falls; the effort required for the MSY has been exceeded.

- Then decrease effort until the yield falls again; the effort is then below that required for the MSY.

- Iterate until E^*, and hence the MSY, is attained.

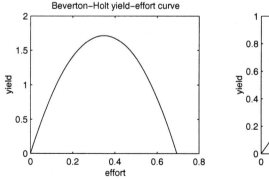
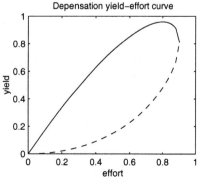

Figure 1.7   Steady state yield-effort curve (1.5.13) for (a) the Beverton–Holt Equation (1.5.12) and (b) a depensatory stock-recruitment model.

Unfortunately the policy outlined here depends on having a smooth yield-effort curve, which in turn depends on the particular model taken for stock recruitment. If the stock-recruitment curve exhibits depensation , an increase in per capita recruitment with density, then a yield-effort curve such as that in Figure 1.7(b) occurs. Here an effort only just above that required for maximum sustained yield may lead to population crashes. Fish that form large closely packed schools, such as the *clupeids* (anchovies, sardines and herrings), typically exhibit depensatory stock recruitment, and are therefore particularly vulnerable to over-exploitation. A spectacular crash in the Peruvian anchovy fishery from about 10 million tons per annum in the 1960s, 15% by weight of the total global catch of all marine fisheries, to around 2 million tons per annum in the 1970s, was due to a combination of over-fishing (through a loss of political control and against the advice of biologists) and the warm El Niño current.

EXERCISES

1.9. *Beverton–Holt equation.* The Beverton–Holt stock-recruitment curve, given by equation (1.5.12), is derived from a sub-model in continuous time that describes the dynamics of the larval population between hatching and recruitment to the adult population. Let $L(t)$ be the number of larvae at time t, and assume that they are subject to density-dependent mortality through intraspecific competition or other effects between time $n + t_1$ and $n + t_2$, where $0 \le t_1 < t_2 \le 1$. In this time, they die according to

$$\frac{dL}{dt} = -(\mu_1 + \mu_2 L)L,$$

where μ_1 and μ_2 are positive constants. If $L(n + t_1)$ is proportional to N_n, the number of adults in the population at time n, and N_{n+1} is proportional to $L(n + t_2)$, derive the Beverton–Holt Equation (1.5.12).

1.10. A fishery population is modelled by

$$N_{n+1} = f(N_n) - Y_n$$

where Y_n is the catch taken in year n, the *yield*. and f is given by the Ricker model $f(N) = Ne^{r(1-N/K)}$, a model for over-compensatory competition, where r and K are positive parameters. Sketch Y^* as a function of N^*.

1.11. Models of fisheries in continuous time, where growth and harvesting occur continuously and simultaneously, are easier to analyse than the discrete time stock-recruitment models of this section. A population of fish that would otherwise grow according to the logistic law $\dot{N} = rN(1 - N/K)$ is fished at constant effort E, leading to a harvest at constant rate qEN, where q is the catchability coefficient.

  a) Show that the steady state in the presence of fishing is given by $N^* = K(1 - qE/r)$.

  b) Find the steady state yield-effort relationship.

  c) Find the maximum sustainable yield.

1.12. Consider a population of fish that would grow without harvesting at a rate $\dot{N} = NF(N)$ that is depensatory, so that F increases, for small values of N. It is then fished at constant effort E, leading to a harvest at constant rate qEN.

  a) Show that, at steady state, $qE^*N^* = N^*F(N^*)$.

  b) Sketch qE^*N^* and $N^*F(N^*)$ on the same graph, and deduce the yield-effort relationships for critical and non-critical depensation.

1.6 Metapopulations

Studies of patchy environments such as archipelagos have shown that species or ecological communities often persist through a combination of local extinctions and re-colonisations. It is difficult to make a mathematical model that includes such spatial effects explicitly and yet is tractable. However, there is a way of including space implicitly which captures the essence of the process and which can be analysed reasonably easily.

Consider a large number K of potentially habitable sites, each of which is either occupied or unoccupied by an individual (or a population) of a particular species. Each of these sites is assumed to be identical and to be isolated from the others in an identical way. Let the fraction occupied at time t be $p(t)$. Let the probability of an occupied site becoming unoccupied in the next interval of time δt be $e\delta t$, so that e is a local extinction rate. Then the mean fraction of sites that become unoccupied in the next interval of time δt is $ep(t)\delta t$. Let the probability of an unoccupied site becoming occupied in the next interval of time δt be $cp(t)\delta t$, so that c represents a colonisation rate. Then the mean fraction of sites that are colonised in the next interval of time δt is $cp(t)(1 - p(t))\delta t$.

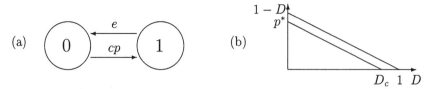

Figure 1.8   (a) The transitions between the vacant and the occupied state in the archetypal metapopulation model given by Equation (1.6.14). Extinction for a given occupied site is assumed to be independent of how many other sites are occupied, whereas colonisation of a given vacant site is linearly dependent on the number of sites available to provide colonists. (b) The bifurcation diagram plotting non-negative steady state solutions p^* of Equation (1.6.16) as a function of the fraction D of habitat removed. There is a transcritical bifurcation at $(D_c, 0)$. Also shown is the fraction $1 - D$ of patches remaining. The difference between these is the number of empty patches remaining at steady state, which stays constant as D increases until the species becomes extinct at D_c.

We have, on average, taking the limit as $\delta t \to 0$,

$$\frac{dp}{dt} = f(p) = cp(1 - p) - ep. \tag{1.6.14}$$

The true behaviour can be expected to track this mean behaviour closely if K is large, and tracks it exactly in the idealised situation of an infinite number of patches. The critical parameter is the *basic reproductive ratio*, the number of sites an occupied site can expect to colonise before going extinct when the species is rare, given by $R_0 = \frac{c}{e}$, analogous to $R_0 = \frac{b}{d}$ for the continuous Malthus Equation (1.3.6).

Example 1.2

The function f in Equation (1.6.14) is shown in Figure 1.9 for $R_0 > 1$ and for $R_0 < 1$. For $R_0 > 1$ show that if $p(0) > 0$ then $p(t) \to p^*$ as $t \to \infty$, where

$$p^* = 1 - \frac{e}{c} = 1 - \frac{1}{R_0}. \tag{1.6.15}$$

For $R_0 < 1$ show that $p(t) \to 0$ as $t \to \infty$.
    The results follow immediately on considering the sign of $\dot{p}$.

As intuition suggests, there is a threshold at $R_0 = 1$; if the colonisation rate when rare is greater than the local extinction rate, $R_0 > 1$, then the population

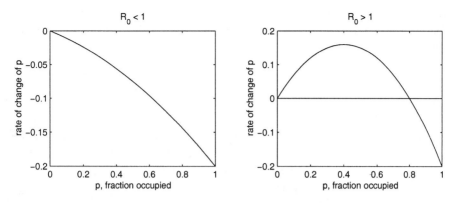

Figure 1.9   The function f in Equation (1.6.14), for (a) $R_0 < 1$ and (b) $R_0 > 1$. The fraction of sites colonised increases if $f > 0$, decreases if $f < 0$.

persists, whereas if the local extinction rate is greater than the colonisation rate when rare, $R_0 < 1$, the population as a whole goes extinct. The dispersal, here represented by the colonisation parameter c, must be sufficiently large $(c > e)$ for persistence.

In Chapter 3 we shall discuss the Kermack and McKendrick model for an SIS disease, i.e. a disease where all infected individuals eventually recover and become susceptible again. There is a parallel here, with occupied patches analogous to infected individuals, and unoccupied patches to susceptible individuals. There the basic reproductive ratio R_0 is the number of infectious contacts an infected individual can expect to make when the disease is rare, and there is again a threshold at $R_0 = 1$, with $R_0 > 1$ necessary for the disease to remain endemic.

What is the effect of habitat destruction on the species? Of course it leads to smaller populations and hence greater chances of extinctions, but the effects are more far-reaching than that. We shall investigate the effect of removal of some habitat by modifying Equation (1.6.14). Removal of a fraction D of habitat leads to the equation

$$\frac{dp}{dt} = cp(1 - D - p) - ep, \qquad (1.6.16)$$

where it has simply been assumed that any attempted colonisation of a removed patch is unsuccessful and leads to immediate death of the propagules. The basic reproductive ratio has now changed from $R_0 = c/e$ to $R_0' = c(1 - D)/e$, since only a fraction $1 - D$ of the original colonisations are now successful. The threshold is $R_0' = 1$, or $R_0 = 1/(1 - D)$, and extinction occurs when $D = D_c = 1 - 1/R_0$. (The epidemiological analogy of this result is that only a fraction of susceptibles needs to be vaccinated to eradicate a disease.) The critical fraction D_c of sites to be deleted is the value of the steady state p^*

when $D = 0$. When the fraction of sites removed becomes equal to the original fraction occupied, the population becomes extinct (despite the fact that the sites are removed at random).

EXERCISES

1.13. In a mainland-island metapopulation model there is supposed to be a mainland in addition to the island patches, where there is a large population with negligible risk of extinction. The equation for the fraction of occupied patches is modified to

$$\frac{dp}{dt} = (m + cp)(1 - p) - ep.$$

a) Explain the model.

b) Find the (biologically realistic) steady states of the model, and discuss their stability.

A multispecies version of this is the basis of MacArthur and Wilson's dynamic theory of *island biogeography*.

1.7 Delay Effects

If a system is controlled by a feedback loop in which there is a delay, then oscillations may result. We have looked at one example which may be thought of as a delay effect, when there are discrete breeding seasons and the size of the population at time $n + 1$ depends on what it was at time n, but there are many others. One of the first models to include delay was by Volterra (1926), who took into account the time taken for pollutants produced by a population to build up, eventually increasing the death rate of the population. Delays may also be due to development time, of the population itself or of its resources. Let us consider population development time. An increase in the amount of resources available to an insect may result in an (almost) immediate increase in the number of eggs laid, but the number of adults will not increase until these have passed through the larval and pupal stages. Nicholson showed that a laboratory population of blowflies may exhibit regular large amplitude oscillations when resources for the adults are limited, in contrast to the irregular behaviour considered in Section 1.2.2 when resources for the larvae were limited. We investigate the hypothesis that the periodic behaviour arises because of delay effects. Let $N(t)$ be the number of adult blowflies at time t, and w the rate at which they are supplied with food. Let m be the rate at which an individual needs to feed in order to maintain itself. Then $(w - mN)_+$ is the excess rate of food supply, defined by

$$(w - mN)_+ = \begin{cases} w - mN & \text{if } w > mN, \\ 0 & \text{otherwise.} \end{cases}$$

Assume this is put towards reproduction, and results in a proportionate number of adults with a delay of τ. Let us also assume constant per capita mortality at rate c. Let τ be the amount of time it takes for an egg to become an adult. Then

$$\frac{dN}{dt}(t) = k(w - mN(t - \tau))_+ - cN(t). \tag{1.7.17}$$

There is a steady state of this equation at N^*, where $N^* = kw/(c + km)$. If this steady state is unstable by growing oscillations, then this indicates that the nonlinear equation may have oscillatory solutions. We investigate stability by linearising about the steady state, defining $n = N - N^*$ and neglecting higher order terms. The perturbation n satisfies the approximate equation

$$\frac{dn}{dt}(t) = -cn(t) - mkn(t - \tau). \tag{1.7.18}$$

This is a linear equation for n, so we look for solutions of the form $n(t) = n_0 \exp(st)$, which satisfies Equation (1.7.18) as long as the *characteristic equation* or *eigenvalue equation*

$$s = -c - mk \exp(-s\tau) \tag{1.7.19}$$

holds. The roots of this equation are eigenvalues of the problem. This is a typical example of an equation that arises from linearising about the steady state of a differential equation with discrete delay. Note that if $\tau = 0$ there is only one eigenvalue, $s = -c - mk < 0$, and so if there were no delay the steady state would be stable. We think of τ as a bifurcation parameter. For $\tau > 0$ an infinite number of eigenvalues, real and complex, bifurcate from the root at $s = -c - mk$. As τ increases, do any of these eigenvalues cross the imaginary axis into the right half of the complex plane, leading to instability? If so, are they real, passing through the origin and leading to exponential divergence from the steady state, or complex, crossing the imaginary axis away from the origin and leading to growing oscillations? First, since $s = 0$ can never be a solution of Equation (1.7.19), there can be no bifurcations with real eigenvalues. Looking for a bifurcation with complex eigenvalues, we put $s = u + iv$ and equate real and imaginary parts. We obtain

$$u = -c - mke^{-u\tau} \cos v\tau, \quad v = mke^{-u\tau} \sin v\tau,$$

so that

$$(u + c)^2 + v^2 = m^2 k^2 e^{-2u\tau}.$$

The eigenvalues appear in complex conjugate pairs, so we may take $v > 0$ without loss of generality. For fixed $v > 0$ and $\tau > 0$, a plot of the two functions $f_1(u) = (u+c)^2 + v^2$ and $f_2(u) = m^2 k^2 e^{-2u\tau}$ (see Figure 1.10) shows that $u < 0$ whenever $m^2 k^2 < c^2 + v^2$. Hence no instability is possible if $mk < c$, and we shall assume henceforth that $mk > c$.

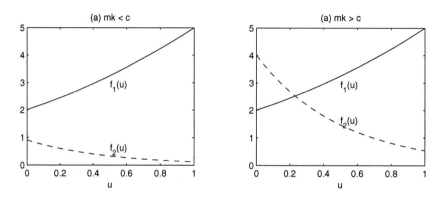

Figure 1.10   No instability is possible if $mk < c$.

We now look for points in (c, mk)-space where stability changes can occur, so we look for eigenvalues of the form $s = iv$. Stability boundaries in the parameter space are given by

$$c = -mk \cos v\tau, \quad v = mk \sin v\tau.$$

For $v > 0$, we must have $\cos v\tau < 0$ and $\sin v\tau > 0$, so that $2n\pi + \frac{1}{2}\pi < v\tau < 2n\pi + \pi$, for $n = 0, 1, 2, \ldots$. These equations may be solved to obtain $v = \sqrt{m^2 k^2 - c^2}$, $\tau = \tau_n = \tau_0 + 2n\pi$, $\tau_0 = \arccos(-c/mk)/\sqrt{m^2 k^2 - c^2}$, where arccos is the main branch of the inverse cosine function. It is easy to check that $\frac{du}{d\tau}(\tau_n)$ is positive, so that two eigenvalues cross into the right half complex plane every time τ passes through τ_n. The first bifurcation to instability via growing oscillations occurs at τ_0. More generally, the effect of delay is often to destabilise steady states, leading to periodic or even chaotic behaviour of the system under consideration.

EXERCISES

1.14. One of the best-studied delay equations is Hutchinson's (1948) equation, a modification of the logistic equation, given by

$$\frac{dN}{dt}(t) = rN(t)\left(1 - \frac{N(t - \tau)}{K}\right).$$

One way to interpret this is that the per capita growth rate depends on the availability of a resource, which in turn depends on the population size a time τ earlier. The population takes a time τ to respond to the resource. Find the steady state of this equation, and investigate its stability.

1.15. Hutchinson's equation may be generalised to

$$\frac{dN}{dt}(t) = rN(t)\left(1 - \frac{1}{K}\int_0^\infty N(t-u)k(u)du\right), \qquad (1.7.20)$$

where $k(u)$ denotes the weight given to the population size a time u earlier, normalised so that $\int_0^\infty k(u)du = 1$. (If $k(u) = \delta(u-\tau)$, we recover the original equation.) In this question we shall take $k(u) = \frac{1}{\tau}\exp\left(-\frac{u}{\tau}\right)$, which has an average delay $\int_0^\infty uk(u)du = \tau$. This is called the *weak generic delay kernel*.

a) Show that $N = 0$ and $N = K$ are steady states of Equation (1.7.20).

b) Show that the equation linearised about $N = K$ is given by

$$\frac{dn}{dt}(t) = -r\int_0^\infty n(t-u)k(u)du. \qquad (1.7.21)$$

c) This equation is linear, so that we expect solutions like $n(t) = n_0\exp(st)$. Derive the characteristic equation for s.

d) Show that the steady state is linearly stable.

e) Show that $P(t) = \int_0^\infty N(t-u)k(u)du$ satisfies $\frac{dP}{dt} = \frac{P-N}{\tau}$. The equation is equivalent to the system $\dot{N} = rN(1 - P/K)$, $\dot{P} = (P-N)/\tau$, and may be analysed by methods to be discussed in Chapter 2.

1.16. Analyse Equation (1.7.20) with strong generic delay, $k(u) = \frac{u}{\tau^2}\exp\left(-\frac{u}{\tau}\right)$. Derive a condition for instability of its non-trivial steady state K.

1.8 Fibonacci's Rabbits

We have so far neglected any population structure, by implicitly assuming that any member of the population is equally likely to give birth or to die. This is not true for a real population. Very old or very young individuals may be more likely to die, female individuals may be more likely to give birth, large or healthy individuals may be more likely to survive, and so on. We now look at one of the earliest models of population growth, as an introduction to incorporating age structure into our models. Leonardo of Pisa, known as Fibonacci, was born in Italy in about 1170 but educated in North Africa, where his father was a diplomat, and died in 1250. The book for which he is most famous is the *Liber abaci*, published in 1202, which brought decimal or Hindu-Arabic numerals into general use in Europe. In the third section of the book he posed the following question.

> A certain man put a pair of rabbits in a place surrounded on all sides by a wall. How many pairs of rabbits can be produced from that pair in a year if it is supposed that every month each pair begets a new pair which from the second month on becomes productive?

This is one of the first mathematical models for population growth. It states most of its assumptions clearly, and although it has some obvious defects its analysis provides some insights into the process considered. We shall assume that each pair of rabbits consists of one male and one female. To translate the words into equations, define

- $y_{k,n}$ to be the number of k-month-old pairs of rabbits at time n;

- y_n to be the total number of pairs of rabbits at time n.

Time n is n months after the man puts the first pair of rabbits into the place surrounded by a wall, but we have to decide whether to census the population just before or just after the births for that month have taken place. We shall census before births, leaving the post-birth census as an exercise. This may seem somewhat counterintuitive, but it is a direct generalisation of models for populations that are not age-structured, and it simplifies the calculations if we never have to consider new-borns. Thus $y_n = \sum_{k=1}^{\infty} y_{k,n}$.

Then, since no rabbits ever die, all those that were k months old at time n are $k + 1$ months old at time $n + 1$, $y_{k+1,n+1} = y_{k,n}$ for $n \geq 0$, $k \geq 0$. There are as many one-month-old pairs at time $n + 1$ as there were pairs at least two months old at time n, $y_{1,n+1} = y_{2,n} + y_{3,n} + \cdots$.

Let us assume the initial pair is adult, two months old without loss of generality. The initial conditions are $y_{1,0} = 0$, $y_{2,0} = 1$, $y_{k,0} = 0$ for $k > 2$. So for $n \geq 0$ we may write

$$y_{n+2} = y_{1,n+2} + y_{2,n+2} + y_{3,n+2} + \cdots$$
$$= (y_{2,n+1} + y_{3,n+1} + \cdots) + (y_{1,n+1} + y_{2,n+1} + \cdots)$$
$$= (y_{1,n} + y_{2,n} + \cdots) + y_{n+1} = y_n + y_{n+1},$$

i.e.

$$y_{n+2} = y_{n+1} + y_n, \tag{1.8.22}$$

Our single initial condition for the total population is $y_0 = 1$, which is not enough to solve this second order equation. However, it is easy to see that $y_{1,1} = 0$, $y_{2,1} = 0$, $y_{3,1} = 1$, $y_{k,1} = 0$ for $k > 3$, so that $y_1 = 1$. We have initial conditions

$$y_0 = y_1 = 1. \tag{1.8.23}$$

The resulting sequence is 1, 1, 2, 3, 5, 8, 13, 21, 34, 55, $\cdots$, and the answer to the original question is 144 pairs of rabbits after a year (233 if you count newborns). This sequence, in which each number is the sum of the two preceding numbers, has proved extremely fruitful and appears in many different areas of mathematics and science. The *Fibonacci Quarterly* is a modern journal devoted to studying mathematics related to this sequence.

Example 1.3

Solve Equation (1.8.22) with initial conditions (1.8.23). Show that the solution grows geometrically as $n \to \infty$, and find the geometric ratio (the *growth ratio*). Equation (1.8.22) is a homogeneous linear difference equation. A matrix method for solving such problems is given in Section A.3 of the appendix, but here we use a method which is simpler in this case. The general solution of the difference equation (1.8.22) is $y_n = A_1\lambda_1^n + A_2\lambda_2^n$, where each λ_i is a solution of the auxiliary equation. The constants A_i are then determined by the initial conditions. The auxiliary equation is obtained by substituting λ^n into Equation (1.8.22), to obtain

$$\lambda^{n+2} = \lambda^{n+1} + \lambda^n,$$

or dividing through by λ^n,

$$\lambda^2 - \lambda - 1 = 0.$$

This is a quadratic equation and has solutions $\lambda = \frac{1}{2}(1 \pm \sqrt{5})$, and the solution of the difference equation (1.8.22) satisfying the initial conditions (1.8.23) has $A_1 = (1 + \sqrt{5})/(2\sqrt{5})$, $A_2 = -(1 - \sqrt{5})/(2\sqrt{5})$, so

$$y_n = \frac{1}{\sqrt{5}}\left(\frac{1 + \sqrt{5}}{2}\right)^{n+1} - \frac{1}{\sqrt{5}}\left(\frac{1 - \sqrt{5}}{2}\right)^{n+1}.$$

The model has geometrically growing solutions, $y_n \approx A_1\lambda_1^n$ as $n \to \infty$ (see also Figure 1.11), with growth ratio $\lambda_1 = \frac{1}{2}(1 + \sqrt{5})$, the so-called golden ratio.

We have seen geometric growth in the Malthusian Equation (1.2.1), a linear equation without structure. Geometric growth or decay occurs in *almost every* linear difference equation model, even when population structure is included. It is not necessarily a problem if we are treating a finite time period, as in

Fibonacci's question, but it clearly will not be possible to extrapolate such a solution indefinitely. The growth ratio λ_1 is clearly of prime importance. Here it is found as a solution of a quadratic equation, or equivalently as an eigenvalue of a matrix. An important part of population dynamics lies in determining the equivalent parameter in different circumstances.

EXERCISES

1.17. Give at least three criticisms of Fibonacci's rabbit model.

1.18. What difference does it make to the model and the analysis if the census takes place just after the births instead of just before?

1.9 Leslie Matrices: Age-structured Populations in Discrete Time

Fibonacci's rabbits are *age-structured*, i.e. their age is important in determining their vital parameters. In this case the only vital parameter is the birth rate, as none of the rabbits ever die. If we define

- $z_{1,n}$ to be the number of juvenile (one-month-old) pairs of rabbits at time n, and

- $z_{2,n}$ to be the number of adult pairs of rabbits at time n,

we have $z_{1,n+1} = z_{2,n}$, $z_{2,n+1} = z_{1,n} + z_{2,n}$. Hence

$$\begin{pmatrix} z_{1,n+1} \\ z_{2,n+1} \end{pmatrix} = \begin{pmatrix} 0 & 1 \\ 1 & 1 \end{pmatrix} \begin{pmatrix} z_{1,n} \\ z_{2,n} \end{pmatrix}, \qquad (1.9.24)$$

or, in matrix notation,

$$\mathbf{z}_{n+1} = L\mathbf{z}_n, \qquad (1.9.25)$$

where $\mathbf{z}_n = (z_{1,n}, z_{2,n})^T$.

How do we generalise all this? Rather than looking at pairs, it is usual to keep track of each sex separately. From now on we shall restrict attention to females only. If there is an oldest age class ω, and if no individual can stay in an age class for more than one time period, which is *not* the case in Fibonacci's model, then $\mathbf{z}_{n+1} = L\mathbf{z}_n$, where $\mathbf{z}_n = (z_{1,n}, \ldots, z_{\omega,n})^T$ and L is an $\omega \times \omega$ matrix of the general form

$$
L = \begin{pmatrix}
s_1 m_1 & s_1 m_2 & \cdots & s_1 m_{\omega-1} & s_1 m_\omega \\
s_2 & 0 & \cdots & 0 & 0 \\
0 & s_3 & \cdots & 0 & 0 \\
0 & 0 & \cdots & 0 & 0 \\
\vdots & \vdots & & \vdots & \vdots \\
0 & 0 & \cdots & s_\omega & 0
\end{pmatrix}. \tag{1.9.26}
$$

Here s is the survival fraction and m the age-dependent *maternity function*, i.e. the s_i are the probabilities of surviving from age $i - 1$ to age i, and each individual of age i produces m_i offspring on average. Then $s_1 m_i$ is the expected number of offspring produced by each individual of age i that survive from birth to their first census. In the Fibonacci model the element 0 in the bottom right-hand corner of the matrix was replaced by 1, representing unit probability of survival in the oldest age class, and in general it may be replaced by $s_{\omega+1}$, the probability of an individual in the oldest age class surviving to the next census. Usually, $m_i = 0$ unless $\alpha \leq i \leq \beta$, where $[\alpha, \beta]$ is the fertile period. The census is taken just before the offspring have been produced.

In general, the non-negative matrix L is known as a Leslie matrix or population projection matrix. If the process is stationary, L constant, then for almost all initial conditions the population approaches a *stable age structure* (age distribution) and growth ratio λ_1 as $n \to \infty$, $\mathbf{z}_n \approx A\mathbf{v}_1\lambda_1^n$, where λ_1 is the eigenvalue of L of greatest modulus, called the *principal eigenvalue* of L, $\mathbf{v}_1$ is the corresponding (left) eigenvector ($L\mathbf{v}_1 = \lambda_1\mathbf{v}_1$), and A is a constant. It is easy to see that if we restrict attention to the pre-reproductive and reproductive age-classes, then all age-classes contribute at some future date to all others, so the corresponding restriction of L is irreducible. In the usual case that L is also primitive, the Perron–Frobenius theorem (Section D.1 of the appendix) then guarantees that the principal eigenvalue λ_1 of L is real and positive and that the components of $\mathbf{v}_1$ are positive. The caveat that this is only true for *most* initial conditions is annoying but generally necessary. For example, the result would not hold if all members of the initial population were too old to produce offspring.

Example 1.4

Find the principal eigenvalue and eigenvector of the matrix in Equation (1.9.24). Hence show that the stable age structure of Fibonacci's rabbits is $1 : \lambda_1 - 1 : 1$ newborns:juveniles:adults, where λ_1 is the golden ratio $\frac{1}{2}(1 + \sqrt{5})$.

The eigenvalues satisfy $\det(L - \lambda I) = \lambda^2 - \lambda - 1 = 0$, which has solutions λ_1 and λ_1^{-1}. The principal eigenvalue is the greater of these, λ_1, and $L\mathbf{v}_1 =$

$\lambda_1 \mathbf{v}_1$ (where $\mathbf{v}_1 = (v_1, v_2, v_3)^T$) as long as $v_2 + v_3 = \lambda_1 v_1$, $v_1 = \lambda_1 v_2$, and $v_2 + v_3 = \lambda_1 v_3$. The solution of these (up to an arbitrary multiplicative constant) is $\mathbf{v}_1 = (1, \lambda_1 - 1, 1)^T$, and the result follows.

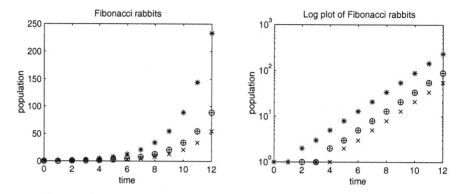

Figure 1.11   Two plots, the second logarithmic, of the solution of Fibonacci's rabbit problem. Newborn rabbits are marked $+$, juvenile rabbits $\times$, adult rabbits $\circ$, and the total $*$. The geometric growth and the stable age structure can be seen.

Similar equations arise if vital parameters depend on the *stage* that an organism has reached rather than its age. Its stage may be a stage in the life-cycle, such as the larval stage, or it may be an attribute such as size. Fibonacci's model may be seen as stage-structured since age is immaterial once the rabbits mature.

EXERCISES

1.19. Consider a population of annual plants with the following characteristics. Seeds are produced at the end of the summer. A proportion survive one winter, and a proportion of these germinate the following spring. Of the remainder, a proportion survive a second winter, and a proportion of these germinate the spring following this second winter, but none can germinate later than this.

a) Justify the model

$$N_{n+2} = \alpha\sigma\gamma N_{n+1} + \beta(1-\alpha)\sigma^2\gamma N_n$$

for the population, and interpret the parameters biologically.

b) Write this in terms of a Leslie matrix.

c) Use an eigenvalue equation to show that the condition for the plant population to thrive is

$$\gamma > \frac{1}{\alpha\sigma + \beta(1-\alpha)\sigma^2}.$$

d) Show that this condition is equivalent to the condition $R_0 > 1$, where R_0 is the basic reproductive ratio, interpreted here to be the expected number of offspring produced by an individual during its lifetime that survive to breed, in the absence of any other mortality.

1.20. In the circulatory system, the red blood cells (RBCs) are constantly being destroyed and replaced. Assume that the spleen filters out and destroys a certain fraction f of the cells daily and that the bone marrow produces a number proportional to the number lost on the previous day. If

- C_n is the number of RBCs in circulation on day n,

- M_n is the number of RBCs produced by the marrow on day n,

then

$$C_{n+1} = (1-f)C_n + M_n$$
$$M_{n+1} = \gamma f C_n$$

where γ is a constant.

a) Explain these equations.

b) Find the principal eigenvalue λ_1.

c) The definition of homeostasis is that $C_n \to C^*$, a non-zero constant, as $n \to \infty$. How may the parameters be chosen to achieve this?

d) Why cannot this be a good model for homeostasis?

1.21. Let the matrix L have distinct eigenvalues λ_i with corresponding eigenvectors $\mathbf{v}_i$. If $\mathbf{u}_{n+1} = L\mathbf{u}_n$ and $\mathbf{u}_0 = \sum_{i=0}^{\omega} A_i \mathbf{v}_i$, where the A_i are constants, show that

$$\mathbf{u}_n = \sum_{i=0}^{\omega} A_i \lambda_i^n \mathbf{v}_i.$$

1.22. If L is given by Equation (1.9.26), show that

$$\det(\lambda I - L) = \lambda^\omega - \sum_{i=1}^\omega f_i \lambda^{\omega-i},$$

where $f_i = l_i m_i$, $l_i = s_i s_{i-1} \ldots s_1$.

1.23. Consider the population process described by equation (1.9.25), where L is the Leslie matrix (1.9.26). Show that the stable age structure is given by

$$\mathbf{v}_1 = (\lambda_1^{\omega-1} l_1, \lambda_1^{\omega-2} l_2, \cdots, \lambda_1 l_{\omega-1}, l_\omega)^T,$$

where $l_i = s_i s_{i-1} \ldots s_1$, λ_1 is the principal eigenvalue of L and ω is the oldest age-class.

1.10 Euler–Lotka Equations

Leonhard Euler, one of the greatest mathematicians of all time, published a paper on the mathematics of the age structure of human populations in 1760. His work was extended by Alfred Lotka, an influential figure in Mathematical Biology after the publication of his book in 1925, whose name we shall meet again when we look at interacting populations. Euler worked in discrete time, and we shall present this approach in discrete time first, and then look at the analogous equations in continuous time. Although Euler considered males only, we shall keep track of females only.

1.10.1 Discrete Time

Let $u_{i,n}$ be the number of females in age class i at time n, and let $\mathbf{u}_n = (u_{0,n}, u_{1,n}, \cdots, u_{\omega,n})^T$ satisfy the equation

$$\mathbf{u}_{n+1} = L\mathbf{u}_n, \tag{1.10.27}$$

where L is the Leslie matrix (1.9.26). There are two questions we could ask.

– What is the general solution of Equation (1.10.27), for arbitrary initial conditions?

– What is the solution $\mathbf{u}_n$ satisfying given initial conditions?

We shall consider the second question, so that $\mathbf{u}_0$ is given.

Then, using Equation (1.10.27) with L given by Equation (1.9.26), we have

$$u_{i,n} = \begin{cases} u_{i-n,0} s_i s_{i-1} \cdots s_{i-n+1} = u_{i-n,0} l_i / l_{i-n} & \text{for } i > n, \\ u_{0,n-i} s_i s_{i-1} \cdots s_1 = b_{n-i} l_i & \text{for } i \leq n, \end{cases} \quad (1.10.28)$$

where we have defined the *survival function* $l_i = s_i s_{i-1} \cdots s_1$ to be the chance of surviving from birth to age i, and $b_n = u_{0,n}$ to be the number of births at time n. The expression for $i > n$ represents those who were initially of age $i - n$ and who then survived from age $i - n$ to age i in time n, and the expression for $i < n$ represents those who were born at time $n - i$ and then survived to reach age i at time n.

This looks like the solution that we wanted for $\mathbf{u}_n$, but of course the birth rate b is unknown, and depends on $\mathbf{u}$, so we need to find an equation for b. Multiplying Equation (1.10.28) by m_i and summing, we have

$$b_n = \sum_{i=1}^{\infty} m_i u_{i,n} = \sum_{i=1}^{n} l_i m_i b_{n-i} + \sum_{i=n+1}^{\infty} u_{i-n,0} \frac{l_i}{l_{i-n}} m_i = \sum_{i=1}^{n} f_i b_{n-i} + g_n, \tag{1.10.29}$$

where $f_i = l_i m_i$ is the *net maternity function*, the number of offspring produced at time n by each female born at time $n - i$, i seasons earlier. The function g_n is determined by the initial conditions. The equation is known as the *Euler renewal equation*. It may be used to determine b_n (and hence by Equation (1.10.28) $u_{i,n}$) uniquely, but we shall only use it here to answer the following question.

– How does b_n (and hence $\mathbf{u}_n$) behave asymptotically for large n?

The series defining g_n will always be finite, assuming there is a maximum age class ω, and will be zero for $n \geq \omega$. Now let us take $n \geq \omega$. The equation becomes

$$b_n = \sum_{i=1}^{\omega} f_i b_{n-i}. \tag{1.10.30}$$

This is also the equation that we would have obtained if we had looked for the general solution of Equation (1.10.27), so we are answering this question as well. It is a linear equation with constant coefficients for b, and we look for a solution $b_n = b_0 \lambda^n$. This satisfies Equation (1.10.30) as long as

$$\lambda^{\omega} = f_1 \lambda^{\omega-1} + f_2 \lambda^{\omega-2} + \cdots + f_{\omega}, \tag{1.10.31}$$

an ωth order equation for λ. It has ω roots λ_i, and the general solution of Equation (1.10.30) is given by

$$b_n = A_1 \lambda_1^n + A_2 \lambda_2^n + \cdots + A_{\omega} \lambda_{\omega}^n, \tag{1.10.32}$$

assuming all the roots are distinct. The general solution of Equation (1.10.27) for $n \geq \omega$ is $u_{i,n} = b_{n-i}l_i$.

It may be shown that λ multiplied by Equation (1.10.31) is the equation $\det(L - \lambda I) = 0$, so that its roots and zero are the eigenvalues of L. Let us assume that L is *primitive*, i.e. there are at least two non-zero coefficients f_i and f_j of Equation (1.10.31) with i and j co-prime. Then the Perron–Frobenius theorem and the particular form of L tell us that L has a real positive principal eigenvalue λ_1, and all other eigenvalues satisfy $|\lambda| < \lambda_1$. Then, as $n \to \infty$, $b_n \approx A_1\lambda_1^n$, and we again have geometric growth (or decay). Moreover, the population tends to a stable age distribution.

Equation (1.10.31) is often written in an alternative form by putting $\lambda = e^r$, to obtain

$$1 = \sum_{i=1}^{\omega} l_i m_i e^{-ri} = \sum_{i=1}^{\omega} f_i e^{-ri}, \qquad (1.10.33)$$

the *Euler–Lotka equation*. The real root $r_1 = \log \lambda_1$ of this equation is known as *Lotka's intrinsic rate of natural increase*. The equation has to be solved numerically, using e.g. Newton's method.

The number of female offspring that a given female produces in her lifetime, the basic reproductive ratio R_0, is given by

$$R_0 = \sum_{i=1}^{\omega} l_i m_i = \sum_{i=1}^{\omega} f_i. \qquad (1.10.34)$$

We would expect the population to grow if and only if $R_0 > 1$, and this intuition is correct. But it is r_1 or λ_1 that tells us *how fast* the population is growing.

EXERCISES

1.24. a) Show that the discrete Euler–Lotka function $\hat{f}(r) = \sum_{i=1}^{\omega} f_i e^{-ri}$ is a monotonic decreasing function of r, satisfying $\hat{f}(r) \to \infty$ as $r \to -\infty$, $\hat{f}(r) \to 0$ as $r \to \infty$.

   b) Deduce that $R_0 > 1$ if and only if $r_1 > 0$, or $\lambda_1 > 1$.

   c) Show that $r_1 \approx (1 - \hat{f}(0))/\hat{f}'(0)$ if r_1 is small.

1.25. A life table for the vole *Microtus agrestis* reared in the laboratory is given on the right. Each age class is eight weeks long.

i	l_i	m_i
1	1.0000	0.0000
2	0.8335	0.6504
3	0.7313	2.3939
4	0.5881	2.9727
5	0.4334	2.4662
6	0.2928	1.7043
7	0.1813	1.0815
8	0.1029	0.6683
9	0.0535	0.4286
10	0.0255	0.3000

a) Calculate R_0.

b) Calculate an approximation to Lotka's intrinsic rate of natural increase r_1.

c) How would you improve this approximation?

1.26. *Generating function method for discrete renewal equations* . Let the sequences b_n, f_n and g_n satisfy the Euler renewal equation (1.10.29). Define generating functions by

$$b(s) = \sum_{i=1}^{\infty} b_i s^i, \quad f(s) = \sum_{i=1}^{\infty} f_i s^i, \quad g(s) = \sum_{i=1}^{\infty} g_i s^i.$$

a) Show that $b(s) = g(s)/(1 - f(s))$.

b) Show that the roots of $1 - f(s) = 0$ are the reciprocals of the roots of the Euler eigenvalue equation (1.10.31).

c) If these roots are distinct, show that

$$1 - f(s) = \prod_{i=i}^{\omega}(1 - \lambda_i s),$$

where the λ_i are the non-zero eigenvalues of L.

d) Use partial fractions to deduce that

$$b_n = \sum_{i=1}^{\omega} \frac{g(s_i)}{f'(s_i)} \lambda_i^{n+1},$$

where $s_i = \frac{1}{\lambda_i}$. This determines the constants A_i in (1.10.32), and so gives the solution of the initial value problem (1.10.27).

1.27. Let r_1 be the real root of the Euler–Lotka Equation (1.10.33), and let R_0 be given by Equation (1.10.34).

a) Let $\bar{a} = (\sum_{i=1}^{\omega} i f_i)/(\sum_{i=1}^{\omega} f_i)$ be the average age of giving birth. If r_1 is small, show that

$$r_1 \approx \frac{R_0 - 1}{\bar{a} R_0}.$$

b) The average age of giving birth $\bar{a}$ is sometimes used as a definition of generation time. An alternative definition is $g = \frac{\log R_0}{r_1}$.
   Show that $g \approx \bar{a}$ if r_1 is small.

1.10.2 Continuous Time

The approach in continuous time is analogous. Let the number of females in
the age interval $(a, a + \delta a)$ at time t be given by $u(a, t)\delta a + O(\delta a^2)$. Let $l(a)$ be
the *survival function*, the probability that an individual survives to age a. Let
the probability that an individual who survives to age a gives birth in the age
interval $(a, a + \delta a)$ be $m(a)\delta a + O(\delta a^2)$; $m(a)$ is called the *maternity function*.
Let $f(a) = l(a)m(a)$, called the *net maternity function*, be the probability
density that an individual both survives to age a and gives birth then. We shall
derive an equation for $u(a, t)$, the *age density* of the female population, if the
survival and maternity functions are known. We shall again consider the initial
value problem; let the initial age density of the female population be given by
$u(a, 0) = u_0(a)$. Let $b(t)$ be the female birth rate at time t. Then we have,
analogously to (1.10.28),

$$u(a, t) = \begin{cases} u_0(a - t)l(a)/l(a - t) & \text{for } a > t, \\ b(t - a)l(a) & \text{for } a < t. \end{cases} \qquad (1.10.35)$$

The expression for $a > t$ represents those who were initially of age $a - t$ and
who then survived from age $a - t$ to age a in time t, and the expression for
$a < t$ represents those who were born at time $t - a$ and then survived to reach
age a at time t.

Again the birth rate b is unknown, and depends on u, so we need to find an
equation for it. Multiplying Equation (1.10.35) by $m(a)$ and integrating,

$$\begin{aligned} b(t) &= \int_0^\infty u(a, t)m(a)da \\ &= \int_0^t b(t - a)l(a)m(a)da + \int_t^\infty u_0(a - t)\frac{l(a)}{l(a - t)}m(a)da \\ &= \int_0^t b(t - a)f(a)da + g(t), \end{aligned} \qquad (1.10.36)$$

say, where $g(t)$ is a known function. This is an integral equation for b, called the
continuous Euler renewal equation. It may be used to determine b explicitly,
but as for the discrete case we shall use it only to answer the following question.

– How does $b(t)$ behave asymptotically for large t?

From this we can find the asymptotic behaviour of u by $u(a,t) = b(t-a)l(a)$. Let us assume that $m(a) = 0$ unless $a \in (\alpha, \beta)$. For $t > \beta$, $g(t) = 0$, we can replace the top limit in $\int_0^t b(t-a)l(a)m(a)da$ by ∞, and we obtain the homogeneous equation

$$b(t) = \int_0^\infty b(t-a)l(a)m(a)da = \int_0^\infty b(t-a)f(a)da. \qquad (1.10.37)$$

It is important to notice that the equation is *linear* in b, as its discrete analogue was. Guided by this, we try $b(t) = b_0 \exp(rt)$. This satisfies (1.10.36) as long as

$$1 = \int_0^\infty e^{-ra}l(a)m(a)da = \int_0^\infty e^{-ra}f(a)da, \qquad (1.10.38)$$

the *continuous Euler–Lotka* or *characteristic* equation for the growth rate r.

It may be shown that this equation has a unique real root r_1. As in the discrete case, r_1 is known as Lotka's intrinsic rate of natural increase. Often, when we are looking at (1.10.38), we shall omit the subscript and assume that the unique real root is meant. The population almost always settles down to grow exponentially at this rate, so $b(t) \to Be^{rt}$, and the age structure settles down to a stable one, $u(a,t) \to Ae^{rt}v(a)$, where

$$v(a) = \frac{\exp(-ra)l(a)}{\int_0^\infty \exp(-ra)l(a)da}.$$

Here, the *basic reproductive ratio* R_0 is given by

$$R_0 = \int_0^\infty l(a)m(a)da = \int_0^\infty f(a)da,$$

the integral of the net maternity function. Again, it is intuitively obvious that the population will grow if and only if $R_0 > 1$, but again it is r_1 that gives us the *rate* of increase.

The stationarity assumption, that l and m depend only on age a, not on time t, is unlikely to be valid for current human populations given the recent changes in birth rates and mortality rates. If this is not true we can still derive a renewal equation for $b(t)$, but it no longer has an exponential solution.

EXERCISES

1.28. Let the net maternity function f on $[0, \omega]$ be continuous, non-negative (at each point) and not identically zero. Show that the Euler–Lotka function

$$\tilde{f}(r) = \int_0^\infty \exp(-ra)f(a)da, \qquad (1.10.39)$$

(which is the Laplace transform of the net maternity function), is monotonic decreasing and satisfies $\tilde{f}(r) \to \infty$ as $r \to -\infty$, $\tilde{f}(r) \to 0$ as $r \to \infty$. Deduce that the Euler–Lotka Equation (1.10.38) has a unique real root r_1.

1.29. Show that the Euler–Lotka function $\tilde{f}$ given by (1.10.39) crosses the vertical axis at R_0, and deduce that $R_0 > 1$ if and only if $r_1 > 0$.

1.30. *Laplace transform method for continuous renewal equations, for readers with a knowledge of Laplace transforms only.* Define the Laplace transform $\tilde{h}(s)$ of a function $h(t)$ by

$$\tilde{h}(s) = \int_0^\infty h(t)e^{-st}dt.$$

The inverse Laplace transform of $\tilde{h}(s)$ is defined by

$$h(t) = \frac{1}{2\pi i}\int_{c-i\infty}^{c+i\infty} e^{st}\tilde{h}(s)ds,$$

where the contour of integration is to the right of all the roots of $\tilde{h}(s)$. Let the functions b, f and g satisfy the Euler renewal equation (1.10.36).

a) Show that $\tilde{b}(s) = \tilde{g}(s)/(1 - \tilde{f}(s))$.

b) The characteristic (Euler–Lotka) equation is given by

$$1 = \int_0^\infty f(a)e^{-sa}da = \tilde{f}(s),$$

and $\tilde{f}$ is the Euler–Lotka function. Show that the characteristic equation has one real root s_1, and that all the other roots have real part less than s_1.

c) If the characteristic equation only has simple roots, denoted by s_i, show by using Cauchy's integral formula that

$$b(t) = \sum_{i=0}^\infty \frac{g(s_i)}{f'(s_i)}\exp(s_i t).$$

1.11 The McKendrick Approach to Age Structure

This approach was discussed by Lieut.-Col. McKendrick (sometimes written M'Kendrick), whose name we shall meet again in Chapter 3, in a wide-ranging paper published in 1926, and rediscovered by von Foerster in 1959. It is most easily visualised using a Lexis diagram, as shown below. We shall assume as usual that the process is stationary, mortality and maternity depending on age but not explicitly on time.

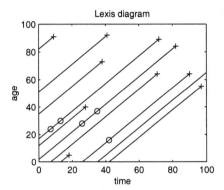

Figure 1.12   A Lexis diagram for a population with age structure. Each line represents a life, either of an individual already alive at $t = 0$ or of an individual born after $t = 0$. A circle represents giving birth, a cross death. A cohort lifetable is calculated by following all those born between t_1 and t_2, and recording the ages at which they give birth and die, while a static lifetable is calculated by recording all such events that occur between t_1 and t_2. They are the same (on average) for stationary processes.

Let $u(a, t)$ be the density of females of age a at time t. Then a time δt later, all individuals who are still alive will have aged by an amount $\delta a = \delta t$,

$$u(a + \delta a, t + \delta t) = u(a, t) - d(a)u(a, t)\delta t + O(\delta t^2),$$

where d is the mortality function, $d(a)\delta t + O(\delta t^2)$ giving the probability of an individual of age a dying in the next small interval of time δt. Then, expanding by Taylor series, and using $\delta a = \delta t$,

$$\frac{\partial u}{\partial a} + \frac{\partial u}{\partial t} = -du. \tag{1.11.40}$$

This is McKendrick's partial differential equation. It is to be solved for $0 \leq a \leq \omega$, $t \geq 0$, where ω is the greatest possible age in the population, with initial conditions given at $t = 0$,

$$u(a, 0) = u_0(a). \tag{1.11.41}$$

Births enter as a boundary condition at age zero,

$$b(t) = u(0, t) = \int_\alpha^\beta u(a, t) m(a) da,$$

where m is the maternity function, and α and β are the youngest and oldest ages for child-bearing.

Example 1.5

Assume temporarily that $u(0, t) = b(t)$ is known. Solve McKendrick's equation (1.11.40) with this condition and the initial condition (1.11.41), to obtain (1.10.35), with $l(a) = \exp(- \int_0^a d(b) db)$. (You may wish to refer to Section C.1 of the appendix.) McKendrick's formulation may therefore be reduced to a continuous Euler renewal equation, as in Section 1.10.2.

Let the characteristic curves of Equation (1.11.40) be given by $a = \hat{a}(s)$, $t = \hat{t}(s)$. On these curves, we define $\hat{u}(s) = u(\hat{a}(s), \hat{t}(s))$, and (by the definition of a characteristic curve) the partial differential equation reduces to an ordinary differential equation. We have

$$\frac{d\hat{u}}{ds} = \frac{\partial u}{\partial a} \frac{d\hat{a}}{ds} + \frac{\partial u}{\partial t} \frac{d\hat{t}}{ds} = -d\hat{u}.$$

Comparing this with Equation (1.11.40), we have

$$\frac{d\hat{a}}{ds} = 1, \quad \frac{d\hat{t}}{ds} = 1, \quad \frac{d\hat{u}}{ds} = -d\hat{u}.$$

The characteristic curves are straight lines of unit slope in the positive quadrant of the (a, t)-plane, and the initial $s = 0$ conditions that we need to apply depend on whether they intersect the a-axis or the t-axis, as in the Lexis diagram of Figure 1.12. For $t > a$, they intersect the t-axis when $s = 0$. If we parametrise the t-axis by $t = \tau$, $a = 0$, then the conditions for $\hat{t}$, $\hat{a}$ and $\hat{u}$ at $s = 0$ are $\hat{t}(0) = \tau$, $\hat{a}(0) = 0$, and $\hat{u}(0) = b(\tau)$. Integrating the equations for $\hat{a}$, $\hat{t}$, and $\hat{u}$ with these initial conditions, we have $a = \hat{a}(s) = s$, $t = \hat{t}(s) = s + \tau$, $\hat{u}(s) = b(\tau) \exp(- \int_0^s d(s') ds') = b(\tau) l(s)$, which defines $l(a) = \exp(- \int_0^a d(b) db)$. In terms of the original variables a and t, $u(a, t) = b(t - a) l(a)$. A similar argument gives $u(a, t) = u_0(a - t) l(a)/l(a - t)$ for $a > t$, and the result follows.

The survivorship function is therefore given by

$$l(a) = \exp\left(- \int_0^a d(b) db \right)$$

(see also Exercise 1.32). Survivorship functions have been investigated empirically for a large number of species. They are usually described in terms of three archetypal curves, sketched in Figure 1.13.

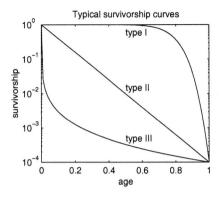

Figure 1.13 Typical survivorship curves, with survivorship on a logarithmic scale. In the figure, unit age is defined arbitrarily to be the age at which 10^{-4} of the original cohort remain alive, but a sensible definition will depend on the organism.

Type I is typical for human populations, where individuals tend to survive until a maximum age and then die (although infant mortality distorts the true curve from the ideal), type II is typical for small birds (although again true curves are distorted down from the ideal near age zero), and type III is typical for organisms such as fish that lay millions of eggs, the vast proportion of which never make it to adulthood. The easiest to deal with mathematically is type II survivorship, where the mortality $d(a)$ is a constant d independent of a. Better approximations for humans are (i) to assume that everyone survives to age ω and then dies, or (ii) to assume so-called Gompertz mortality, $d(a) = d_0 \exp(a/a_0)$, where d_0 and a_0 are positive constants.

EXERCISES

1.31. Consider a female producing offspring in a population growing steadily at rate r. To maximise her contribution to the population as a whole, she would like not only to produce many children but to produce them early, when they will make up a greater fraction of the population of their age. A birth delayed by time τ will be worth $e^{-r\tau}$ times as much as a birth not subject to such a delay. The *reproductive value* of an individual of age a relative to the reproductive value of a new-born individual is defined by

$$v(a) = \frac{\int_a^\infty \exp(-rb)l(b)m(b)db}{\exp(-ra)l(a)}.$$

a) Interpret this definition biologically.

b) What qualitative characteristics would you expect v to have as a function of a?

c) The deaths from malaria and heart disease and reproductive values in the Philippines in 1959 and 1960 were as follows.

Age	Malaria	Heart disease	Reproductive value
All ages	913	918	
$0-4$	251	12	1.21
$5-14$	156	7	1.64
$15-24$	133	37	2.00
$25-44$	186	198	0.76
$45-64$	138	322	0
65+	45	333	0
Unknown	4	9	

Calculate the total reproductive value of the deaths from each cause, and hence quantify (as far as possible) the different effects of each cause on subsequent population growth.

1.32. a) Explain why the survival function $l(a)$ must satisfy McKendrick's equation (1.11.40).

b) Deduce that $l(a) = \exp(-\int_0^a d(b)db)$.

1.33. Consider a population growing at Lotka's intrinsic rate of increase r.

a) Show that the *life expectancy* of a cohort, defined to be the mean age at death of individuals born at the same time, is given by $\int_0^\infty ad(a)l(a)da$.

b) Show that the mean age of those dying simultaneously is given by

$$\frac{\int_0^\infty ad(a)e^{-ra}l(a)da}{\int_0^\infty d(a)e^{-ra}l(a)da}.$$

c) Show that these two quantities are different in general.

1.12 Conclusions

- Linear models are often explicitly soluble by elementary techniques. In particular, geometric or exponential solutions are the norm.

- Linear models cannot be applied to a growing population indefinitely, as solutions tend to infinity.

- Whether a model population grows or decays is determined by the eigenvalues of the problem. For a difference equation formulation, stability occurs when all eigenvalues have modulus less than unity, whereas for a differential equation formulation it occurs when all eigenvalues have real part less than zero.

- An important parameter in difference equation models for populations without age structure is R_0, the basic reproductive ratio. In an unchecked annual population this is identical to λ, the Malthusian parameter or geometric growth ratio.

- High values of R_0 tend to destabilise the steady state of a nonlinear equation, leading to periodic solutions or chaos.

- Intraspecific competition may increase mortality or decrease fertility, or both. If it increases mortality, an extra individual at time n may lead to exactly one, more than one, or less than one extra death at time $n+1$, known as exact, over- and under-compensation respectively. These may be discriminated by the asymptotic behaviour $F(N) \sim c/N^b$ as $N \to \infty$, with $b = 1$, $b > 1$ and $b < 1$ respectively.

- Over-compensation tends to destabilise the steady state of a nonlinear equation.

- It is still controversial whether chaotic systems exist in population biology, although there is good laboratory data that suggest that they do.

- The logistic is a model of limited population growth that gives an S-shaped growth curve, in line with many experimental results. Other such models exist, but the logistic is often used because of its simplicity.

- Evolutionary arguments that neglect inhomogeneity and unpredictability lead to the conclusion that all organisms should be K-selected, but there are many r-selected organisms in nature.

- Mathematical models can be used to give resource management policies, but care has to be taken that they are robust to the particular model chosen.

- In a metapopulation model, populations survive through a combination of local extinctions and re-colonisations.

- The crucial parameter in metapopulations is the basic reproductive ratio R_0, the number of sites an occupied site can expect to colonise before going extinct when the species is rare.

- Habitat destruction has parallels to vaccination. If a fraction of sites equal to the original fraction occupied is destroyed at random, then the population is predicted to go extinct.

- Delay can destabilise an otherwise stable steady state. Instability via growing oscillations is typical.

- Age structure can be dealt with by a variety of approaches. Importantly, though, if the model is linear we still get geometric or exponential growth or decay, and the age structure settles down to a stable limit, for almost all initial conditions.

- The Leslie matrix equation $\mathbf{z}_{n+1} = L\mathbf{z}_n$ gives the numbers in each age class in a population at time $n+1$ in terms of the numbers at time n. The Leslie matrix has non-negative components. Its eigenvalue of greatest modulus is real and non-negative, and has a real non-negative (left) eigenvector. Both the eigenvalue and the eigenvector are usually positive (see Appendix D).

- Important parameters for linear difference equations for age-structured populations are r_1, Lotka's intrinsic rate of natural increase, $\lambda_1 = e^{r_1}$, the asymptotic growth ratio, and R_0, the basic reproductive ratio. Conditions for the population to grow are $R_0 > 1$, $\lambda_1 > 1$ or $r_1 > 0$, but it is λ_1 or r_1 that tells us how fast the population grows.

- Important parameters for linear differential equations for age-structured populations are r_1, also called Lotka's intrinsic rate of natural increase, and R_0, the basic reproductive ratio. Growth occurs for $r_1 > 0$, or $R_0 > 1$, but it is r_1 that tells us how fast the growth is.

- The McKendrick approach to age structure essentially involves following a cohort as it ages. It leads to a partial differential equation that may be reduced to a renewal equation by the method of characteristics.

2
Population Dynamics of Interacting Species

- Prey-predator or host-parasitoid interactions have an oscillatory tendency. However the oscillations are often damped out in nature and in more realistic mathematical models. There are many possible stabilising mechanisms.

- Competition, especially for few resources, tends to lead to extinctions. There is a competitive advantage to having a high carrying capacity K and a high rate of consumption of resources relative to competitors. However, extinctions do not always occur where they might be expected; there are many mechanisms that may lead to coexistence.

- Ecosystems modelling includes explicit consideration of mass or energy flows through an ecological system. It is useful when abiotic resources or energy are limiting factors.

- The metapopulation approach is a relatively simple way of including spatial effects in ecological models. Communities persist through a balance between colonisation and local extinction, and the advantages of good dispersal characteristics as a competitive strategy become clear.

2.1 Introduction

In this chapter we shall consider how interactions between small numbers of species affect the population dynamics of those species. As in Chapter 1, we shall choose as variables the sizes of the populations in some sense, e.g. the numbers of individuals in each species, the biomass of the populations, the mass of an essential element incorporated into the population, or the fraction of the total habitat that the population occupies. We shall neglect the age or other structure of the populations.

The terms in the equations that represent interactions are sometimes best obtained by modelling the individual behaviour that leads to the population-level terms. This consideration of individual behaviour is often difficult, but is essential if we are to understand fully the link between behaviour and ecology. In most cases we shall try to give some idea of the individual-based modelling that leads to the equations at the population level.

It is convenient to classify the direct interaction between a pair of species into three categories, as follows.

– Competition, $(-,-)$. Each species has an inhibitory effect on the other.

– Commensalism, or mutualism, $(+,+)$. Each species has a beneficial effect on the other.

– Predation or parasitism, $(+,-)$. One species, the predator or parasite, has an inhibitory effect on the other; the other species, the prey or host, has a beneficial effect on the first.

We shall concentrate on predation (or parasitism) and competition.

2.2 Host-parasitoid Interactions

Parasitoids are creatures, usually wasps or flies, that have at least one free-living and one parasitic stage in their life cycle. The adult usually lays eggs in the larval or pupal stage of its host, e.g. in the caterpillar of a butterfly or moth, which hatch and develop, consuming the host from the inside and eventually killing it before they pupate. These creatures are not uncommon; about 10% of all metazoan (multi-cellular) species on Earth are insect parasitoids. Their use as agents of bio-control of insect pests is increasing. Many of them follow an annual cycle, and we shall therefore assume that they have discrete non-overlapping generations. We shall derive a general class of model for such interactions. We shall then look at the earliest and simplest of these models, derived by Nicholson (a biologist whose later blowfly experiments were mentioned in Chapter 1) and Bailey (a physicist) in 1935. The predictions of this model are not confirmed by observations. Such a situation is often where we learn most from a mathematical model. It focuses attention on assumptions which are suspect, and we consider what modifications might be needed to improve the model.

Define

- H_n to be the number of hosts at generation n;

- P_n to be the number of parasitoids at generation n;

- R_0 to be the host basic reproductive ratio, i.e. the per capita production of the host in the absence of parasitism;

- c to be the average number of eggs laid by an adult parasitoid in a single host that will survive to breed in the next generation.

- $f(H, P)$ is the fraction of hosts *not* parasitised. (This negative definition is standard in the literature.)

The census takes place at the beginning of the season, before parasitism takes place, counting adult parasitoids and the stage of the host that is subject to parasitisation. After parasitism, the number of unparasitised hosts is $Hf(H, P)$, and the number of parasitised hosts is $H(1 - f(H, P))$. Thus

$$H_{n+1} = R_0 H_n f(H_n, P_n), \quad P_{n+1} = cH_n(1 - f(H_n, P_n)) \qquad (2.2.1)$$

We have tacitly assumed that there are no density-dependent effects in the dynamics of the host population alone, e.g. no intra-specific competition, so that in the absence of parasitism the host population would grow exponentially (if $R_0 > 1$, which we shall assume to be true). The model can only be realistic for hosts whose numbers are limited by the parasitoids.

Let us assume that there is a steady state at (H^*, P^*), and denote evaluation at this steady state by an asterisk. Note that $R_0 f^* = 1$, from the first of Equations (2.2.1). The stability of this steady state is determined by the eigenvalues of the linearisation of the system about it, as explained in Section A.4.1 of the appendix. The Jacobian matrix here is given by

$$J^* = \begin{pmatrix} R_0(f^* + H^* f_H^*) & R_0 H^* f_P^* \\ c(1 - f^* - H^* f_H^*) & -cH^* f_P^* \end{pmatrix}, \qquad (2.2.2)$$

where subscripts denote partial differentiation. The trace and determinant of J^* are given by

$$\beta = \operatorname{tr} J^* = R_0 f^* + R_0 H^* f_H^* - cH^* f_P^*,$$

$$\gamma = \det J^* = -R_0 H^* c f_P^*.$$

The steady state is stable if the Jury conditions hold (see Section A.3 of the appendix),

$$|\beta| < \gamma + 1, \quad \gamma < 1. \tag{2.2.3}$$

In the first mathematical model of this class, Nicholson and Bailey modelled the function f by assuming that the parasitoid search is a Poisson process with parameter a, known as the *searching efficiency*. From an individual-based point of view, the reason for this can be modelled as follows. Let searching start at time n, at the beginning of the season, and finish a time τ later. During this search let encounters take place at a rate proportional to both the numbers of unparasitised hosts and the numbers of parasitoids. Then

$$\frac{dH}{dt} = -\alpha P H \tag{2.2.4}$$

for $n < t < n + \tau$. The numbers of parasitoids is constant, $P = P_n$, throughout the search interval. At the beginning of the search, $H(n) = H_n$, and at the end, by the definition of f, $H(n + \tau) = H_n f(H_n, P_n)$. But, solving the differential equation, $H(n + \tau) = H_n \exp(-\alpha P_n \tau) = H_n \exp(-a P_n)$, where $a = \alpha\tau$, so $f(H_n, P_n) = \exp(-a P_n)$. The equations become

$$H_{n+1} = R_0 H_n \exp(-a P_n), \quad P_{n+1} = c H_n (1 - \exp(-a P_n)). \tag{2.2.5}$$

Example 2.1

Find the coexistence steady states of Equations (2.2.5), and determine their stability.

There is only one coexistence steady state (H^*, P^*) given by

$$P^* = \frac{1}{a} \log R_0, \quad H^* = \frac{P^* R_0}{c(R_0 - 1)},$$

biologically realistic as long as $R_0 > 1$. The determinant of the Jacobian of the transformation at the steady state is given by

$$\gamma = \det J(H^*, P^*) = \frac{R_0 \log R_0}{R_0 - 1} > 1$$

for $R_0 > 1$ (see Figure 2.1), violating one of the Jury conditions (2.2.3) for stability. Violation of this condition leads to growing oscillations on perturbation from the steady state.

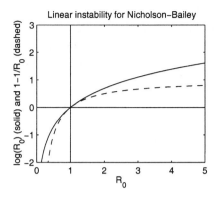

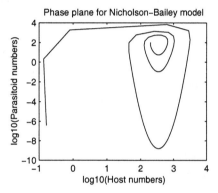

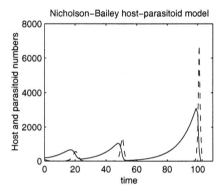

Figure 2.1 The functions $\log R_0$ (solid line) and $1 - 1/R_0$ (dashed line). It can be seen that the ratio $\gamma > 1$ for any value of $R_0 > 1$, so that the steady state is unstable to growing oscillations.

Figure 2.2 Some numerical solutions of the Nicholson–Bailey host-parasitoid model. The numerical scheme fails soon after $t = 100$ as the parasitoid numbers fall to ridiculously low levels. The data is presented in the (H, P)-plane or *phase plane*. The trajectory in the phase plane spirals outwards, leading to higher peaks and lower troughs with each spiral, until effective extinction occurs.

Some numerical simulations of the model are shown in Figure 2.2. The Nicholson–Bailey model predicts that host-parasitoid systems will exhibit oscillations whose amplitude grows without limit. Of course this can never happen in real life. Some such systems do exhibit oscillatory behaviour, but the majority have stable steady states. What unrealistic assumptions have been made in the modelling, and can modifications be made that result in better predictions?

– We may remove the assumption that the host is limited entirely by the predator, by replacing the $R_0 H_n$ in the H-equation by a nonlinear and more realistic function, such as one of those discussed in Chapter 1.

– The *functional response* is defined to be the per capita rate at which parasitoids parasitise hosts during the search period. It is modelled by αH in Equation (2.2.4), linear in the number of hosts, but this is unrealistic at least for large H. More realistic functional responses will be discussed in Section 2.3.

– When a parasitoid comes into contact with another while searching, it may move away or temporarily stop searching. The time available for searching therefore decreases with increasing parasitoid density. This is known as *interference* between parasitoids.

– The parasitoids may not search at random, but tend to search more in places where they have already found a host.

– Some hosts may be easier to find than others. Such heterogeneity of the environment can lead to increased stability.

Modifications of the model have been made to accommodate these points, and can all lead to a stabilisation of the steady state. Which of them is important for any given host-parasitoid system depends on the detailed biology of the interaction, but the one that has received most attention from theorists is that the parasitoids search according to a distribution which is more clumped than a Poisson distribution. This may be represented phenomenologically by the negative binomial distribution

$$f(H, P) = \left(1 + \frac{bP}{k}\right)^{-k},$$

where k is the clumping parameter. As $k \to \infty$, this tends to the Poisson distribution. A comparison of the negative binomial with the Poisson distribution for various values of k is sketched in Figure 2.3.

It may be shown that the steady state is stable if k is a constant satisfying $k < 1$. The graph shows a comparison of observed and predicted densities using Equations (2.2.1) with this f of winter moth larvae before and after biological control with a certain parasitoid. Whether this graph supports the use of the negative binomial distribution is difficult to tell with the limited data available.

EXERCISES

2.1. Modify the Nicholson–Bailey Equations (2.2.5) if the host population satisfies $H_{n+1} = R_0 H_n (1 + aH_n)^{-b}$, Equation (1.2.4), in the absence of parasitism. Will this entail any modification of the P equation? Does it alter the long-term behaviour of the solutions?

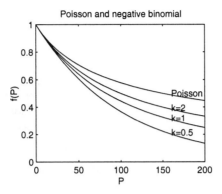

Figure 2.3   Comparison of negative binomial and Poisson distributions.

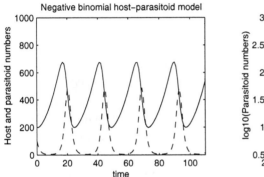 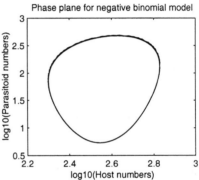

Figure 2.4   Numerical results of simulating Equations (2.2.1) with a negative binomial search function, $k = 1$. For $k < 1$ the steady state is asymptotically stable, and for $k > 1$ the solutions spiral to infinity, as they do with the Poisson distribution (Figure 2.2). Here they reach a closed invariant curve in (H, P)-space. Other parameters are $R_0 = 1.1$, $b = 0.001$, $c = 3$.

2.2. One model proposed for interference between parasitoids is to take f in Equations (2.2.1) to be given by $f(H, P) = \exp(-a_0 P^{1-m})$, where $0 < m < 1$. Why does this have the right character for a model for interference? What difference does it make to solutions of the equations? (This needs to be done numerically.) Measured values of m range from 0.4 to 0.8, so the effect may be important.

2.3. Heterogeneity of the environment (refuges). Let us assume that a certain number $\hat{H}$ of hosts may be accommodated in safe refuges. Modify Equations (2.2.5) to take this into account. Does this stabilise the steady state?

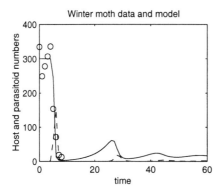

Figure 2.5   Results of simulating Equations (2.2.1) with a negative binomial distribution for winter moth larvae (solid line) and their parasitoids (dashed line). Observations for winter moth larvae are shown as circles.

2.3 The Lotka–Volterra Prey-predator Equations

In 1925 the mathematician Vito Volterra, whose interests were in differential and integral equations, had a conversation with his future son-in-law, the marine biologist Umberto d'Ancona. D'Ancona wondered why, when the fishing effort in the Mediterranean had decreased as a consequence of the First World War, the proportion of predatory fish caught had increased, and asked Volterra if there were any way of studying this mathematically. Volterra published a long paper on the subject, in Italian, in 1926, and a book, pursuing the question in detail, in French, in 1931. Alfred Lotka had proposed the same equations for a prey-predator system in his book, published in 1925. Lotka and Volterra are seen as the instigators of the golden age of theoretical ecology, when attempts were first made to find ecological laws of nature. These are not laws in the same sense as are Mendel's laws of heredity, to be discussed in Chapter 4, being much more qualitative in nature, and the models are consequently more strategic. In other words, they are more geared towards understanding general features of the system being investigated than they are to making quantitative predictions.

In a prey-predator interaction, let U be the numbers of prey and V the numbers of predators. Then, in words, {rate of change of U} = {net rate of growth of U without predation} – {rate of loss due of U to predation}, and {rate of change of V} = {net rate of growth of V due to predation} – {net rate of loss of V without prey}. This assumes implicitly, and unrealistically, that the prey-predator interactions are the only determinants of the population dynamics. Other assumptions made by Lotka and Volterra to model these terms were as follows.

– The prey is limited only by the predator, and in its absence would grow exponentially.

- The *functional response* of the predator is linear. In other words, the predation term is linear in U.

- There is no *interference* between predators in finding prey. In other words, encounters between predators do not take time or otherwise reduce the efficiency of the search for prey. Mathematically, the predation term is linear in V.

- In the absence of the prey, the predator dies off exponentially.

- Every prey death contributes identically to the growth of the predator population. This may be more realistic if we consider U and V as *biomass* rather than numbers.

The first three of these mirror the Nicholson–Bailey assumptions for hosts and parasitoids. The combination of the assumption of a linear functional response and that of no interference between predators leads to terms proportional to UV, analogous to terms in chemical kinetic equations when the chemicals react together according to the *law of mass action*. The phrase *law of mass action* is also used here to describe such terms.

Then, in equations,

$$\frac{dU}{d\tau} = \alpha U - \gamma UV, \quad \frac{dV}{d\tau} = e\gamma UV - \beta V. \tag{2.3.6}$$

The parameters α, β, γ and e are all positive.

In Volterra's application, where U and V are prey and predatory fish, these are the equations in the absence of fishing. Now let us assume that the catchability coefficients (Section 1.5) for prey and predators are p and q, and that fishing takes place with constant effort E. The equations become

$$\frac{dU}{d\tau} = \alpha U - pEU - \gamma UV, \quad \frac{dV}{d\tau} = e\gamma UV - qEV - \beta V. \tag{2.3.7}$$

Let us look for steady states of these equations. There is a trivial steady state at $(0,0)$, and a non-trivial steady state at (U^*, V^*), where

$$U^* = \frac{qE + \beta}{e\gamma}, \quad V^* = \frac{\alpha - pE}{\gamma}.$$

For $E < \alpha/p$, fishing *increases* the prey steady state and decreases the predator steady state. Let us assume for the moment that the system settles down to the non-trivial steady state (U^*, V^*). Then the proportion of predatory fish caught is given by

$$P = \frac{qEV^*}{pEU^*} = \frac{qe(\alpha - pE)}{p(qE + \beta)},$$

a decreasing function of the fishing effort E. If fishing effort decreases, as in the First World War, the fraction of predatory fish caught should increase, as it did.

But is our assumption that the system settles down to (U^*, V^*) justified? Let us go back and analyse Equations (2.3.7). The stability of the steady state is determined by the eigenvalues of the Jacobian matrix

$$J^* = \begin{pmatrix} 0 & -\gamma U^* \\ e\gamma V^* & 0 \end{pmatrix}. \tag{2.3.8}$$

This matrix has zero trace and positive determinant, so that the eigenvalues are purely imaginary. The equations linearised about the steady state have periodic solutions, but the nonlinear equations could have periodic solutions or solutions that spiral into or out from the steady state depending on the nonlinear terms. We are going to have to do some nonlinear analysis and it will pay to simplify the equations.

Define new dependent variables by $u = U/U^*$, $v = V/V^*$, and a new time variable by $t = (\alpha - pE)\tau$. Note that $(\alpha - pE)$ is the rate of increase of prey in the absence of predators. The equations become

$$\frac{du}{dt} = u(1 - v), \quad \frac{dv}{dt} = av(u - 1), \tag{2.3.9}$$

where $a = (qE + \beta)/(\alpha - pE)$. Now we may divide the second of Equations (2.3.9) by the first to obtain

$$\frac{dv}{du} = \frac{av(u - 1)}{u(1 - v)}. \tag{2.3.10}$$

This is an equation in the (u, v)-plane, also known as the *phase plane* from its original application in analysing electrical oscillations.

Example 2.2

Integrate the phase-plane Equation (2.3.10). Deduce that the *nonlinear* system (2.3.9) has periodic solutions.

Equation (2.3.10) is separable, $a(u - 1)du/u + (v - 1)dv/v = 0$, and has integral

$$\Phi(u, v) = a(u - \log u) + v - \log v = A,$$

where A is a constant of integration. These curves in the phase plane may be thought of as the contours of the bowl-shaped surface $w = \Phi(u, v)$ in (u, v, w)-space, and hence are closed curves, and so correspond to periodic solutions.

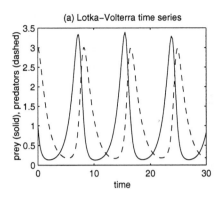
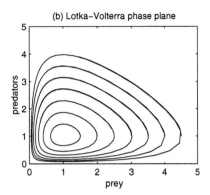

Figure 2.6   Some numerical solutions of the non-dimensional Lotka–Volterra prey-predator model Equations (2.3.9) and the corresponding phase plane Equation (2.3.10).

Some time-dependent solutions of the original system and the corresponding closed curves in the phase plane are shown in Figure 2.6.

Since the populations are periodic, we shall calculate the average population of each kind of fish, which is proportional to the catch, over a period T. Let us define these average populations by

$$\bar{u} := \frac{1}{T}\int_0^T u(t)dt, \quad \bar{v} := \frac{1}{T}\int_0^T v(t)dt.$$

Now, dividing the first equation of (2.3.9) by Tu and the second by Tv and integrating from 0 to T,

$$\frac{1}{T}[\log u(t)]_0^T = 1 - \bar{v}, \quad \frac{1}{T}[\log v(t)]_0^T = a(1 - \bar{u}).$$

But, by periodicity, the left-hand side of each of these equations is zero, so that $\bar{u} = 1$, $\bar{v} = 1$, or translating back into dimensional terms, $\bar{U} = U^*$, $\bar{V} = V^*$. The average population is the steady state population, and our previous conclusion still holds, and explains why the decrease in fishing effort in the Adriatic resulted in an increase in the proportion of predatory fish caught. More generally,

– *Volterra's principle* states that an intervention in a prey-predator system that kills or removes both prey and predators in proportion to their population sizes has the effect of increasing average prey populations.

As a result, caution must be exercised in attempting to achieve additional chemical control of a pest that is already biologically controlled. When the cottony cushion scale insect was accidentally introduced from Australia in 1868, it threatened to destroy the American citrus industry. Its natural predator, a ladybird beetle, was introduced, and reduced the pest population to a low level. When the insecticide DDT was discovered, it was applied in the hope of further control. In agreement with Volterra's principle, the result was an *increase* in the scale insect population.

Let us now focus on another prediction of the Lotka–Volterra model, that there will be oscillations in the abundance of each species. The reason for the cycle is intuitively clear, scarce prey leading to scarce predators, to abundant prey, to abundant predators, and back to scarce prey again. Oscillations are seen in some data sets, particularly those from boreal (Northern) regions where food webs are relatively simple, e.g. those of the lynx and snowshoe hare in Canada (see Figure 2.7).

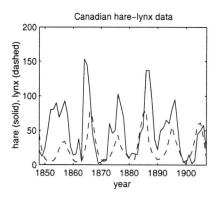 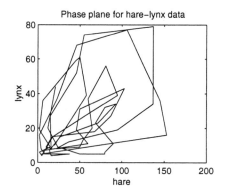

Figure 2.7   The famous hare-lynx data were obtained from the number of pelts sold to the Hudson Bay Company of Canada by hunters between 1848 and 1907. The numbers of pelts brought in are assumed to reflect abundance. An ecological time series over this length of time is very rare. There has been controversy over whether this data is periodic or chaotic, and whether it is caused by the interaction between prey and predator or by exogenous factors. The data are also plotted in the phase plane. They are far from the regular oscillations predicted by Lotka–Volterra, and indeed occasionally move clockwise rather than anti-clockwise, suggesting that it is the hares that are eating the lynx!

Can the Lotka–Volterra model explain these oscillations? Unfortunately, there is a fundamental problem here. No model can reflect nature exactly, and this one in particular has some unrealistic assumptions. We would like to be sure that if we made some small changes to the equations in order to patch up these problems, then the qualitative predictions would remain the same; in particular, oscillations would still occur. However, this is not so, as shown in Exercise 2.4 below; the system is *structurally unstable*, and the addition of another term to the equations, however small, can change the qualitative predictions of the model. This structural instability is connected to the fact that the phase plane Equation (2.3.10) is separable; there is in general no conserved quantity corresponding to $\Phi(u, v)$ after a small change in the equations. There is a family of oscillatory solutions, while we require an isolated oscillatory solution, known as a *limit cycle*, if the system is to be structurally stable. In the next section we shall try and modify the system to remove some of the unrealistic assumptions and obtain a structurally stable system that can explain oscillatory prey-predator behaviour.

EXERCISES

2.4. The Lotka–Volterra prey-predator equations with self-limitation of prey are given in non-dimensional form by

$$\frac{du}{dt} = u(1 - \epsilon u - v), \quad \frac{dv}{dt} = av(u - 1).$$

a) Show that the coexistence steady state (u^*, v^*) is a stable focus for ϵ small and positive.

b) Use the Dulac criterion (Section B.2.3 of the appendix) with $B(u, v) = (uv)^{-1}$ to show that there are no periodic solutions of the system in the positive quadrant.

c) Lyapunov functions (Section B.3.2 of the appendix) for Lotka–Volterra systems may often be found in the form $\Phi(\mathbf{u}) = \sum c_i(u_i - u_i^* \log u_i)$, where the c_i are positive constants to be determined. Show that the function Φ given by $\Phi(u, v) = a(u - \log u) + v - v^* \log v$ is a Lyapunov function for this system.

d) Sketch the phase plane for the system.

e) $\dot{\Phi}$ is *not* negative definite. Show that nevertheless the coexistence steady state is globally asymptotically stable for solutions with initial conditions in the positive quadrant.

2.5. Let a prey and predator satisfy the Lotka–Volterra equations, but with a refuge for the prey.

a) Explain the basis for the new model

$$\frac{du}{dt} = u - (u - k)v, \quad \frac{dv}{dt} = av(u - k - 1).$$

b) What effect does this modification have on the solutions of the system?

2.4 Modelling the Predator Functional Response

In this section we remove two of the unrealistic assumptions of the Lotka–Volterra prey-predator model (2.3.6). First we replace the constant α, the per capita growth rate of the prey in the absence of predation, by the more realistic form $\Theta(U) = r(1 - U/K)$, already derived in Chapter 1. Next we replace γU, the linear functional response of the predator to the prey which was chosen for its simplicity, by the general functional response $\Phi(U)$. We shall derive an *individual-based* form for this response. The equations become

$$\frac{dU}{d\tau} = rU\left(1 - \frac{U}{K}\right) - V\Phi(U), \quad \frac{dV}{d\tau} = eV\Phi(U) - \beta V, \qquad (2.4.11)$$

known as the Rosenzweig–MacArthur model.

To model the functional response, we start by partitioning the time budget of an individual predator. We assume that the number of prey caught by a predator is proportional to the prey density and to the time spent in actual search. Then we note that the time spent searching is less than the total amount of time allocated to food-gathering activities by the time needed to handle individual prey items. So, if N denotes the number of prey caught during a food-gathering period, T the duration of that period, U the prey density, s the effective search rate and h the handling time, then

$$N = sU(T - hN)$$

and therefore

$$\Phi(U) = \frac{N}{T} = \frac{sU}{1 + shU}.$$

This is known as *Holling's disc equation*, so called because of a class experiment in which students picked up discs from the floor of the laboratory and deposited them in a distant waste-paper basket. It is not however based on this experiment but on field data for an invertebrate predator. This so-called type II functional response gives a good fit with most data sets, although qualitatively different type I and type III responses do occur.

With this response function, the prey-predator equations become

$$\frac{dU}{d\tau} = rU\left(1 - \frac{U}{K}\right) - \frac{sUV}{1 + shU}, \quad \frac{dV}{d\tau} = \frac{esUV}{1 + shU} - mV. \qquad (2.4.12)$$

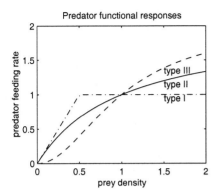

Predator functional responses

Figure 2.8   The three qualitatively different types of functional response of predators to prey. The type I response may be observed in cases where handling times are effectively zero, for example in filter feeders, and the type III response when, for example, the predator switches to an alternative prey at low densities.

We would like to simplify the equations by non-dimensionalising. An obvious non-dimensionalisation for the prey is $u = U/K$. For the predator, note that at steady state we have $V^* = (r/s)(1 - U^*/K)(1 + shU^*)$, proportional to r/s. Based on this, we take $v = sV/r$. For the time, there are at least three non-dimensionalisations we could choose, based on

− r, the per capita growth rate for the prey at low densities,

− e/h, the per capita growth rate for the predator at high prey densities, or

− m, the per capita death rate for the predator.

We choose to define $t = e\tau/h$. After some algebra the equations become

$$\frac{du}{dt} = a\left(u(1 - u) - v\frac{bu}{1 + bu}\right),$$
$$\frac{dv}{dt} = v\left(\frac{bu}{1 + bu} - c\right),$$

(2.4.13)

where

$$a = \frac{rh}{e}, \quad b = shK, \quad c = \frac{mh}{e}.$$

This is the non-dimensional version of the Rosenzweig–MacArthur model with type II functional response. The parameters a and c represent two ratios of the time scales above, and b is the ratio of handling time to search time for the predator when the prey is at carrying capacity. Steady states of these equations are at $(0,0)$, $(1,0)$ and (u^*, v^*), where $u^* = \frac{c}{b(1-c)}$, $v^* = \frac{b-c(1+b)}{b^2(1-c)^2}$. The last of these is only biologically relevant if $b/(1 + b) > c$. The interpretation of this is clear from the second of Equations (2.4.13), that the growth rate of the predator must exceed its death rate when the prey is at carrying capacity. We shall assume that this is the case. If not, the prey will never be able to sustain the predator, which will go to extinction.

We are particularly interested in the stability of the coexistence steady state (u^*, v^*), because its instability would suggest that a stable limit cycle oscillation could occur. In fact, the Hopf bifurcation theorem described in Section B.4.2 of the appendix guarantees that a limit cycle occurs close to the borderline between stability and oscillatory instability, although it does not guarantee its stability.

Example 2.3

Equations (2.4.13) may be written as

$$\frac{du}{dt} = uf(u,v), \quad \frac{dv}{dt} = vg(u,v),$$

and are said to be in *Kolmogorov form*. For such equations, show that the Jacobian matrix at a coexistence state (u^*, v^*) is given by

$$J^* = \left(\begin{array}{cc} u^* f_u^* & u^* f_v^* \\ v^* g_u^* & v^* g_v^* \end{array} \right),$$

where asterisks denote evaluation at (u^*, v^*).

The result follows directly from partial differentiation on noting that $f^* = g^* = 0$ at a coexistence steady state.

In this case, $g_v^* = 0$, so that $\beta = \operatorname{tr} J^* = u^* f_u^*$, $\gamma = \det J^* = -u^* v^* f_v^* g_u^* > 0$. Since $\gamma > 0$, the coexistence state is unstable if $\beta > 0$, stable if $\beta < 0$, and if $\beta = 0$ there are two purely imaginary eigenvalues, by the Routh–Hurwitz criteria (Section B.2.2 of the appendix).

So for instability we require $\beta = u^* f_u^* > 0$, i.e. $f_u^* > 0$. This condition for instability makes intuitive sense, since $u^* f_u^*$ is the per capita growth rate of perturbations of prey density from the steady state. But

$$f_u^* = a\left(-1 + (b - (b+1)c)\right), \tag{2.4.14}$$

so instability occurs if $b - (b+1)c > 1$. Oscillations are possible if the positive terms $b - (b+1)c$ arising from the functional response outweigh the negative term -1 from the logistic growth function. But why does the type II functional response give a positive contribution to f_u^*? The reason is that it is a saturating function, so that an increase in prey density results in a smaller proportion of the prey being taken by predators. Mathematically, the function $\psi(u) = \phi(u)/u$ satisfies $\psi'(u) < 0$. Factors that destabilise the system and may lead to oscillations are those that increase the importance of this saturating effect, allowing u to move from a steep to a shallow part of the curve. This may be done (i) by making the functional response curve steeper, (ii) by reducing the prey steady state in the presence of predation further into the steep region, or (iii) by increasing the prey steady state in the absence of predation further into the saturating region. Some destabilising factors are therefore

- an increase in the effectiveness s of the predators in capturing prey,

- an increase in the efficiency e of the predators in converting prey to predator biomass,

- a decrease in predator death rate m, or

- an increase in prey carrying capacity K, e.g. by enrichment of the system. (Early researchers had assumed that enrichment would be stabilising, and referred to this result as the *paradox of enrichment*.)

What does the phase plane for this system look like? For large values of u we will in general have $f_u < 0$, because of the compensatory effects of intraspecific competition, so for oscillations to be possible ($f_u > 0$) the u-nullcline f must have a hump. Two possible configurations for the phase plane are shown in Figure 2.9(a) and (b). For instability the v-nullcline must lie to the left of the hump, as in (b).

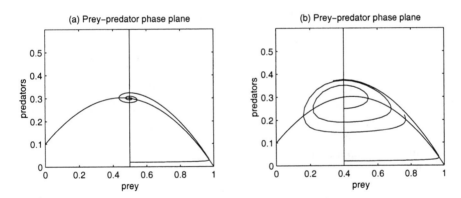

Figure 2.9 Two possible configurations for the phase plane of the prey-predator equations (2.4.13). The oscillatorily stable steady state of (a) becomes unstable through a Hopf bifurcation (Section B.4.2 of the appendix), leading to a stable periodic limit cycle solution in (b).

We have shown that a nonlinear saturating functional response $\phi(u)$ may lead to prey-predator oscillations, but an alternative mechanism is that of depensation (Section 1.3). By definition, depensation (or an Allee effect) occurs when per capita fertility increases (or per capita mortality decreases) with density; in a population growing according to $\dot{u} = u\theta(u)$, we have $\theta'(u) > 0$. Recall the condition $f_u^* > 0$ for instability of the steady state, from Equations (2.4.14), which may be written $\theta'(u^*) - v^* \psi'(u^*) > 0$. So far we have concentrated on the possibility that $\psi'(u^*) < 0$, but now we consider $\theta'(u^*) > 0$. If the prey is driven to the depensatory part of the growth curve by the predator, its recovery at a per capita rate $\theta'(u)$ may lead to divergent oscillations. This model can therefore lead to oscillatory behaviour either because of a nonlinear functional response or because of depensation. It gives insight into when we may expect oscillations in real prey-predator systems.

EXERCISES

2.6. *Holling type II functional response.* Let U be prey and V predator, and let V exist in one of two states, V_0 searching and V_1 handling prey.

   a) Explain the equations

$$\frac{dU}{dt} = rU - aUV_1,$$

$$\frac{dV_0}{dt} = aUV_1 - \beta V_0 + \text{births} - \text{deaths},$$

$$\frac{dV_1}{dt} = -aUV_1 + \beta V_0 + \text{births} - \text{deaths},$$

   where births and deaths are to be specified later.

   b) Assume that the births and deaths take place on a much longer time scale than the switching between V_0 and V_1. Then, to a good approximation, $aUV_1 - \beta V_0 = 0$. Show that the fraction of predators in the searching state is given by

$$\frac{V_1}{V} = \frac{\beta}{\beta + aU} = \frac{1}{1 + abU},$$

   where $b = 1/\beta$.

   c) Explain the equations

$$\frac{dU}{dt} = rU - \frac{aUV}{1 + abU}, \quad \frac{dV}{dt} = h\frac{aUV}{1 + abU} - cV.$$

2.7. *Gause's prey-predator model.* Let u and v satisfy

$$\frac{du}{dt} = ug(u) - vp(u), \quad \frac{dv}{dt} = v(-d + q(u)),$$

   where

- $g(u) > 0$ for $u < K$, $g(u) < 0$ for $u > K$, $g(K) = 0$,

- $p(0) = 0$, $p(u) > 0$ for $u > 0$,

- $q(0) = 0$, $q'(u) > 0$ for $u > 0$.

a) Interpret the conditions on g, p and q biologically.

b) Sketch two possible configurations for the nullclines in the phase plane, depending on whether there is a coexistence state or not.

c) Show that the steady states $(0,0)$ and $(K,0)$ are saddle points, if a coexistence steady state exists.

d) In the case that there is a coexistence state, derive a condition for it to be unstable.

e) Sketch the phase plane in this case.

2.8. *Kolmogorov's conditions for a limit cycle.* For a predator-prey system in Kolmogorov form,

$$\frac{du}{dt} = uf(u,v), \quad \frac{dv}{dt} = vg(u,v),$$

let the functions f and g satisfy the following conditions.

- $f_v < 0$, $g_u > 0$.

- For some $v_1 > 0$, $f(0,v_1) = 0$, and for some $u_1 > 0$, $g(u_1,0) = 0$.

- There exists $u_2^* > 0$ such that $f(u_2^*,0) = 0$.

- $f_u > 0$ for small u, $f_u < 0$ for large u, and $g_v < 0$.

- There is an unstable coexistence state (u^*,v^*), and the slope of $f(u,v) = 0$ at (u^*,v^*) is positive.

a) Interpret the conditions above.

b) Sketch the phase plane for the system.

c) Show that the system has a limit cycle in the positive quadrant.

2.9. *Leslie's prey-predator model.* Consider the equations

$$\frac{du}{dt} = au - bu^2 - cuv, \quad \frac{dv}{dt} = dv - e\frac{v^2}{u},$$

where a, b, c, d and e are positive parameters. The predator equation is logistic, with carrying capacity proportional to the prey population.

a) Criticise the model.

b) Sketch the phase plane.

c) What is the behaviour of solutions of the model?

2.5 Competition

— *The principle of competitive exclusion* states that if two species occupy the same *ecological niche*, then one of them will go extinct.

As Dr Seuss puts it,

And NUH is the letter I use to spell Nutches
Who live in small caves, known as Nitches, for hutches.
These Nutches have troubles, the biggest of which is
The fact there are many more Nutches than Nitches.
Each Nutch in a Nitch knows that some other Nutch
Would like to move into his Nitch very much.
So each Nutch in a Nitch has to watch that small Nitch
Or Nutches who haven't got Nitches will snitch.

The principle is often named after the Russian biologist Gause, who did experimental work on competitive systems in the 1930s in an attempt to test the theoretical ideas of Lotka and Volterra, although the concept did not originate with him.

Two species occupy the same niche if their ecology is the same, i.e. if they interact in the same way with the same other species, require the same nutrients, live in the same habitat at the same time, and so on. This principle is not very useful in practice, as we can always find a way in which two species differ. What is more, it is not even always true; among other things, interactions with other species at different trophic levels can lead to stable coexistence. However, if two competitors coexist there must be a reason for them to do so, and the principle of competitive exclusion focuses attention on this fact.

The most straightforward analysis of competition between two species is as follows. Let U_1 and U_2 grow in each other's absence according to the logistic equation

$$\frac{dU_i}{d\tau} = r_i U_i (1 - \frac{U_i}{K_i})$$

for $i = 1, 2$. The term $\frac{U_i}{K_i}$ represents the effects of *intraspecific* competition. Now assume that they compete interspecifically in exactly the same way as they compete intraspecifically. The equations become

$$\frac{dU_1}{d\tau} = r_1 U_1 (1 - \frac{U_1 + \alpha U_2}{K_1}), \quad \frac{dU_2}{d\tau} = r_2 U_2 (1 - \frac{U_2 + \beta U_1}{K_2}),$$

where α is the *competition coefficient* of species 2 on species 1, and β is the reverse competition coefficient. Suppose that species 1 and 2 were competing flour beetles, and that one individual of species 2 ate two grains of flour for every one eaten by an individual of species 1. Then $\alpha = 2$, and conversely $\beta = \frac{1}{2}$. Generally, for species in the same ecological niche, $\alpha\beta = 1$. On the other hand, for species not in the same niche it is difficult to interpret the competition coefficients, and the model must be interpreted with caution.

Simplifying the equations by defining

$$u = \frac{U_1}{K_1}, \quad v = \frac{U_2}{K_2}, \quad t = r_1\tau,$$

we obtain

$$\frac{du}{dt} = u(1 - u - av), \quad \frac{dv}{dt} = cv(1 - bu - v),$$

where $a = \alpha\frac{K_2}{K_1}$, $b = \beta\frac{K_1}{K_2}$, $c = \frac{r_2}{r_1}$. There are steady states at $(0,0)$, $(1,0)$, $(0,1)$, and (u^*, v^*), where

$$u^* = \frac{1-a}{1-ab}, \quad v^* = \frac{1-b}{1-ab}.$$

The coexistence state is only in the positive quadrant and therefore biologically realistic if either $a < 1$ and $b < 1$, or $a > 1$ and $b > 1$. Phase planes for the four possible configurations are shown in Figure 2.10. We have the following outcomes.

- If $a < 1$ and $b < 1$, there is stable coexistence.

- If $a > 1$ and $b > 1$, the system is *bistable*; the ultimate winner depends on the initial conditions.

- If $a < 1 < b$, then species 1 is the *superior competitor*, and always wins.

- If $a > 1 > b$, then species 2 is the superior competitor, and always wins.

Let us look at this from the evolutionary point of view. As species 1, your best evolutionary strategy must be based on increasing b or decreasing a, i.e. increasing your carrying capacity K_1 relative to your competitor's, a so-called K-strategy, or increasing your competition coefficient relative to your competitor's. There is no advantage in increasing your growth rate r_1, a so-called r-strategy. As we have seen in Chapter 1, an r-strategy is only advantageous in an unpredictable or inhomogeneous environment, and we have included no unpredictability or inhomogeneity in our model.

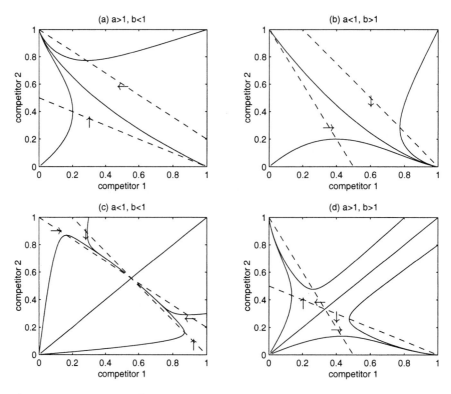

Figure 2.10   The four qualitatively different configurations of the phase plane for the Lotka–Volterra competition equations with two competitors.

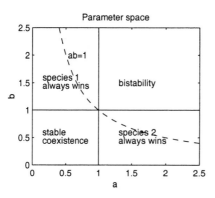

Figure 2.11   The four qualitatively different outcomes plotted in parameter space, the (a, b)-plane.

It can be seen that there can only be a stable coexistence state if the interspecific competition parameters a and b are both less than one. But for competitors occupying the same ecological niche, $ab = \alpha\beta = 1$, so a and b cannot both be less than one, and we have proved the principle of competitive exclusion for this model.

The condition for stable coexistence may be generalised to the competition equations

$$\frac{du}{dt} = uf(u,v), \quad \frac{dv}{dt} = vg(u,v).$$

The Jacobian matrix at the steady state is given by

$$J^* = \begin{pmatrix} u^* f_u^* & u^* f_v^* \\ v^* g_u^* & v^* g_v^* \end{pmatrix}.$$

Let us assume that there are no depensatory (Allee) effects, so that f_u^* and g_v^* are both negative, and $\operatorname{tr} J^* < 0$. The condition for stability of a coexistence state (u^*, v^*) is that $\det J^* > 0$,

$$f_u^* g_v^* > f_v^* g_u^*,$$

i.e. that the product of the intraspecific limitation coefficients is greater than the product of the interspecific ones at the steady state. In this sense, stable coexistence can only occur if intraspecific is stronger than interspecific competition, and therefore the species do not occupy the same ecological niche. This is a re-statement of the principle of competitive exclusion.

EXERCISES

2.10. Show that the condition for stability of a coexistence state is equivalent to the condition that the nullcline $f = 0$ should be steeper than the nullcline $g = 0$ at the steady state, as in Figure 2.10(c).

2.11. Consider two species u and v competing for the same food, and let $f(u,v)$ be the rate at which food is consumed. Let $f(u,v) \to \infty$ as $|(u,v)| \to \infty$. Let u and v satisfy the equations

$$\frac{du}{dt} = r_1 u(1 - \gamma_1 f(u,v)), \quad \frac{dv}{dt} = r_2 v(1 - \gamma_2 f(u,v)),$$

with initial conditions $u(0) = u_0$, $v(0) = v_0$.

a) Show that $r_2 \gamma_2 \frac{d}{dt} \log u - r_1 \gamma_1 \frac{d}{dt} \log v = r_1 r_2 (\gamma_2 - \gamma_1)$.

b) Integrate this equation and deduce a principle of competitive exclusion, giving the condition for u to outcompete v.

c) Interpret this condition biologically.

d) Generalise this result to the case of n species competing for the same food (or for $n - 1$ food resources?).

2.12. The Lotka–Volterra competition equations are given by

$$\frac{du}{dt} = u(1 - u - av), \quad \frac{dv}{dt} = cv(1 - bu - v).$$

a) Use the Dulac criterion (Section B.2.3 of the appendix) to verify that there are no periodic solutions of the system in the positive quadrant.

b) By considering the function Φ given by $\Phi(u, v) = c(u - u^* \log u) + v - v^* \log v$, or otherwise, show that when the coexistence state is stable it is globally asymptotically stable for solutions with initial conditions in the positive quadrant. [*Hint:* Φ is a Lyapunov function (Section B.3.2 of the appendix), and $1 - u - av = -(u - u^*) - a(v - v^*)$.]

2.13. *Disturbance-mediated coexistence.* In some experiments on competition between two species of hydra, it was found that coexistence was only possible if a fraction of the population of each species was removed at regular intervals. A model for the system with this experimental manipulation is given by

$$\begin{aligned}
\frac{dN_1}{dt} &= \frac{r_1 N_1}{K_1} \left(K_1 - m_1 - N_1 - aN_2 \right), \\
\frac{dN_2}{dt} &= \frac{r_2 N_2}{K_2} \left(K_2 - m_2 - bN_1 - N_2 \right).
\end{aligned} \tag{2.5.15}$$

Explain the model. Show that if $K_1 = 100$, $K_2 = 90$, $a = 1.2$, $b = 0.8$, $m_1 = 2$ and $m_2 = 10$, then stable coexistence occurs for this system but not for the corresponding system with $m_1 = m_2 = 0$.

2.14. a) Set up a Lotka–Volterra model for mutualism. That is, use the simplest terms possible for the positive effect of each species on the other. Include self-limitation of each species.

b) Show that if it has a coexistence steady state (u^*, v^*) then this steady state is globally stable in the positive quadrant. (Otherwise all solutions in the positive quadrant are unbounded.)

2.6 Ecosystems Modelling

In all physical or biological systems, the laws of conservation of mass and energy must hold. How can we make use of these powerful principles in Mathematical

Biology? Alfred Lotka, in his book published in 1925 which outlined his programme for a physical biology, was very concerned with the circulation of the elements, and with the energy flows through biological systems. Energy flows through ecosystems are somewhat difficult to measure and to model. However, much of the energy within an ecosystem is chemically bound, and the most successful ecosystems models have concentrated on the balance of essential elements such as carbon, nitrogen and phosphorus rather than explicitly on energy.

Early ecosystem models tended to be extremely complex, with tens or hundreds of state variables. This leads to problems with parameter identification, and testing such models is very difficult. We shall take the approach of formulating strategic models aiming at understanding those aspects of their dynamics that result from constraints on nutrients or essential elements. We focus on *functional groups* of species that play a similar role in the passage of nutrients through the system.

We shall consider an example of plankton in the ocean limited by the essential element nitrogen, and assume for simplicity that the system is closed to nitrogen. Plankton are of two kinds, *phytoplankton*, or plant plankton, which photosynthesise and require essential elements, and *zooplankton*, or animal plankton, which feed on phytoplankton. We shall assume that all the action takes place in a well-mixed surface layer of the ocean. Let N be the concentration of nitrogen available for uptake, measured as mass per unit surface area of the ocean, P the concentration of phytoplankton, and Z the concentration of zooplankton, both measured in the same currency, i.e. as mass of nitrogen incorporated in the plankton per unit surface area of the ocean.

Nitrogen (as dissolved gas or compounds) is taken up from the ocean and incorporated into phytoplankton. It is incorporated into zooplankton through consumption of phytoplankton. It is recycled from the plankton to the ocean through death and excretion. A model of these processes gives

$$\frac{dN}{dt} = aP + bZ - cNP,$$
$$\frac{dP}{dt} = cNP - dPZ - aP, \qquad (2.6.16)$$
$$\frac{dZ}{dt} = dPZ - bZ.$$

We have taken linear functional responses of P to N and Z to P for simplicity. We have also assumed that the nitrogen stored in plankton becomes available for uptake immediately on the death of the plankton, whereas in fact the processes of decay and re-mineralisation take time. It follows from this last assumption that

$$\frac{dN}{dt} + \frac{dP}{dt} + \frac{dZ}{dt} = 0,$$

so that

$$N + P + Z = A,$$

where A is a constant of integration. This is the law of conservation of mass for the nitrogen, and A represents the total concentration of nitrogen, both free (available for uptake) and bound (incorporated in the plankton).

The steady states of the system of equations for (N, P, Z) are $S_0 = (A, 0, 0)$, $S_1 = (\frac{a}{c}, A - \frac{a}{c}, 0)$, and $S_2 = (N^*, P^*, Z^*)$, where

$$P^* = \frac{b}{d}, \quad Z^* = \frac{c}{c+d}\left(A - \frac{c}{a} - \frac{b}{d}\right), \quad N^* = A - P^* - Z^*.$$

Note that P^* does not depend on A. If the supply of the nutrient which limits phytoplankton growth is increased, it is not the phytoplankton but the predatory zooplankton that benefit. It is usual in models of this kind that species at alternate levels in the food chain benefit from an increase in nutrient supply.

As the season progresses, irradiance levels rise, and the phytoplankton become more efficient at fixing nitrogen. How does this affect the dynamics? It causes an increase in the parameter c. For $c < \frac{a}{A}$, the only steady state is the trivial one S_0. Plankton levels are very low. As c increases past $\frac{a}{A}$, S_0 loses its stability through a transcritical bifurcation (Section B.4.1 of the appendix) as an eigenvalue passes through zero, and S_1 becomes stable. There is a bloom of phytoplankton, overshooting its steady state and returning towards it. As c increases still further, there is a second critical value at

$$c = \frac{c}{A - b/d},$$

where S_1 loses its stability as an eigenvalue passes through zero, again through a transcritical bifurcation, and S_2 becomes stable. A zooplankton bloom succeeds the phytoplankton bloom, but this also exceeds its steady state and then returns towards it. A numerical solution of the equations with c changing periodically over the year is shown in Figure 2.12. Similar oscillations in plankton populations also occur in the ocean and in models with more realistic (type II) functional responses of Z to P.

EXERCISES

2.15. In the ecosystem model in the text we assumed that the system was closed to nitrogen, which was a limiting resource for the system. Here we shall assume that the essential resource carbon is freely available. Plants take up carbon through photosynthesis and incorporate it into their biomass. Let the biomass of plants be given by P, and biomass

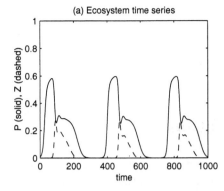

(a) Ecosystem time series

Figure 2.12 A numerical solution of the plankton model given by Equations (2.6.16) with irradiance, and hence c, varying in an annual cycle.

of herbivores feeding on these plants be H. The equations satisfied by P and H are given by

$$\frac{dP}{dt} = \phi - aP - bPH, \quad \frac{dH}{dt} = ebPH - cH.$$

a) Interpret these equations biologically.

b) Find the steady state(s) of these equations and determine their stability.

c) What is the effect of increasing primary production ϕ?

d) Sketch the possible phase planes for the system.

2.16. Analyse an ecosystem model with three trophic levels, phytoplankton P, zooplankton Z, and consumers C of the zooplankton, with nitrogen N as the limiting resource. How does the solution behaviour depend on the total amount of nitrogen?

2.17. Analyse the ecosystem discussed in Section 2.6, but now with type II functional response of the predators to the prey. Can enrichment destabilise the coexistence steady state, as it can in the Rosenzweig–MacArthur model discussed in Section 2.4?

2.18. a) Set up an ecosystem model for two species P_1 and P_2 of phytoplankton in competition for a single abiotic resource N, in a similar way to Exercise 1.5. (Assume that the functional response of each competitor to the resource is linear.)

b) Show that as long as there is sufficient resource, the equations can be non-dimensionalised to obtain the usual Lotka–Volterra competition equations,

$$\frac{du}{dt} = u(1 - u - av), \quad \frac{dv}{dt} = cv(1 - bu - v),$$

where $ab = 1$.

c) Show that the parts of the nullclines that do not run along the axes are parallel.

2.7 Interacting Metapopulations

The metapopulation approach to modelling single species persistence in patchy environments was introduced in Section 1.6. The idea is that populations may persist through a combination of local extinctions and re-colonisations of a large number of potentially habitable sites. In this section we shall apply this approach to interacting populations.

The principle of competitive exclusion of Section 2.5 states that when two species occupy the same ecological niche, one will drive the other to extinction. Generalising, it states that n species cannot coexist if there are fewer than n niches available. But near the surface of the ocean there are few limiting resources and many planktonic species. Are there more niches available than limiting resources, or does the principle break down? It has been suggested that this *paradox of the plankton* could be resolved if different species occupied different regions of space. This spatial heterogeneity could be and to some extent would have to be generated by the species themselves through their interactions.

Spatial heterogeneity may also promote persistence in prey-predator systems. In a famous experiment in the 1950s, Huffaker studied herbivorous and predatory mites in a patchy environment consisting of oranges and rubber balls in a tray. He showed that the herbivorous mites survived in the absence of predators, but if predators were added they were liable to drive the prey to extinction and then to become extinct themselves, rarely completing more than a single prey-predator cycle. However, if more oranges were included and the dispersal rates between the oranges were artificially reduced, especially for the predatory mites, then the populations could coexist through a mosaic of unoccupied patches, prey-predator patches heading for extinction, and thriving prey patches. This is a clear candidate for modelling with the metapopulation approach.

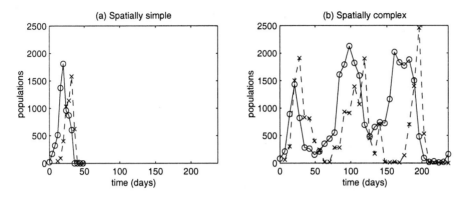

Figure 2.13 Some of the results of Huffaker's experiments. In (a) the spatial lay-out of the experiment is simple, while in (b) it is more complex. The increased complexity allows more prey-predator cycles.

2.7.1 Competition

The metapopulation model (1.6.14) for a single population may be extended to one for two competing populations by considering that patches may be in one of four states, empty or occupied by one or other or both competitors. It is usual to ignore local dynamics and to assume that the effects of competition are to lower colonisation rates, or to increase extinction rates, or both. It may be shown that one species can exclude the other if it reduces its chances of colonisation or increases its chances of extinction sufficiently. A limiting case occurs when the superior competitor can immediately displace the inferior in a patch, so that there are no patches with both competitors present.

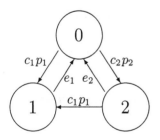

Figure 2.14 State transitions for patches in the metapopulation model with two competitors, species 1 and 2, where species 1 immediately displaces species 2 from a contested patch.

The equations are

$$\frac{dp_1}{dt} = c_1 p_1 (1 - p_1) - e_1 p_1, \quad \frac{dp_2}{dt} = c_2 p_2 (1 - p_1 - p_2) - e_2 p_2 - c_1 p_1 p_2. \quad (2.7.17)$$

Species 1, whose occupancy fraction is p_1, is the superior competitor, and species 2 has no impact on it whatever. In contrast, species 2 can only colonise empty sites (the term $c_2 p_2 (1 - p_1 - p_2)$), and loses sites when species 1 colonises them (the term $-c_1 p_1 p_2$).

EXERCISES

2.19. a) Find the steady states of the system (2.7.17).

b) Show that species 1 goes extinct if $c_1 < e_1$, and species 2 goes extinct if $c_2 < e_2$.

c) Assuming that $c_1 > e_1$ and $c_2 > e_2$, sketch the phase plane in two cases, and find the condition for species 2 to survive.

d) Deduce that the poorer competitor can survive by being a good disperser.

In the metapopulation paradigm, in contrast to the Lotka–Volterra approach, we can see the importance of having a high growth rate, or being an r-strategist, in the sense of being quick to colonise newly vacated patches. Paradoxically, this mechanism for coexistence, known as fugitive coexistence, relies on local extinctions.

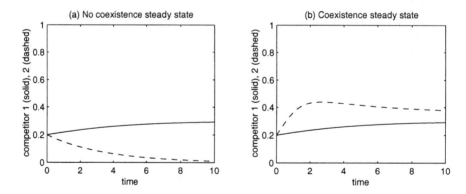

Figure 2.15   Numerical results for a metapopulation model with competition. Parameter values are $c_1 = 1$, $e_1 = 0.7$, $e_2 = 0.7$, and (a) $c_2 = 1.2$, (b) $c_2 = 3$. In (b) the better dispersal characteristics of the inferior competitor allow it to coexist with the superior one.

2.7.2 Predation

Here we present a spatially implicit metapopulation model for a prey-predator system, such as Huffaker's experiments of herbivorous and predatory mites on oranges. We used the subscript 2 to denote a competitor above and will continue to do so in the following; we use the subscript 3 here and later to denote a predator. The system consists of K sites (oranges), and at time t a fraction p_1 is in state 1, occupied by prey only, a fraction p_{13} in state 13, occupied by both prey and predators, and a fraction $p_0 = 1 - p_1 - p_{13}$ in state 0, empty. We must specify the possible moves that an orange can make from state to state, and the probability of each in the next interval of time δt. For movement from state 0 to state 1 let this probability be $c_1 p_1 \delta t$, so that c_1 is the prey colonisation rate. (Colonisation by the prey is assumed to take place only from oranges in state 1, which are predator-free and therefore able to produce colonisers.) For movement from state 1 to state 13 let it be $c_3 p_{13} \delta t$, so that c_3 is the predator colonisation rate. From state 13 to state 0 let it be $e_{13} \delta t$, so that e_{13} is the local extinction rate of the prey in the presence of the predators, followed immediately by the extinction of the predators themselves. From state 1 to state 0 let it be $e_1 \delta t$, so that e_1 is the local extinction rate of the prey in the absence of predators.

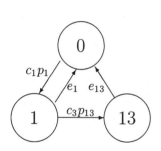

Figure 2.16　　State transitions for patches in the metapopulation model for predation, with species 1 the prey and species 3 the predator. A patch in state 13 contains prey and predators, but if the prey in such a patch go extinct then it is assumed that the predators immediately do so as well, so that there are no pure predator patches.

The equations for the fraction of sites in each state are given by

$$\frac{dp_1}{dt} = c_1 p_0 p_1 - c_3 p_1 p_{13} - e_1 p_1,$$

$$\frac{dp_{13}}{dt} = c_3 p_1 p_{13} - e_{13} p_{13},$$

(2.7.18)

EXERCISES

2.20. Assume that $c_1 > e_1$, so that the prey can persist in the absence of predators.

a) Sketch the phase plane for the system in the two cases (i) $e_{13}/c_3 > 1 - e_1/c_1$ and (ii) $e_{13}/c_3 < 1 - e_1/c_1$, and deduce that the predators die out if c_3 is too small or e_{13} is too large.

b) Show that in case (ii) the coexistence steady state is stable.

The model therefore predicts stable coexistence for c_3 sufficiently large. However, as c_3 increases, the fraction of occupied patches becomes smaller. There are two ways in which this could lead to extinction in a real system, although we shall not consider them further here. First, a delay in the system could lead to growing oscillations about the steady state, and population crashes to zero. Second, small population sizes could lead to stochastic effects becoming important, invalidating the deterministic assumptions of the model.

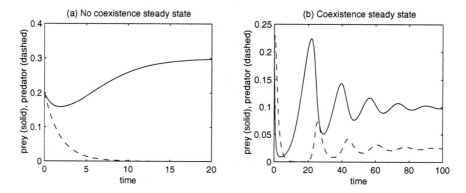

Figure 2.17   Numerical results for a metapopulation model of a predator-prey interaction, in the cases (a) $e_{13}/c_3 > 1 - e_1/c_1$ and (b) $e_{13}/c_3 < 1 - e_1/c_1$.

2.7.3 Predator-mediated Coexistence of Competitors

We shall combine models (2.7.17) and (2.7.18) to demonstrate how a predator can mediate coexistence between two competitors.

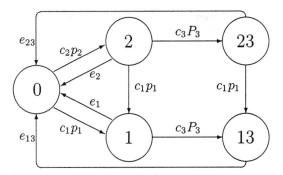

Figure 2.18 State transitions for patches in the model with two competing prey, species 1 and 2, and one predator, species 3. Prey-predator patches include species 3 and either species 1 (type 13) or species 2 (type 23). Upper case P_3 denotes *all* patches containing the predator, $P_3 = p_{13} + p_{23}$.

The model becomes

$$\frac{dp_1}{dt} = c_1(p_0 + p_2)p_1 - c_3 p_1(p_{13} + p_{23}) - e_1 p_1,$$

$$\frac{dp_2}{dt} = c_2 p_0 p_2 - c_1 p_1 p_2 - c_3 p_2(p_{13} + p_{23}) - e_2 p_2,$$

$$\frac{dp_{13}}{dt} = c_3 p_1(p_{13} + p_{23}) + c_1 p_1 p_{23} - e_{13} p_{13},$$

$$\frac{dp_{23}}{dt} = c_3 p_2(p_{13} + p_{23}) - c_1 p_1 p_{23} - e_{23} p_{23}.$$

Again, species 1 is a better competitor than species 2, and can immediately displace it from a patch. With certain parameter values all three species can coexist. Essentially, the mechanism is fugitive coexistence, with the predator providing the necessary empty patches which the better disperser, species 2, can colonise and exploit. With no predators species 2, the poorer competitor, goes extinct in competition with species 1, the better.

2.7.4 Effects of Habitat Destruction

Assemblies of competing populations have been looked at using the spatially implicit metapopulation method, and assumed that poorer competitors are better dispersers. It has been shown that when habitat is destroyed, and under certain conditions on the parameters, it is the better competitors that go extinct first. As in the single species case considered in Section 1.6, the critical fraction D_c of sites to be deleted, when the best competitor goes extinct, is the

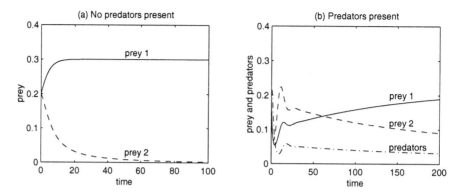

Figure 2.19   Numerical results for a metapopulation model of predator-mediated coexistence. The only difference between the two panels is in the initial conditions. In panel (a), initial conditions are chosen with each species of competitor occupying a fraction 0.2 of the patches. The inferior prey species soon goes extinct. In panel (b), the initial numbers of competitor-only patches are the same, but there is also a fraction 0.1 of 13 patches and a fraction 0.1 of 23 patches. The system settles down to a coexistence steady state. Parameter values are $c_1 = 1$, $c_2 = 1.4$, $c_3 = 2.5$, $e_1 = 0.7$, $e_2 = 0.7$, $e_{13} = 0.7$, $e_{23} = 0.7$.

fraction of sites occupied by the best competitor at steady state. Very complicated effects may occur with only slightly more complex systems, with many possibilities for the sequence of extinctions and possibly recolonisations as the amount of destruction increases.

EXERCISES

2.21. Investigate the effect of habitat destruction on the metapopulation prey-predator system, Equations (2.7.18). Assume that the species coexist when $D = 0$.

2.22. Investigate the effect of habitat destruction on the metapopulation competition system, Equations (2.7.17). Assume that there is a trade-off between competitive ability and R_0, so that the worse competitor, species 2, has the higher R_0, $c_2/e_2 > c_1/e_1$. Assume also that the competitors coexist when $D = 0$.

2.8 Conclusions

- Host-parasitoid and prey-predator relationships are essentially equivalent.

- However, the existence of an annual cycle for many host-parasitoid relationships means that they are often modelled as a system of difference rather than differential equations.

- Simple models often assume that the host or prey is limited by the parasitoid or predator. Some *self-limitation* must always be present, but if the parasitoid or predator limits the population well below its carrying capacity, it may *sometimes* be safely neglected.

- The *functional response* is the per capita rate at which parasitoids parasitise hosts or predators eat prey, as a function of host or prey density. In simple models it is often taken to be linear, but must realistically be a *saturating* (increasing but bounded) function.

- *Interference* between parasitoids or predators in searching for hosts or prey is widespread in nature, but often neglected in models.

- Simple host-parasitoid and prey-predator models tend to oscillate.

- The simplest (Nicholson–Bailey) host-parasitoid model has unrealistic behaviour in that its oscillations increase in amplitude without limit.

- The simplest (Lotka–Volterra) prey-predator model has a family of periodic solutions. This is a sign of an underlying weakness of the model, its structural instability, and the periodic solutions are easily destroyed by small changes to the equations. Self-limitation of the prey, however small, *cannot* be neglected.

- Volterra's principle states that an intervention in a prey-predator system that kills or removes both prey and predators in proportion to their population sizes has the effect of increasing prey populations. Caution must be exercised in attempting to achieve additional chemical control of a pest that is already biologically controlled.

- Host-parasitoid or prey-predator oscillations are sometimes seen in nature but are the exception rather than the rule, stable steady states being more common. Possible stabilising mechanisms for the real system and its model are:

  - self-limitation or interference, as discussed above;

  - heterogeneity of the environment, e.g. some cover for the prey;

  - non-random search patterns.

- When a prey-predator oscillation does occur in nature it must be modelled by a structurally stable oscillation, such as a limit cycle. Limit cycle oscillations in continuous-time prey-predator systems may occur under certain conditions, such as

  - predators capable of high birth-rate to death-rate ratios,

  - high environmental carrying capacity for the prey,

  - a depensatory (Allee) effect for the prey.

- The *principle of competitive exclusion* states that if two species occupy the same ecological niche, then one of them will go extinct.

- However, the principle is only valid under restrictive conditions. Coexistence of competitors can occur for any of the following reasons, which do not rely on niche differentiation.

  - Predator-mediated coexistence in a patchy environment.

  - Fugitive coexistence in a patchy environment. Poor competitors can survive if they are sufficiently good dispersers.

  - Predator-mediated coexistence in a homogeneous environment is also possible, although not in such a simple system, as is resource-mediated coexistence.

- Stable coexistence of two competitors in a Lotka–Volterra or similar system is possible if intra-specific is stronger than inter-specific competition, in the sense that $f_u^* g_v^* > f_v^* g_u^*$ (or $\det J^* > 0$). This can only happen if the species occupy different (but possibly overlapping) ecological niches.

- Ecosystem modelling concentrates mainly on the balance of essential elements in an ecological system.

- An increase in the amount of nutrient available may not lead to an increase in the population size of the consumers of this nutrient, but may benefit the consumers of these consumers instead.

- Metapopulation modelling includes spatial effects, particularly patchy environments, in a tractable way.

- In the metapopulation paradigm it is easy to see the advantage of an r-strategy, whereas the Lotka–Volterra approach emphasises a K-strategy.

- Patchy environments can allow persistence of competitive and prey-predator communities.

- Assuming that poorer competitors are better dispersers, destruction of habitat tends to drive better competitors to extinction before poorer.

3
Infectious Diseases

In this chapter, questions such as the following about infectious diseases will be investigated by looking at simple deterministic models.

– Will there be an epidemic?

– If so, how many individuals will be affected?

– If the disease is endemic, what is the prevalence of the infection?

– Can the disease be eradicated or controlled?

– What is the effect of population age structure?

3.1 Introduction

Despite the importance of diseases in human communities, there was little work on mathematical models for them until the beginning of the last century. An interesting exception is a paper by Daniel Bernoulli, written in 1760 and published in 1766, which analyses deaths from smallpox. It was aimed at influencing public policy towards variolation, a technique of injecting a mild strain of the smallpox virus to induce immunity against the full disease. This paper has similarities with Euler's work on demography, published in 1760 and discussed in Chapter 1. The Euler and Bernoulli families were both from Basel, and Leonhard (born 1707) and Daniel (born 1700) knew each other as children. They both obtained chairs at the Russian Academy in St Petersburg under Catherine the Great in the 1720s and lived there together from 1727 until 1733, when Bernoulli returned to Basel, so it is possible that they corresponded about the mathematical modelling of human populations.

More systematic work on modelling disease was done in the early 20th century by Hamer, who was interested in the regular recurrence of measles epidemics, and Ross, who obtained a Nobel prize in 1902 for showing that malaria was transmitted by mosquitoes. They put forward hypotheses about transmission of infectious disease and investigated their consequences through mathematical modelling. Based on their work, Kermack and McKendrick published a classic paper in 1927 that discovered a threshold condition for the spread of a disease and gave a means of predicting the ultimate size of an epidemic.

Kermack and McKendrick and other early authors assumed that the population mixed homogeneously, and much has been done since their paper was published to investigate the effect of removing this unrealistic assumption. Their threshold theory has been extended to more complex models. Mechanisms of spatial spread have been analysed, and control theory has been applied to optimise public health policies. The mechanisms of recurrent epidemics have been elucidated. The development of stochastic models has been very important, although we shall not deal with these in this book.

In modelling an epidemic process we need to make assumptions about

- the population affected,

- the way the disease is spread, and

- the mechanism of recovery from the disease or removal from the population.

With regard to the population, we model

- the population dynamics: whether the population is *closed*, so that immigration, emigration, and birth and disease-unrelated death can be neglected, or *open*;

- the disease status structure of the population: a mutually exclusive and exhaustive classification of individuals according to their disease status, and

- possibly other population structure, such as age or sex.

With regard to disease status, an individual is in one of the following classes.

- *Susceptible.*

- *Latent*, or *exposed*. Infected by the disease, but not yet infectious.

- *Infective*, or *infectious*. An individual may become infectious before symptoms appear. The period before symptoms appear is the *incubation period*.

- *Removed.* No longer infectious, whether by acquired immunity, isolation or death.

- *Carrier.* In some diseases, there may be individuals who remain infectious for long periods, maybe for life, but do not show any symptoms of the disease themselves. They may be important for the progress of the disease.

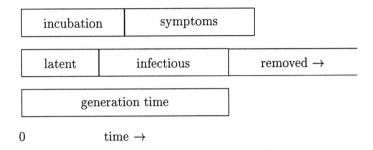

Figure 3.1 Diagrammatic representation of the progress of a disease within an individual. The point of infection is at $t = 0$.

The infective class may be split up further depending on whether the disease is

- *microparasitic*, caused by a virus (e.g. measles), a bacterium (e.g. TB), or a protozoon (e.g. malaria), where (to a first approximation) an individual either has the disease or does not have it, or

- *macroparasitic*, caused by a helminth (e.g. a tapeworm), or an arthropod (e.g. a tick), when the degree of infestation may be important.

We shall distinguish between

- *epidemic* diseases, which are prevalent in a population only at particular times or under particular circumstances, and

- *endemic* diseases, which are habitually prevalent.

(The *prevalence* of a disease in a population is the fraction infected. The *incidence* is the rate at which infections occur.)

In the next three sections we shall consider a homogeneous and homogeneously mixing population suffering from a microparasitic infection with various characteristics, in the epidemic and the endemic cases. We shall discuss eradication and control. We shall then consider the effects of inhomogeneity in mixing and age. This will include populations that are stratified geographically, i.e. split into a number of homogeneously mixing groups with possibilities of infection passing between groups, and we shall consider diseases where the age at infection is important. Spatial heterogeneity with continuous spatial variation will be discussed in Chapter 5. Then we shall consider criss-cross infection, which includes diseases carried by a vector, such as the malarial mosquito, and sexually transmitted diseases. We shall then introduce models that include the degree of infection, important for macroparasitic diseases. Finally we shall discuss some evolutionary questions.

3.2 The Simple Epidemic and SIS Diseases

In a *simple epidemic*, the population consists only of susceptibles and infectives. A few infectives are assumed to be introduced into a population of susceptibles. The disease is *contagious*, i.e. spread by contact between a susceptible and an infective. A susceptible, once infected, becomes infectious immediately and remains so indefinitely:

$$S \to I.$$

This is a reasonable approximation to the initial stages of many diseases. We shall assume that the population is closed, so that

$$S(\tau) + I(\tau) = N,$$

where $S(\tau)$ and $I(\tau)$ are the numbers of susceptible and infectious individuals at time τ, and N is the constant population size. The differential equations satisfied by S and I are given by

$$\frac{dS}{d\tau} = -f(S,I), \quad \frac{dI}{d\tau} = f(S,I),$$

where $f(S,I)$ is *incidence* of the disease, i.e. the rate at which infections occur. Clearly f is an increasing function of both S and I, and the simplest model is

$$f(S,I) = \lambda(I)S = \beta IS.$$

The function $\lambda(I)$ is called the *force of infection,* an important epidemiological concept. It is defined to be the *probability density* of a given susceptible contracting the disease, i.e. the probability that a given susceptible will contract the disease in the next small interval of time $\delta\tau$ is given by $\lambda(I)\delta\tau + O(\delta\tau^2)$. For $\lambda(I) = \beta I$, the parameter β is called the *(pairwise) infectious contact rate*, i.e. the rate of infection per susceptible and per infective. This is the *law of mass action*, which we have seen before in Chapter 2. As before, there are problems with it.

– The *functional response* βS of the infectives to the susceptibles (analogous to the functional response of predators to prey in Chapter 2) is linear, whereas we would expect a saturating response.

– The force of infection $\lambda(I) = \beta I$ is linear; we would again expect this to be a saturating function.

However, just as for the Lotka–Volterra equations, the law of mass action $f(I,S) = \beta IS$ is a good starting point.

The disease is represented diagrammatically as shown:

$$S \quad \xrightarrow{\beta IS} \quad I.$$

We can analyse the system by eliminating S from the differential equation for I, giving

$$\frac{dI}{d\tau} = \beta I(N - I).$$

We recover the logistic equation, familiar from Chapter 1, with $r = \beta N$ and $K = N$. This simple epidemic will always spread and will eventually infect all susceptibles, conflicting with data.

The SI disease does not include recovery; the SIS disease does, with recovered individuals again being susceptible. It is possible that such a disease remains *endemic* in the population:

$$S \rightarrow I \rightarrow S.$$

Most microparasitic infections confer some measure of immunity, but some, especially bacterial infections like TB, meningitis and gonorrhoea, confer little or none, in which case this is a reasonable approximation. We assume that the modelling time scale is short compared to the lifetime of its hosts, so that we can neglect birth and death. We again have a closed population,

$$S(\tau) + I(\tau) = N,$$

with

$$\frac{dS}{d\tau} = -f(S,I) + g(I), \quad \frac{dI}{d\tau} = f(S,I) - g(I).$$

We shall model the incidence function f as for the simple epidemic, and need a model of the term $g(I)$ representing recovery from the disease. The simplest is

$$g(I) = \gamma I,$$

where γ is the *rate of recovery*. From an individual-based point of view, the argument is that each infective has a probability $\gamma \delta\tau + O(\delta\tau^2)$ of leaving the infective class in the next small interval of time $\delta\tau$. You are asked to show below that the amount of time an infective spends in the I class is exponentially distributed with mean $1/\gamma$. In reality, we would expect the probability of leaving the infective class to depend on the length of time that the individual had already spent there, and later we shall discuss a class of models that allows for this.

The disease is represented diagrammatically as shown:

$$S \xrightarrow{\beta IS} I \xrightarrow{\gamma I} S.$$

Let us define non-dimensional variables by

$$u = \frac{S}{N}, \quad v = \frac{I}{N}, \quad t = \gamma \tau.$$

We obtain

$$\frac{du}{dt} = -(R_0 u - 1)v, \quad \frac{dv}{dt} = (R_0 u - 1)v. \tag{3.2.1}$$

The equations are to be solved on the one-dimensional simplex (or line segment) $S_1 = \{(u,v)|0 \le u \le 1, 0 \le v \le 1, u + v = 1\}$. Since $u + v = 1$, we could substitute for either u or v, but it is helpful to retain both equations. In the equations,

$$R_0 = \frac{\beta N}{\gamma} \tag{3.2.2}$$

is the *basic reproductive ratio*. Interpreting the formula, βN is the rate at which a single infective introduced into a susceptible population of size N makes infectious contacts, and $\frac{1}{\gamma}$ is the expected length of time such an infective remains infectious, so R_0 is the expected number of infectious contacts made by such an infective. Thus it is analogous to the basic reproductive ratio in population dynamics or metapopulation theory. It is the most important concept in the chapter, and will recur in each section.

Theorem 3.1 (Threshold for SIS Epidemic)

If $R_0 < 1$, the disease dies out, but if $R_0 > 1$, it remains endemic in the population.

Since $\dot{v} < (R_0 - 1)v$, then if $R_0 < 1$ the infected fraction decays exponentially to zero. If $R_0 > 1$, then the second of Equations (3.2.1) written in the form $\dot{v} = (R_0(1 - v) - 1)v$ implies that $v(t) \to v^* = 1 - R_0^{-1}$ as $t \to \infty$.

The dimensionless per capita growth rate $R_0 u - 1$ of infectives is reduced by the disease from $R_0 - 1$ when the population is wholly susceptible to $R_0 u^* - 1 = 0$ when steady state is reached. The system is analogous to a prey-predator system, with u the prey and v the predators; the predators limit the otherwise exponential growth of the prey. Here, the infectives satisfy the logistic equation with $r = R_0 - 1$ and $K = 1 - R_0^{-1}$. In dimensional terms, the initial per capita growth rate of the disease is given by

$$r = \gamma(R_0 - 1), \tag{3.2.3}$$

analogous to the relationship $r = d(R_0 - 1) = b - d$ in Malthusian growth (Section 1.3). The parameter R_0 tells us whether the disease will spread, but r tells us how fast it will do so.

EXERCISES

3.1. The *disease age* of an individual is the time τ since that individual was infected. For a certain disease, let each infective have probability $\gamma \delta \tau$ of leaving the infective class in the next small interval of time $\delta \tau$.

    a) Show that the probability that an individual of disease age τ is still infective is $\exp(-\gamma \tau)$, i.e. the time spent in the infective class is exponentially distributed.

    b) Show that the mean time spent in the infective class is $1/\gamma$.

3.2. Consider the initial phase of an epidemic, where there are so few infectives that $S \approx N$, and the rate at which infectious contacts are made may be approximated by $\beta I N$. Let the probability that an individual of disease age σ is still in the infective class be $f(\sigma)$.

    a) The *incidence* i of a disease is the rate at which new cases occur. Show that

$$i(\tau) = \beta N \int_0^\infty f(\sigma) i(\tau - \sigma) d\sigma.$$

    b) For the usual exponentially distributed time in the infective class $f(\sigma) = \exp(-\gamma \sigma)$. Try a solution for the incidence of the form $i(\tau) = i_0 \exp(r\tau)$, and find r.

    c) What is the relationship between r and R_0 for this disease?

3.3. *SEI diseases.* Consider the simple epidemic with an exposed (latent) state E between susceptible and infective, whose members have contracted the disease but are not yet infectious. Explain the equations

$$\frac{dS}{d\tau} = -\beta IS, \quad \frac{dE}{d\tau} = \beta IS - \delta E, \quad \frac{dI}{d\tau} = \delta E$$

for the SEI disease.

3.4. Consider an SIS disease with infective period exactly τ_I and incidence function $i(\tau)$.

a) Argue that

$$I(\tau) = \int_{\tau-\tau_I}^{\tau} i(\sigma)d\sigma,$$

(or equivalently $\frac{dI}{d\tau}(\tau) = i(\tau) - i(\tau - \tau_I)$).

b) If $i(\tau) = \beta I(\tau)S(\tau)$, deduce that the steady state I^* satisfies $I^* = \beta I^*(N - I^*)\tau_I$, so that $I^* = N - 1/(\beta\tau_I)$.

c) What is R_0 in this situation?

3.3 SIR Epidemics

In the SIR disease individuals leaving the infective class play no further role in the disease. They may be immune, or dead, or removed by an isolation policy or otherwise. Most childhood diseases, such as measles, have such a *removed* class R. The disease can be represented diagrammatically by

$$S \to I \to R.$$

In this section we shall look at the simplest model in this class, which dates back to a classic paper of 1927 by Kermack and McKendrick. The results are basic to mathematical epidemic modelling. We assume that the duration of the epidemic is short compared to the lifetime of its hosts, so that we can neglect birth and disease-unrelated death. The population is therefore closed, of constant size N, and

$$N = S(\tau) + I(\tau) + R(\tau).$$

We model the movement between the classes as follows:

$$\frac{dS}{d\tau} = -\beta IS, \quad \frac{dI}{d\tau} = \beta IS - \gamma I, \quad \frac{dR}{d\tau} = \gamma I. \qquad (3.3.4)$$

This may be represented diagrammatically by

$$S \quad \xrightarrow{\beta IS} \quad I \quad \xrightarrow{\gamma I} \quad R.$$

Let us non-dimensionalise the equations by defining

$$u = \frac{S}{N}, \quad v = \frac{I}{N}, \quad w = \frac{R}{N}, \quad t = \gamma\tau.$$

The equations become

$$\frac{du}{dt} = -R_0 uv, \quad \frac{dv}{dt} = (R_0 u - 1)v, \quad \frac{dw}{dt} = v, \qquad (3.3.5)$$

where $R_0 = \beta N/\gamma$ is the basic reproductive ratio. This system is to be solved on the two-dimensional simplex (or triangle) $S_2 = \{(u, v, w) | 0 \leq u \leq 1, 0 \leq v \leq 1, 0 \leq w \leq 1, u + v + w = 1\}$. However the first two equations do not involve w, and we shall look at the flow determined by these equations on the projection of this simplex onto the (u, v)-plane, the triangle Δ bounded by the axes and the line $u + v = 1$. The u-axis is a nullcline for both equations, so that any point on it is a steady state. If $R_0 > 1$ the nullcline $R_0 u - 1 = 0$ intersects this triangle, but if $R_0 < 1$ it does not. The phase plane in the two cases is sketched below. Equations for the trajectories may be found explicitly, by integrating Equations (3.3.6) below, but we are only interested in qualitative behaviour at the moment.

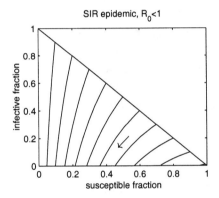
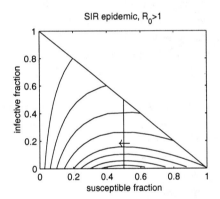

Figure 3.2   Phase plane for the SIR epidemic, in the cases $R_0 < 1$ and $R_0 > 1$.

Now let us confirm the stability properties of the disease-free state $(u, v) = (1, 0)$ suggested by Figure 3.2.

Example 3.2

Find the eigenvalues of the Jacobian of the first two equations of (3.3.5) at $(1, 0)$.

The Jacobian at $(1, 0)$ is given by

$$J = \begin{pmatrix} -R_0 v & -R_0 u \\ R_0 v & R_0 u - 1 \end{pmatrix}\Bigg|_{(1,0)} = \begin{pmatrix} 0 & -R_0 \\ 0 & R_0 - 1 \end{pmatrix},$$

which has eigenvalues $R_0 - 1$ and zero.

The steady state is stable but not asymptotically stable to perturbations along the u-axis, corresponding to the zero eigenvalue, and stable if $R_0 < 1$, unstable if $R_0 > 1$ to perturbations along the line $u + v = 1$, corresponding to the eigenvalue $R_0 - 1$ and representing the infection of a small number of susceptibles in an otherwise disease-free population. We have the following result.

Theorem 3.3 (Threshold for SIR Epidemic)

The disease-free steady state is stable (but not asymptotically stable) if $R_0 < 1$, so that the disease dies out, unstable if $R_0 > 1$, so that an epidemic may potentially occur.

Again the crucial parameter is the basic reproductive ratio $R_0 = \frac{\beta N}{\gamma}$, the average number of new cases produced by a single infective introduced into a purely susceptible population of size N. Again the initial per capita growth rate of the infectives is $r = \gamma(R_0 - 1)$ in dimensional terms.

It is important to find the size of the epidemic, the total number who will suffer from the disease. This is given by the number who are eventually in the removed class. We can find this by noting that the system (3.3.5) is separable in (u, v, w)-space. We have

$$\frac{dw}{du} = -\frac{1}{R_0 u}, \quad \frac{dv}{du} = -1 + \frac{1}{R_0 u}. \tag{3.3.6}$$

Trajectories in the simplex S_2 in (u, v, w)-space are found by integrating these equations. We wish to discover where the trajectory T that starts at the disease-free steady state $(1, 0, 0)$ ends up. Integrating the first of Equations (3.3.6) and applying the condition that $(1, 0, 0)$ is on T, we have

$$u = \exp(-R_0 w). \tag{3.3.7}$$

This equation is satisfied everywhere on T. Since, from Equations (3.3.5), u and w are monotonic bounded functions of t,

– they tend to limits, $u(t) \to u_1$, $w(t) \to w_1$, say, as $t \to \infty$, and

– $\frac{dw}{dt}(t) \to 0$, so $v(t) \to 0$, as $t \to \infty$.

Hence $(u(t), v(t), w(t)) \to (u_1, 0, w_1) = (1 - w_1, 0, w_1)$ as $t \to \infty$, and taking the limit of (3.3.7),

$$1 - w_1 = \exp(-R_0 w_1). \tag{3.3.8}$$

These two functions are shown in Figure 3.3, and it is clear that there is a unique positive root of Equation (3.3.8) for $R_0 > 1$. In contrast to the simple (SI) epidemic, susceptibles remain after the epidemic has passed.

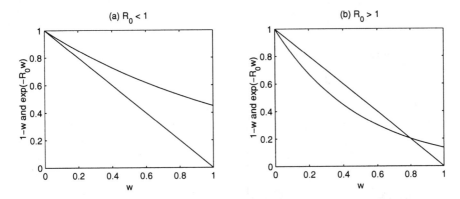

Figure 3.3   The functions $1 - w_1$ and $\exp(-R_0 w_1)$, for $R_0 < 1$ and $R_0 > 1$. The intersection point determines the final size of the epidemic, if any.

We have

– if $R_0 > 1$, the total number infected by the epidemic is $N w_1$, where w_1 is the unique positive root of Equation (3.3.8).

We can also analyse the time course of the epidemic. From Equations (3.3.5) and (3.3.7), we have

$$\frac{dw}{dt} = 1 - w - \exp(-R_0 w). \tag{3.3.9}$$

There is no closed form solution of this equation, and a numerical solution is often required, but analytical progress may be made for a small epidemic.

EXERCISES

3.5. Verify the qualitative features of Figure 3.3 for $R_0 < 1$ and $R_0 > 1$. Show that if $R_0 > 1$ then $R_0 u_1 < 1$, and interpret this biologically.

3.6. Let $R_0 w$ be small throughout the course of the epidemic.

a) Show that Equation (3.3.9) may be approximated by

$$\frac{dw}{dt} = (R_0 - 1) w \left(1 - \frac{w}{w_1} \right),$$

where $w_1 = 2(R_0 - 1)/R_0^2$.

b) Deduce Kermack and McKendrick's second threshold theorem, that if $(R_0 - 1)/R_0$ is small, the size of the epidemic is given approximately by $w_1 = 2(R_0 - 1)/R_0^2$.

c) Show that the incidence of death (the rate at which deaths occur) in a small epidemic should follow a sech^2 curve.

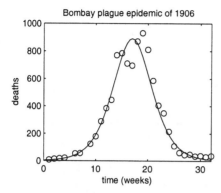

Figure 3.4   Bombay plague of 1906, showing typical disease progression. The epidemic was small (in the sense that its final size was a small fraction of the population size), and a fit is shown with a sech^2 curve.

3.7. Find the equations of the trajectories in Figure 3.2. [Hint: integrate the second of Equations (3.3.6).]

3.8. If the probability of death given infection is p, how many deaths does the epidemic cause?

3.9. In the Spanish invasion of South America, the Aztecs were devastated by a smallpox epidemic introduced by one of Cortez' men. Smallpox is an immunity-conferring or lethal disease. Take $N = 1000$, $\beta = 0.1$, and assume that the infective period for smallpox is two weeks.

a) Find R_0.

b) Compare the final size of the epidemic in the two cases (i) that the population is initially wholly susceptible and (ii) that the population is initially 70% immune. In case (i), show that more than 750 would die, and in case (ii) estimate the number.

3.10. *SIRS diseases.* Some diseases confer only temporary immunity, so that there is a flow of individuals from the R class back to the S class. For obvious reasons, these are known as SIRS diseases. Like SIS diseases, they may die out or remain endemic in the population.

a) Write down a model for such a disease if the time spent in the R class is exponentially distributed.

b) Find R_0 for this disease.

c) What is the condition for the disease to remain endemic?

3.11. *SEIR diseases.* Suppose that the population structure according to disease status is SEIR, where E represents an exposed (latent) state.

a) Explain the equations

$$\frac{dS}{d\tau} = -\beta IS, \quad \frac{dE}{d\tau} = \beta IS - \delta E, \quad \frac{dI}{d\tau} = \delta E - \gamma I, \quad \frac{dR}{d\tau} = \gamma I$$

for the SEIR disease.

b) What is R_0 in this situation?

c) Does a threshold theorem hold for this disease?

d) Consider the epidemic case, where the infection is introduced into an initially susceptible population of size N. Show that the final size of this epidemic is the same as the final size for the SIR epidemic.

3.12. The SIR and SEIR diseases are idealisations of the real situation, where the *infectivity* of an individual who has contracted the disease is a function f, called the *infectivity function*, of the time σ since the disease was contracted, the *disease age*. The easiest interpretation of the infectivity of an individual is the probability of transmission of the disease, given a contact between that individual and a susceptible.

a) For an SIR disease, let this probability be p if the individual is in the infective class, zero otherwise. Let the rate at which (not necessarily infectious) contacts take place between susceptibles and infectives be given by $\beta' IS$. Argue that $p\beta' = \beta$.

b) Argue that the basic reproductive ratio is given by

$$R_0 = \beta' N \int_0^\infty f(\sigma) d\sigma \qquad (3.3.10)$$

c) For an SIR disease with exponentially distributed time in the infective class, the probability that an individual with disease age σ is still in the infective class is given by $\exp(-\gamma\sigma)$. Show that the infectivity function f in this case is given by $f(\sigma) = p\exp(-\gamma\sigma)$.

d) Check that Equation (3.3.10) holds in this case.

e) Show that the infectivity function in the SEIR disease is given by

$$f(\sigma) = p\frac{\delta(\exp(-\delta\sigma) - \exp(-\gamma\sigma))}{\gamma - \delta}.$$

[Hint: in order to be infective at disease age σ, the exposed individual must enter the infective class at some disease age $s < \sigma$ and then remain there in the interval (s, σ). Calculate this probability for arbitrary s, and then integrate.]

f) Use Equation (3.3.10) to find the basic reproductive ratio in this case.

g) Why is R_0 independent of δ?

3.4 SIR Endemics

In this section we consider an endemic disease, habitually prevalent in a population. In the epidemic case we assumed that the duration of the disease was short compared to the life expectancy of the host, so that we could neglect any birth and disease-unrelated death. For an endemic we are interested in long-term behaviour, and this is no longer reasonable. It is also no longer sensible to lump together immune and dead people into the same (removed) class, as their differences are now important! R should now be considered the *immune* class.

With births and deaths included, the population is no longer closed, and the total population size N will only be constant under additional assumptions on the birth and death rates. We consider an open population with birth rate (*not* per capita birth rate) B and per capita disease-related and disease-unrelated death rates c and d. For simplicity we shall take c and d to be constant, but we shall make different assumptions about B. There is assumed to be no *vertical transmission*, so that all births are assumed to enter the susceptible class. Vertical transmission is transmission from parent to foetus or new-born offspring, and occurs for example in AIDS and BSE. The disease is represented diagrammatically below:

$$
\begin{array}{ccccc}
B & & \beta IS & \uparrow cI \quad \gamma I & \\
\longrightarrow & S & \longrightarrow & I & \longrightarrow & R. \\
& \downarrow dS & & \downarrow dI & & \downarrow dR
\end{array}
$$

There are two ways the population can approach an endemic steady state. First, we could have $B = bN$, $b = d$ and $c = 0$. Second, we could have the size of the population controlled by the disease. We shall investigate the possibilities in turn.

3.4.1 No Disease-related Death

Let $B = bN$, $b = d$ and $c = 0$. The equations are

$$\frac{dS}{d\tau} = bN - \beta IS - bS, \quad \frac{dI}{d\tau} = \beta IS - \gamma I - bI, \quad \frac{dR}{d\tau} = \gamma I - bR. \quad (3.4.11)$$

The total population size N is constant, and we can write $u = S/N$, $v = I/N$, $w = R/N$ as usual. However, we use a different non-dimensionalisation for time, $t = (\gamma + b)\tau$, since individuals leave the infective class at an increased rate because of the chance of disease-unrelated death in that class. We obtain

$$\frac{du}{dt} = \frac{b}{\gamma + b}(1 - u) - R_0 uv, \quad \frac{dv}{dt} = (R_0 u - 1)v, \quad \frac{dw}{dt} = \frac{\gamma}{\gamma + b}v - \frac{b}{\gamma + b}w. \quad (3.4.12)$$

The basic reproductive ratio R_0 of the disease is defined to be the expected number of infectious contacts made by a single infective in an otherwise totally susceptible population, as before, but now

$$R_0 = \frac{\beta N}{\gamma + b}, \quad (3.4.13)$$

since the mean time in the I class is reduced from $1/\gamma$ to $1/(\gamma + b)$. Normally $b \ll \gamma$, since the mean infectious period is much less than the life expectancy of the host, so this makes very little difference to the numerical value of R_0.

Equations (3.4.12) are to be solved on the simplex $S_2 = \{(u, v, w) | 0 \le u \le 1, 0 \le v \le 1, 0 \le w \le 1, u + v + w = 1\}$, but as before w is uncoupled from the system and we can consider them on the projection Δ of S_2 onto the (u, v)-plane. There is a disease-free state at $(1, 0)$; the eigenvalues of the Jacobian there are given by $-b/(\gamma + b)$ and $R_0 - 1$, so $(1, 0)$ is asymptotically stable (not just neutrally stable) if $R_0 < 1$, unstable if $R_0 > 1$. An endemic steady state (u^*, v^*) exists and is stable as long as $R_0 > 1$, where

$$u^* = \frac{1}{R_0}, \quad v^* = \frac{b}{\gamma + b}\left(1 - \frac{1}{R_0}\right).$$

Figure 3.5 shows a numerical solution of the equations with initial conditions $(u(0), v(0)) = (u_0, v_0)$, where $v_0 = 1 - u_0$ is small and positive. The initial epidemic phase is almost identical to the epidemic analysed in Section 3.3, and there is then a fast oscillatory approach to the endemic steady state. punctuated by smaller and smaller epidemics.

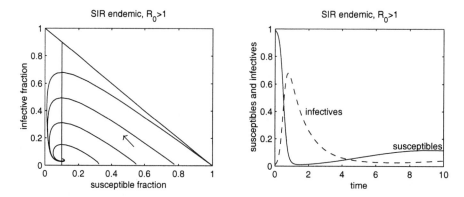

Figure 3.5   Numerical solution of Equations (3.4.12), for $R_0 > 1$ (in fact $R_0 = 10$). We have taken $\gamma/b = 0.1$, much larger than is realistic for most diseases, so that the final infective fraction is large enough to be easily seen. For $R_0 < 1$ the disease dies out.

Note that the equation for the infective class is given by

$$\frac{dv}{dt} = (R_0 u - 1)v,$$

so that the initial per capita growth rate $(R_0 - 1)$ is reduced to zero as $u(t) \to u^*$, and the disease controls the susceptible population so that this is so, just as in the SIS endemic. But u^* may undershoot this value markedly and recovery to it may be slow, as Figure 3.5 makes clear.

EXERCISES

3.13. Analyse the system of Equations (3.4.12) for an SIR epidemic with no disease-related death, and verify the results given in the text on existence and stability of the disease-free and the endemic steady states.

3.14. This exercise extends the model of this subsection to include the infectivity function f, where $f(\sigma)$ is the infectivity (i.e. the probability of transmission of the disease between the infective and a susceptible) of an individual who caught the disease a time σ ago.

   a) Explain why a death rate d from disease-unrelated causes is expressed as a factor $\exp(-d\sigma)$ in $f(\sigma)$.

   b) Deduce that a latency (exposed) period E of fixed length τ_E is reflected as a factor $\exp(-d\tau_E)$ in $f(\sigma)$.

c) Derive Equation (3.4.13) from part (a) of this question.

d) Consider the SEIR model of Exercise 3.11 with the addition of a disease-unrelated death rate d. During the exposed period there are competing risks, to die or to become infectious. If the latter occurs, we are again in the situation of part (c). Argue that

$$R_0 = \frac{\delta}{\delta + d} \frac{\beta N}{\gamma + d}.$$

3.4.2 Including Disease-related Death

If we let $B = bN$ with $b > d$ and $c > 0$, the equations are

$$\frac{dS}{d\tau} = bN - \beta IS - dS, \quad \frac{dI}{d\tau} = \beta IS - \gamma I - cI - dI, \quad \frac{dR}{d\tau} = \gamma I - dR. \quad (3.4.14)$$

There is a disease-free solution, $(S, I, R) = (N, 0, 0)$, with $S(t) = N(t) = N_0 \exp((b - d)t)$, which cannot be realistic indefinitely. The existence of this solution makes analysis awkward without the compensation of a realistic model, and it is normal to take constant birth rate B instead of constant per capita birth rate. The equations then become

$$\frac{dS}{d\tau} = B - \beta IS - dS, \quad \frac{dI}{d\tau} = \beta IS - \gamma I - cI - dI, \quad \frac{dR}{d\tau} = \gamma I - dR. \quad (3.4.15)$$

Adding these, we obtain

$$\frac{dN}{d\tau} = B - cI - dN. \quad (3.4.16)$$

Any three of these four equations, together with $N = S + I + R$, are the equations we need to analyse. We shall choose the (N, S, I) equations. Since the population is open, we cannot reduce the system to two equations as we have done before.

There is a disease-free steady state $(N, S, I) = (N_0^*, N_0^*, 0)$, with $N_0^* = B/d$, and an infective introduced into this steady state expects to make $R_0 = \frac{\beta B}{d(\gamma + c + d)}$ infectious contacts. There is an endemic steady state $(N, S, I) = (N_1^*, S_1^*, I_1^*)$, with $S_1^* = (\gamma + c + d)/\beta = B/(dR_0)$, $I_1^* = (B - dS_1^*)/(\beta S_1^*) = d(R_0 - 1)/\beta$, and $N_1^* = (B - cI_1^*)/d$, as long as $R_0 > 1$.

The endemic steady state can be shown to be stable whenever it exists (in the positive octant), i.e. whenever $R_0 > 1$. The usual threshold behaviour with threshold $R_0 = 1$ holds for this disease. The approach to the endemic steady state is a damped oscillation, with the frequency of the oscillation given approximately by $\omega = \sqrt{\beta^2 I_1^* S_1^*}$, so that the period is

$$T = \frac{2\pi}{\omega} = \frac{2\pi}{\sqrt{(\gamma + c + d)d(R_0 - 1)}}. \quad (3.4.17)$$

In the absence of vaccination, diseases such as measles often exhibit periodic behaviour, with a period close to T but driven by exogenous influences such as the start of the school year. This counteracts the damping effect. A typical time series for measles is shown in Figure 3.6.

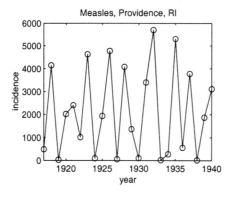

Figure 3.6   A typical time series for measles without vaccination, from Providence, Rhode Island, USA.

EXERCISES

3.15. Consider Equations (3.4.16) and the first two of (3.4.15), together with $N = S + I + R$, for an SIR endemic.

a) Confirm the results on existence and stability of the disease-free and the endemic steady states given in the text.

b) Confirm the result (3.4.17) for the period of the damped oscillations approaching the endemic steady state.

c) Measles has an infective period of twelve days, and $R_0 \approx 12 - 13$. Estimate the period of outbreaks of the disease in a population with a life expectancy of 70 years.

3.5 Eradication and Control

One of the reasons for making models of infectious diseases is to enable us to design policies aimed at eradicating or at least controlling them. Such control policies might aim to reduce R_0, the basic reproductive ratio, below 1. In the simplest (closed population) models above $R_0 = \frac{\beta N}{\gamma}$, so there are three strategies available to us.

- Increase γ, the rate of removal of infectives.

- Decrease β, the pairwise infectious contact rate.

- Decrease the effective value of N, which should be interpreted as decreasing the initial susceptible population.

Control measures in the recent foot-and-mouth epidemic in the UK involved slaughtering infected animals (increasing γ), disinfection and movement controls to prevent the spread of the virus (reducing β), and slaughtering potential contacts of the infected animals (reducing N). One method which was not employed was vaccination, which also reduces N.

Example 3.4 (Vaccination Against an SIR Epidemic)

Imagine that there is a threat of infection being introduced to the closed population described in Section 3.3, and assume that we have a perfect vaccine against the disease. What proportion p of the population do we have to vaccinate in a one-off programme to remove the threat of an epidemic?

Vaccination essentially moves a proportion p of the initial susceptible population from the susceptible class to the removed class, leaving a proportion $q = 1 - p$ there. An epidemic will potentially occur if the new steady state $(u, v) = (q, 0)$ is unstable. The Jacobian has eigenvalues $qR_0 - 1$ and zero, so the steady state is stable for $qR_0 < 1$. We have to vaccinate a proportion $p \geq \hat{p} = 1 - R_0^{-1}$ to remove the threat of an epidemic.

What happens in the endemic situation? In this case, rather than a one-off vaccination of the population, we vaccinate a proportion of susceptibles as they join the population. (In fact we do not vaccinate new-born babies, partly because their maternal anti-bodies imply that the vaccination will not be effective, but we assume that vaccination takes place early enough that this approximation is reasonable.) In the case discussed in Section 3.4.1 Equations (3.4.11) are replaced by

$$\frac{dS}{d\tau} = bqN - \beta IS - bS, \quad \frac{dI}{d\tau} = \beta IS - \gamma I - bI, \quad \frac{dR}{d\tau} = bpN + \gamma I - bR, \quad (3.5.18)$$

where again $q = 1 - p$, the fraction unvaccinated. It may be shown that there is no steady state with disease present, so that the disease will die out, if $qR_0 - 1 \leq 0$, $p \geq \hat{p} = 1 - R_0^{-1}$, just as in the epidemic case.

The same result may be shown to hold more generally, and we have the following result.

- Vaccination of a fraction $\hat{p} = 1 - R_0^{-1}$ of susceptibles is enough to prevent a disease spreading.

Although there are still susceptibles in the population, the disease does not spread, so the population as a whole is in a sense immune. This is called *herd immunity*.

Estimates of R_0 and hence $\hat{p}$ for various diseases are given below. Most of these data are from England and Wales, USA, or other developed countries; values of R_0 in developing countries tend to be higher, especially in densely populated countries such as those of the Indian subcontinent. The values for smallpox are from the developing world but are still very low. It is partly these low R_0 values, and the consequent low level of coverage required for elimination, that enabled the success of the global campaign for the eradication of smallpox. The $\hat{p}$ values here assume a vaccine that is 100% effective; even higher coverage would be required for less effective vaccines such as that for pertussis (whooping cough).

Infection	R_0	$\hat{p}$, %
Smallpox	3–5	67–80
Measles	12–13	92
Pertussis (whooping cough)	13–17	92–94
Rubella (German measles)	6–7	83–86
Chickenpox	9–10	89–90
Diphtheria	4–6	75–83
Scarlet fever	5–7	80–86
Mumps	4–7	75–86
Poliomyelitis	6	83

EXERCISES

3.16. Show that the disease of Equations (3.5.18) dies out if $p \geq \hat{p} = 1 - 1/R_0$, as claimed in the text.

3.17. Consider a population of size N with per capita birth rate $b(N)$ and death rate $d(N)$. Assume that it reaches a stable steady state N^* in the absence of disease, with $b(N^*) = d(N^*)$.

   a) Explain the equations

$$\frac{dS}{d\tau} = b(N)N - \beta IS - d(N)S,$$

$$\frac{dI}{d\tau} = \beta IS - \gamma I - cI - d(N)I,$$

$$\frac{dR}{d\tau} = \gamma I - d(N)R$$

in the presence of disease.

b) What is R_0 when this disease is introduced into a population at the disease-free steady state?

c) Analyse the stability of the disease-free steady state with $N = N^*$. Show that the disease can invade if $\beta N^* > \gamma + c + d(N^*)$.

d) What fraction of the population must be vaccinated at birth to drive the disease to extinction?

3.6 Age-structured Populations

If infectious contact rates are different at different ages, a classic example being that of measles in developed countries before vaccination programmes were set up being driven by the start of school at age 5 or so, then it becomes important to include age structure in the models. Alternatively, age at infection may be important, especially if the disease is more serious at some ages than others. We have already looked at age structure in single populations in Chapter 1, and we shall use the McKendrick approach that we introduced in Section 1.11, but with a somewhat different notation to conform to that used in this chapter. Discrete time Leslie-matrix-type models could also be used.

3.6.1 The Equations

We define the densities

- $s(a,t)$ of susceptibles of age a at time t,
- $i(a,t)$ of infectives of age a at time t,
- $r(a,t)$ of removed individuals of age a at time t,
- $n(a,t) = s(a,t) + i(a,t) + r(a,t)$ of hosts of age a at time t,

so that for example $s(a,t)\delta a$ is the number of susceptibles aged between a and $a + \delta a$ at time t. Then the equations satisfied by these variables are given by

$$\frac{\partial s}{\partial t}(a,t) + \frac{\partial s}{\partial a}(a,t) = -(\lambda(i)(a,t) + d(a))s(a,t),$$

$$\frac{\partial i}{\partial t}(a,t) + \frac{\partial i}{\partial a}(a,t) = \lambda(i)(a,t)s(a,t) - (\gamma(a) + c(a) + d(a))i(a,t), \quad (3.6.19)$$

$$\frac{\partial r}{\partial t}(a,t) + \frac{\partial r}{\partial a}(a,t) = \gamma(a)i(a,t) - d(a)r(a,t).$$

Here $\lambda(i)$, the force of infection, is a functional, i.e. it depends on the function i which gives the density of infectives of all ages. It will in general also depend on the pairwise infectious contact rate between a susceptible of age a and an infective of age a'. Let us call this rate $\beta(a, a')$. Then a straightforward generalisation of the model $\lambda(I) = \beta I$ used in previous sections gives

$$\lambda(i)(a, t) = \int_0^\infty \beta(a, a')i(a', t)da'. \tag{3.6.20}$$

Equations (3.6.19) need to be supplemented by initial conditions at $t = 0$, and age-zero conditions at $a = 0$ specifying the birth rate $B(t)$, as in Section 1.11. We assume as usual that there is no vertical transmission, so that all individuals are born into the susceptible class. If $b(a)$ is the maternity function, then we have

$$B(t) = s(0, t) = \int_0^\infty b(a)n(a, t)da. \tag{3.6.21}$$

(In Section 1.11, the notation $m(a)$ rather than $b(a)$ was used for the maternity function, and the notation $b(t)$ rather than $B(t)$ for birth rate.)

Equations (3.6.19) are a set of coupled nonlinear partial integro-differential equations, and not surprisingly a general analysis is rather complicated. We shall now make some drastic simplifying assumptions, in search of some insight into the average age at infection, its connection with R_0, and how it is affected by a vaccination programme. We shall assume that the pairwise infectious contact rate $\beta(a, a')$, the disease-unrelated death rate $d(a)$, the maternity function $b(a)$, the disease-related death rate $c(a)$, and the recovery rate $\gamma(a)$ are all independent of age.

If β is constant, Equation (3.6.20) reduces to

$$\lambda(i)(t) = \beta \int_0^\infty i(a', t)da' = \beta I(t). \tag{3.6.22}$$

The force of infection is proportional to the total number of infectives, as before. (We shall generally use upper case letters to denote the total number of susceptibles, infectives or removed individuals.)

EXERCISES

3.18. With β independent of a and a', c, d and γ independent of a, show that Equations (3.6.19) with (3.6.20) may be integrated with respect to a from 0 to ∞ to recover Equations (3.4.15), with $S(t) = \int_0^\infty s(a, t)da$, etc, and that Equation (3.6.21) reduces to $B(t) = bN(t)$.

3.6.2 Steady State

Guided by the results of Chapter 1, let us assume that the system settles down to a steady age distribution, which we wish to find. In the steady state $\frac{\partial}{\partial t} = 0$, and we no longer require initial conditions. We shall take the disease-related death rate $c = 0$, and assume that the birth rate and the disease-unrelated death rate are age-independent and equal (to b), and the pairwise infectious contact rate is independent of the ages of the susceptible and the infective, $\beta(a, a') = \beta$, constant. The basic reproductive ratio is $R_0 = \frac{\beta N}{\gamma + b}$. The equations become

$$\frac{ds}{da}(a) = -(\lambda(i) + b)s(a) = -(\beta I + b)s(a),$$

$$\frac{di}{da}(a) = \lambda(i)s(a) - (\gamma + b)i(a) = \beta I s(a) - (\gamma + b)i(a), \qquad (3.6.23)$$

$$\frac{dr}{da}(a) = \gamma i(a) - br(a),$$

with

$$s(0) = bN, \quad i(0) = 0, \quad r(0) = 0, \qquad (3.6.24)$$

so that $n(0) = bN$. Adding the equations together and solving with the birth conditions (3.6.21), $n(a) = n(0)l(a) = bN \exp(-ba)$. The function $l(a) = \exp(-ba)$ is the *survivorship function* of Chapter 1, exponentially distributed because of our assumption of type II mortality, i.e. that the per capita death rate d (equal to b) is independent of age. Since I and hence the force of infection λ is constant, we can also solve the other equations to obtain

$$s(a) = bN e^{-(\lambda + b)a},$$

$$i(a) = bN \frac{\lambda}{\lambda - \gamma}(e^{-(\gamma + b)a} - e^{-(\lambda + b)a}), \qquad (3.6.25)$$

$$r(a) = n(a) - s(a) - i(a).$$

Integrating,

$$S = \frac{bN}{\lambda + b}, \quad I = \frac{\lambda bN}{(\lambda + b)(\gamma + b)}, \quad R = \frac{\lambda \gamma N}{(\lambda + b)(\gamma + b)}.$$

Since the force of infection is $\lambda = \beta I$, the second of these leads to an equation we can solve for λ, to give

$$\lambda = \frac{\beta bN}{\gamma + b} - b = b(R_0 - 1). \qquad (3.6.26)$$

Our motivation for introducing age-dependent models has mainly been to find the average age at infection, which is of interest epidemiologically for reasons that will become clear. The number acquiring the infection between ages a

and $a + \delta a$ is given by $\lambda s(a)\delta a$ to first order, so that the average age at infection is

$$\bar{a} = \frac{\int_0^\infty a\lambda s(a)da}{\int_0^\infty \lambda s(a)da} = \frac{1}{\lambda + b} = \frac{1}{bR_0} \qquad (3.6.27)$$

In public health $\bar{a}$ is easy to measure, and so may be used to estimate the force of infection and the basic reproductive ratio, which are not so easily measured directly. Defining life expectancy by $\bar{\omega} = 1/b$, we have

$$R_0 = \frac{\bar{\omega}}{\bar{a}}. \qquad (3.6.28)$$

This is still approximately true for other mortality schedules than than the type II mortality that we have analysed here.

Now assume that a vaccination programme is put in place to reduce the incidence of a disease, vaccinating a fraction p of susceptibles soon after birth. The effect is to change the age-zero (birth) conditions to

$$s(0) = qbN, \quad i(0) = 0, \quad r(0) = pbN, \qquad (3.6.29)$$

where $q = 1 - p$ is the fraction unvaccinated, while keeping everything else unchanged. Thus R_0 becomes $R_0' = qR_0$, and the average age at infection $\bar{a}$ becomes $\bar{a}' = \bar{a}/q$. Vaccination increases average age at infection.

This increased age at infection may be important from the public health point of view. Vaccination against a disease which is more serious in younger individuals leads to greater gains than would be estimated simply by counting cases, whereas vaccination against a disease which is more serious in older individuals leads to smaller gains, or even perhaps to a net loss. Rubella (german measles) is usually a mild infection, but there is a high risk of foetal damage through congenital rubella syndrome if the disease is contracted in the first trimester of pregnancy, leading to possible cataracts, mental retardation, deafness and cardiac defects. The benefits of rubella vaccination in reduced incidence have to be carefully weighed against the increased chance that a given case leads to this syndrome. A policy of vaccination before entry into kindergarten (as practised in every state in the USA) may be inferior to a policy of vaccinating girls only just before puberty (as was practised in the UK until the recent introduction of the triple MMR vaccine against measles, mumps and rubella). A policy of early vaccination could lead to a 10-fold increase in cases of congenital rubella syndrome in a country like The Gambia, where the average age of infection without vaccination is only 2–3 years, unless unrealistically high levels of coverage were attained.

3.7 Vector-borne Diseases

Malaria and other vector-borne diseases are not contagious (spread by contact), but are caused by a parasite whose life-cycle is spent partly in an intermediate host. They exhibit *criss-cross infection*, the vector infecting the host and the host then infecting another vector. In the case of malaria the vector is the *Anopheles* mosquito. The infectious agent is a protozoan parasite, which is injected into the human bloodstream by a female mosquito when she is taking a blood meal, necessary for the development of her eggs. The parasite develops in the human and eventually produces gametocytes, which may be taken up by a biting mosquito. The gametocytes fuse to form zygotes, which develop and eventually make their way to the salivary glands of the mosquito, ready to start the life cycle over again. To model malaria we have to track the disease status of both the human and the mosquito population. We assume that the human population is closed, of size N_1, and that the mosquitoes do not die of malaria and have equal birth and death rates, so that the mosquito population size N_2 is also constant.

Is malaria an SIS or an SIR disease? Certainly reinfections of the same individual human can occur, but the rate of infection decreases with exposure. However, for the simplest model, we shall assume that no acquired immunity occurs either for humans or for mosquitoes. Then we have

$$\frac{dS_1}{d\tau} = -f_1(S_1, I_2) + \gamma_1 I_1, \quad \frac{dI_1}{d\tau} = f_1(S_1, I_2) - \gamma_1 I_1,$$

$$\frac{dS_2}{d\tau} = -f_2(S_2, I_1) + \gamma_2 I_2 + b_2 N_2 - d_2 S_2, \quad \frac{dI_2}{d\tau} = f_2(S_2, I_1) - \gamma_2 I_2 - d_2 I_2,$$

with $b_2 = d_2$. The incidence function for the humans is given by

$$f_1(S_1, I_2) = ap_1 \frac{S_1}{N_1} I_2.$$

The parameter a is the biting rate of the mosquitoes, i.e. each female mosquito takes on average $a\delta\tau + O(\delta\tau^2)$ blood meals from humans in a small interval of time $\delta\tau$. This functional response of the mosquitoes to humans is assumed constant, so that we are on the saturating part of the response curve. This is equivalent to assuming that there are always enough humans for the mosquitoes to find the blood meals that they require. A fraction $\frac{S_1}{N_1}$ of these meals are taken from susceptibles, and each leads to infection with probability p_1.

The incidence function for mosquitoes is given by

$$f_2(S_2, I_1) = ap_2 S_2 \frac{I_1}{N_1}.$$

This is made up in the same way as f_1, with p_2 denoting the probability that a susceptible insect will become infected on biting an infectious human.

Let us define R_0 to be the number of secondary cases in humans that we expect to be produced by a single primary case (in humans) introduced into a wholly susceptible population (of mosquitoes and humans). The human case leads to $ap_2 N_2/(\gamma_1 N_1)$ cases in mosquitoes, each of which leads to $ap_1/(\gamma_2 + b_2)$ cases in humans. Thus

$$R_0 = \frac{a^2 p_1 p_2 N_2}{\gamma_1 N_1 (\gamma_2 + b_2)}. \tag{3.7.30}$$

Simplifying the equations by defining

$$u_1 = \frac{S_1}{N_1}, \quad v_1 = \frac{I_1}{N_1}, \quad u_2 = \frac{S_2}{N_2}, \quad v_2 = \frac{I_2}{N_2},$$

we obtain

$$\frac{dv_1}{d\tau} = \gamma_1(\alpha_1 v_2 u_1 - v_1), \quad \frac{dv_2}{d\tau} = (\gamma_2 + b_2)(\alpha_2 v_1 u_2 - v_2),$$

with $u_1 + v_1 = 1$, $u_2 + v_2 = 1$, where $\alpha_1 = ap_1 N_2/(\gamma_1 N_1)$, $\alpha_2 = ap_2/(\gamma_2 + b_2)$. There is a steady state of the system at $(v_1, v_2) = (0, 0)$, and a non-trivial one at (v_1^*, v_2^*), where

$$v_1^* = \frac{\alpha_1 \alpha_2 - 1}{\alpha_1(\alpha_2 + 1)}, \quad v_2^* = \frac{\alpha_1 \alpha_2 - 1}{\alpha_2(\alpha_1 + 1)}.$$

This is biologically realistic if $\alpha_1 \alpha_2 > 1$, i.e. if

$$R_0 = \frac{a^2 p_1 p_2 N_2}{\gamma_1 N_1 (\gamma_2 + b_2)} > 1.$$

In this case it is stable, so malaria remains endemic if and only if $R_0 > 1$.

EXERCISES

3.19. *Sexually transmitted diseases.* (Sexually transmitted diseases in heterosexual populations and vector-borne diseases both exhibit crisscross infection.) Gonorrhoea is a bacterial infection and confers very little if any immunity. A large proportion of infectious females are asymptomatic.

   a) Explain the following model for the spread of gonorrhoea in a heterosexual population:

$$\frac{dS_1}{d\tau} = -\beta_{21} I_2 S_1 + \gamma_1 I_1, \quad \frac{dS_2}{d\tau} = -\beta_{12} I_1 S_2 + \gamma_2 I_2,$$

$$\frac{dI_1}{d\tau} = \beta_{21} I_2 S_1 - \gamma_1 I_1, \quad \frac{dI_2}{d\tau} = \beta_{12} I_1 S_2 - \gamma_2 I_2.$$

Why should β_{12} be different from β_{21}? Why should γ_1 be different from γ_2?

b) Show that this fourth-order system may be reduced to the non-dimensional second-order system given by

$$\frac{dv_1}{d\tau} = \gamma_1 R_{01}(1 - v_1)v_2 - v_1, \quad \frac{dv_2}{d\tau} = \gamma_2 R_{02}(1 - v_2)v_1 - v_2,$$

giving expressions for R_{01} and R_{02}.

c) For what values of the parameters does an epidemic occur if the disease is introduced into an otherwise wholly susceptible population? Interpret this in terms of an overall basic reproductive ratio.

3.20. Consider the introduction of an immunity-conferring sexually transmitted disease into an initially wholly susceptible heterosexual population.

a) Make a mathematical model of this disease, modifying that of Exercise 3.19 appropriately.

b) Derive a threshold theorem in this case.

c) Obtain the transcendental equations that determine the final size of an epidemic.

3.8 Basic Model for Macroparasitic Diseases

Many tropical and other diseases are caused by macroparasites, multicellular organisms such as helminths (worms), including flukes, tapeworms and round-worms, and arthropods, including lice, fleas and ticks. They differ from the microparasites that we have considered so far in generally having much longer generation times and more complex life-cycles, which often involve more than one host species. For simplicity, we shall assume that the parasite infects a single host species and does so at the adult stage, while the larvae live freely in the environment, maturing when they encounter and infect a host.

In such diseases the degree of infestation, i.e. the number of parasites harboured, is often an important indicator of its severity. Mathematical models therefore need to take the degree of infestation into account. We define H_i to be the number of hosts harbouring i parasites. The dynamics of the interaction are shown diagrammatically below. Here λ is the force of infection (the probability density of a host acquiring a (or another) parasite), and δ the parasite death rate.

$$H_0 \xrightarrow[\lambda H_0]{\overset{\delta H_1}{\longleftarrow}} H_1 \xrightarrow[\lambda H_1]{\overset{2\delta H_2}{\longleftarrow}} H_2 \xrightarrow[\lambda H_2]{\overset{3\delta H_3}{\longleftarrow}} H_3 \quad \cdots$$

We shall consider the epidemic situation, analogous to Section 3.3, where the birth and death rates of the hosts are neglected. The equations for the hosts are

$$\frac{dH_0}{dt} = -\lambda H_0 + \delta H_1, \tag{3.8.31}$$

$$\frac{dH_i}{dt} = -(\lambda + i\delta)H_i + (i+1)\delta H_{i+1} + \lambda H_{i-1} \tag{3.8.32}$$

for $i \geq 1$. Let us define the total number of hosts by $N = \sum_{i=0}^{\infty} H_i$. Adding the equations together, we obtain

$$\frac{dN}{dt} = \sum_{i=0}^{\infty} \frac{dH_i}{dt} = 0,$$

so that N is constant, as expected.

The equations so far are not sufficient to determine H_i, since the force of infection λ depends on the parasite population, so that we need to derive equations for the parasites. Let L be the number of larval parasites in the environment, M the number of mature parasites in the hosts. A diagrammatic representation of the parasite dynamics of a macroparasitic disease is given below. Here b is the parasite per capita birth rate, d the larval per capita death rate, and λ and δ are as before.

$$\xrightarrow{bM} L \xrightarrow{\lambda N} M$$
$$\quad\quad dL \downarrow \quad\quad \delta M \downarrow$$

Since M is the total number of mature parasites in the hosts,

$$M = \sum_{i=1}^{\infty} iH_i.$$

The equations for the larvae are

$$\frac{dL}{dt} = bM - \lambda N - dL. \tag{3.8.33}$$

The force of infection λ multiplied by the number of hosts N gives the rate at which larvae infect hosts, and are therefore removed from the larval pool. These immediately become mature, so that

$$\frac{dM}{dt} = \lambda N - \delta M. \tag{3.8.34}$$

Note that this could also have been obtained by multiplying the ith host equation by i and adding.

Finally, we need to model the force of infection λ. Of course, λ depends on L, and we shall take $\lambda = \theta L$, where θ is a constant. We have assumed here that larvae released into the environment immediately become infective, and that as soon as larvae infect a host they become mature. In a real system there will be delays built in to these processes. For example, the nematode *Ascaris lumbricoides*, a roundworm that parasitises humans, releases eggs to the environment that take 10–30 days to become infective larvae, and the time from infection to egg production is 50–80 days. This compares to a lifespan of the infective stage of 4–12 weeks and a lifespan of the mature stage of 1–2 years. This nematode can reach a length of 30 cm and large numbers can block the gut and kill its host. It is estimated that between 800 and 1000 million people are infected with it.

The equations for the larvae are therefore

$$\frac{dL}{dt} = bM - \theta NL - dL, \tag{3.8.35}$$

where b is the per capita rate at which adult parasites produce larvae, and d is the per capita death rate for larvae.

We now have two linear equations with constant coefficients, (3.8.34) and (3.8.35), for the numbers of adult and larval parasites in the system. It is straightforward to show that both populations grow exponentially if the determinant of the corresponding Jacobian matrix is negative, $\delta(\theta N + d) - b\theta N < 0$.

What is the basic reproductive ratio R_0 for such a disease? We interpret it as being the expected number of offspring of an adult parasite that survive to breed. An adult produces larvae at a rate b for expected time $1/\delta$, and each larva has an average lifespan $1/(\theta N + d)$, during which it has probability θN per unit time of maturing successfully. Thus

$$R_0 = \frac{b}{\delta} \frac{\theta N}{\theta N + d}. \tag{3.8.36}$$

The condition for exponential growth of the parasite population, unsurprisingly, is simply $R_0 > 1$.

The lifespan of the infective stage is often an order of magnitude smaller than that of the mature stage, as in *Ascaris lumbricoides*. This suggests that the quasi-steady-state approximation $\dot{L} \approx 0$ (to be discussed in more detail in Chapter 6) will give good results. With this approximation, the force of infection is given by

$$\lambda = \theta L = \frac{\theta b M}{\theta N + d},$$
(3.8.37)

and the equation for M is

$$\frac{dM}{dt} = \delta(R_0 - 1)M.$$
(3.8.38)

Again we can deduce exponential growth for $R_0 > 1$.

EXERCISES

3.21. Schistosomiasis is a disease caused by schistosomes or blood flukes, nematode worms that in their sexual adult stage parasitise humans, and spend the majority of the rest of their life in a molluscan host. It is endemic in many tropical countries, and was so common in ancient Egypt that passing blood in the urine, a symptom of infection, was seen as the male equivalent of menstruation. A model for the disease is given by

$$\frac{dW}{d\tau} = \lambda_1 - \delta W, \quad \frac{dI}{d\tau} = \lambda_2(N - I) - dI,$$

where W is the mean worm burden in human hosts, N the total number and I the number of infected snails, and λ_1 and λ_2 the force of infection for humans and snails respectively.

a) How does the W equation relate to the M Equation (3.8.34)? Explain why λ_2 is multiplied by $N - I$ but λ_1 is not multiplied by an equivalent expression.

b) A model for λ_2 is given by

$$\lambda_2 = \frac{bW^2}{1 + W},$$

quadratic (depensatory) for small W because in this case it may be difficult for the worms to find a mate, linear (zero compensation) for large W. A model for λ_1 is $\lambda_1 = cI$. Find the steady states of the system with these assumptions and determine their stability.

c) What would you expect to happen to the system if c were gradually reduced?

d) Derive a single (approximate) equation for W if the quasi-steady-state approximation $\frac{dI}{d\tau} \approx 0$ holds.

3.9 Evolutionary Aspects

It is in the interests of a parasite population not to over-exploit its hosts, just as it is in the interests of a predator population not to over-exploit its prey. But over-exploitation is rife, as we know only too well in the case of human predators. The reason is that selection does not act on populations but on individuals, and it is therefore the interests of individuals that will decide the course of evolution. Unless there are unusual circumstances so that group selection can take place, the simple argument that over-exploitation will not occur because it is not in the interests of the population is not watertight. We therefore wish to consider whether parasite virulence will be reduced through evolution. Virulence has two meanings in everyday use, but here it is defined as harmfulness rather than infectiousness, and more explicitly as the probability that an infected individual will die of the infection.

Consider a strain of parasite that has reached an equilibrium with its host, and consider a second strain, rare at first, maybe produced by a genetic mutation. Will this second strain invade the equilibrium that has been set up? If so, will it completely displace the first strain or will it coexist with it? We shall consider these questions in the context of an SIR endemic with disease-related death, as in Section 3.4.2.

We shall make the simplifying assumption that each strain provides immunity to the other immediately on infection and indefinitely. Each individual is susceptible (to either strain), or infected (by one strain or the other), or recovered from one strain and immune to both. The equations become

$$
\begin{aligned}
\frac{dS}{d\tau} &= bN - \beta_1 S I_1 - \beta_2 S I_2 - dS, \\
\frac{dI_1}{d\tau} &= \beta_1 S I_1 - \gamma_1 I_1 - c_1 I_1 - dI_1, \\
\frac{dI_2}{d\tau} &= \beta_2 S I_2 - \gamma_2 I_2 - c_2 I_2 - dI_2, \\
\frac{dR}{d\tau} &= \gamma_1 I_1 + \gamma_2 I_2 - dR.
\end{aligned}
\tag{3.9.39}
$$

It may be shown that strain 2 can invade the steady state of strain 1 alone and drive it to extinction as long as its basic reproductive ratio is greater,

$R_{02} > R_{01}$, where

$$R_{0i} = \frac{\beta_i N}{\gamma_i + c_i + d} \tag{3.9.40}$$

for $i = 1, 2$.

The question of which strain will persist reduces to a simple calculation of the parameter R_0, and we shall investigate how R_0 depends on virulence. Let us assume that strain 2 is less virulent than strain 1, in the sense that $c_2 < c_1$. Assuming no other differences, $\beta_1 = \beta_2 = \beta$ and $\gamma_1 = \gamma_2 = \gamma$, then

$$R_{02} = \frac{\beta N}{\gamma + c_2 + d} < \frac{\beta N}{\gamma + c_1 + d} = R_{01}, \tag{3.9.41}$$

and the less virulent strain invades. It looks as if diseases do tend to evolve towards reduced virulence. However, changes in virulence tend to be correlated with other changes in a disease. In myxomatosis, a disease artificially introduced to control the rabbit populations of Australia and the UK in the 1950s, high virulence is correlated with skin lesions which increase transmissibility. The transmission parameter β is an increasing function of virulence, and may be modelled phenomenologically by

$$\beta = \beta_0 c^\alpha, \tag{3.9.42}$$

where $0 < \alpha < 1$ is a parameter. Then the graph of R_0 against c has a maximum, as in Figure 3.7, and we should expect c to evolve to the intermediate level that maximises R_0. This is in good agreement with data for myxomatosis in Australia.

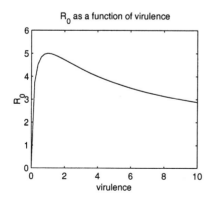

Figure 3.7   The basic reproductive ratio R_0 as a function of the virulence c, if the infectious contact rate β is given by Equation (3.9.42), while γ is constant.

EXERCISES

3.22. Assume that the rate of recovery from a disease decreases with virulence, and may be modelled by $\gamma = \gamma_0 c^{-\alpha}$. How do you expect virulence to evolve, if the transmission parameter β is constant?

3.23. Initially, the system (3.9.39) is at its endemic steady state with strain 1 only, and then a small number of infectives of strain 2 are introduced. Show that strain 2 can invade the steady state and drive strain 1 to extinction as long as its basic reproductive ratio is greater, $R_{02} > R_{01}$. You may assume that the total population size N remains constant.

3.10 Conclusions

- The structure of a population by disease status is an essential element of the models we have considered.

- Different diseases require different disease status classes, depending among other things on whether they are lethal and whether they are immunity-conferring.

- A simple classification of a disease as SIS (neither immunity-conferring nor lethal) or SIR (either immunity-conferring or lethal) is often useful.

- In short-term investigations of epidemics it may be possible to neglect birth and disease-unrelated death, and therefore deal with closed populations.

- In long-term investigations of endemics the populations will normally be open.

- The most important parameter of a disease is R_0, the basic reproductive ratio, defined to be the expected number of infectious contacts made by a single infective introduced into a susceptible population of a given size.

- In many cases there is a threshold theorem. The disease-free steady state is stable if $R_0 < 1$, unstable if $R_0 > 1$.

- Under certain conditions, the total number infected during the course of an epidemic may be calculated. If the assumptions are realistic, some of those who are initially susceptible remain susceptible when the epidemic is over.

- An endemic disease in an open population has a tendency towards oscillations, although under the usual assumptions these are damped. However oscillations can occur if more realistic assumptions are made, and are observed in real diseases, notably childhood diseases like measles.

– The aim of a disease control strategy is often to reduce R_0 below 1, by increasing the rate of removal of infectives, decreasing the infectious contact rate, or decreasing the number of susceptibles.

– Vaccination reduces the number of susceptibles. If it is possible to achieve $R_0 < 1$ in a vaccination programme, the disease will die out, despite the fact that susceptibles may remain; this is known as *herd immunity*.

– The average age at which a disease is acquired is important

  – in itself, if the disease has different effects at different ages,

  – as an indirect method for estimating R_0.

– Vaccination increases the average age at infection according to $\bar{a}' = \bar{a}/q$, where q is the fraction unvaccinated.

– Vector-borne diseases such as malaria require the vector population as well as the host population to be modelled.

– In macroparasitic diseases, the degree of infestation may be important.

– Simple models suggest that the virulence of a disease should decrease through evolution by natural selection. However more realistic models may give different results, e.g. that the disease may evolve to some intermediate virulence that maximises its basic reproductive ratio R_0.

4

Population Genetics and Evolution

- Population genetics considers the genetic basis for evolution in a population, taking into account the effects of sexual reproduction and genetic dominance. It is a quantitative science on a firm theoretical footing.

- Game theory is a much more qualitative approach to the study of evolution. It considers how the effectiveness of strategies depends on the strategies adopted by others, and asks which strategy or mixture of strategies might be expected to be adopted by a population subject to natural selection.

4.1 Introduction

Darwin recognised the importance of the fact that many individuals of a species are destined to die before reaching reproductive age. He argued that as a consequence every slight modification that chanced to arise and that was advantageous to the individual that possessed it would tend to be preserved, and thus change the characteristics of the species as a whole; he called this *evolution by natural selection*. This assumes that such slight modifications are passed on through the generations, but how and under what circumstances this occurs was not known in Darwin's time. The Bohemian monk Gregor Mendel did not publish his experiments on genetics until 1865, six years after *The Origin of Species*, and their significance was not recognised for decades after that. Yet, as Sigmund states in his delightful book *Games of Life*, the basics of inheritance

could have been deduced a hundred years before Mendel by considering qualitative traits like eye-colour. (The inheritance of a quantitative trait like height
tends to be more complex, and so less revealing.) Maupertuis, famous in mathematics for his "principle of least action", tried to find a solution by studying
the occurrence of the rare qualitative trait polydactyly ("multi-fingeredness")
in a family over several generations. Elisabeth Horstmann of Rostock had six
fingers on both hands, and so did her daughter. Four of her daughter's eight
children, among them the famous surgeon Jacob Ruhe of Berlin, had six fingers, and two of his six children also inherited the painful trait. It seemed that
there were certain "factors" causing the trait, that were passed on unchanged
through the generations, and that a child could inherit these factors from either
its father or its mother. Everything is consistent with the hypothesis that each
trait is caused by one factor, which comes with equal probability from either
parent. The rarity of the trait, although it made it obvious that it was inherited, prevented Maupertuis from obtaining a crucial piece of evidence, that two
polydactylous parents may have a five-fingered child (just as two brown-eyed
parents may have a blue-eyed child). The trait is in fact determined by *two*
factors, one inherited from each parent. What happens if you inherit different
factors from your parents?

– Which of the two factors do you pass on to your offspring?

– Which of the two factors is expressed?

Observations suggest that each factor has the same chance of being passed on,
and that, for pairs of contrasting traits (blue eyes *vs* brown eyes, five fingers *vs*
six fingers), one of the two is *dominant* and always overrules the other (*recessive*)
factor. This hypothesis was confirmed by Mendel in an exhaustive series of
experiments on peas. He formulated explicit quantitative laws that have no
counterpart in fields such as ecology, and which have been the basis of very
successful mathematical work, initially in the 1920s and 1930s by Fisher, Wright
and Haldane. This area is still regarded by many as the best example of the
use of mathematics in biology.

EXERCISES

4.1. How many of Elisabeth Horstmann's daughter's eight children would
you have expected to have had six fingers? State any assumptions
you make.

4.2 Mendelian Genetics in Populations with Non-overlapping Generations

Mendel's work applies to sexually reproducing organisms, such as peas and humans. In such organisms adults produce female and male *gametes*, eggs and sperm in humans, which fuse to form *zygotes*, which develop and mature to adulthood. In the absence of genetic mutation, which will be discussed later, factors determining various traits are passed down *unchanged* through the generations. In this chapter we shall make things as easy for ourselves as we can by assuming that generations are discrete and non-overlapping; thus time will be a discrete variable and we shall end up with difference equations. Similar results hold for populations that reproduce in continuous time.

Mendel's first law states that

– each gamete contains only one factor for one of any pair of contrasting traits.

Hence a human gamete contains either a factor for blue eyes or a factor for brown eyes. Zygotes and therefore adults have two factors, one from each parent. In other words, gametes are *haploid* and zygotes are *diploid*. If each of the factors in an individual is the same it is a *homozygote*, if different a *heterozygote*. In heterozygotes, which of the factors is expressed depends on which is *dominant*; the other is *recessive*. More complicated effects than simple dominance may also occur, especially if the trait is not one of a contrasting pair. In modern terminology, the factors are known as *genes*, and may have one of several forms, or *alleles*. The genetic make-up of an individual is its *genotype*; which trait is expressed is its *phenotype*. For example, an individual with phenotype brown-eyes may have the homozygous genotype brown-brown or the heterozygous genotype brown-blue, since brown is dominant over blue. Each gene sits at its own specific *locus* on its own specific *chromosome*.

How does the genetic make-up of a population change over the generations? Suppose that what happens at a given locus is independent of what happens at any other, and focus on changes at a single locus. Suppose that there are two and only two alleles A and B that may sit at this locus. A given individual may then have one of three genotypes: the homozygotes AA or BB or the heterozygote AB.

Let p be the so-called *frequency* of allele A in a population, defined to be the number of alleles A over the total number of alleles in the population, and q the frequency of allele B; clearly, $p + q = 1$. Let us also define x, y and z to be the frequencies of genotype AA, AB and BB respectively; then $p = x + \frac{1}{2}y$, $q = z + \frac{1}{2}y$.

Let us now make the following assumptions:

– expected sex ratio is independent of genotype;

– mating is random, that is the expected number of matings of genotype 1 with genotype 2 is proportional to the product of the frequency of genotype 1 and the frequency of genotype 2;

– fertility is independent of genotype, that is there is no difference in the expected number of gametes produced by genetically different individuals, nor the expected number of viable zygotes produced by genetically different crosses;

– survivorship, the probability of surviving to breed, is independent of genotype;

– there is no mutation or migration.

Then among both male and female gametes, the frequency of allele A is p, and since we have assumed random mating these gametes unite at random. This may be summarised by a *Punnett square*, named after the early 20th century biologist R.C. Punnett.

Punnett square				
			Frequency of female gametes	
			A	B
			p	q
Frequency of	A	p	p^2	pq
male gametes	B	q	pq	q^2

This allows us to read off the frequency of each genetically different union of gametes, e.g. the frequency of the union of a female A gamete with a male B gamete is pq, and the frequency of the union of an A gamete with a B gamete, and hence the frequency of heterozygotes AB in the subsequent generation, is $2pq$. It follows immediately that

$$p_{n+1} = p_n + \frac{1}{2}2p_n q_n = p_n, \quad q_{n+1} = q_n + \frac{1}{2}2p_n q_n = q_n,$$

where n is generation number. Hence p_n and q_n are constants independent of n; let us call them p and q. Moreover, for $n \geq 1$, but not necessarily for $n = 0$, $x_n = p^2$, $y_n = 2pq$, and $z_n = q^2$. Generation 0 is known as the *parental* (P) generation, and generation n as the nth *filial* (F_n) generation for $n \geq 1$.

The *Hardy–Weinberg law* states that, under the assumptions listed above,

– allele frequencies p and q remain unchanged from generation to generation, and are therefore the same in the filial generations as in the parental generation, and

– from generation F_1 onwards the genotype frequencies are $x = p^2$, $y = 2pq$, $z = q^2$.

G.H. Hardy, the pure mathematician, is said to have seen as soon as the question was put to him, at a cricket match in 1908 by Punnett, that the allele frequencies would remain unchanged, and to have been rather dismayed to have his name associated with something so immediately comprehensible.

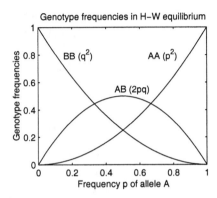

Figure 4.1   Hardy–Weinberg genotype frequencies. It can be seen from this figure that when an allele is rare, it occurs predominantly in heterozygotes. As an illustration, the frequency of the recessive cystic fibrosis gene is about 0.024. About 1 in 1700 newborn Caucasians are homozygous for the gene, and hence suffer from the disease, but about 1 in 21 are heterozygous for it, and are therefore carriers. The disorder is lethal, but the fact that the gene is recessive means that the effects of natural selection on it are drastically reduced.

EXERCISES

4.2. *Mendel's Second Law* states that

> – each one of a pair of contrasted traits may be combined with either of another pair.

Apply this law in the case that one pair of contrasted traits is roundedness and wrinkledness of pea seeds, and the other is yellowness and greenness. Given that roundedness is dominant over wrinkledness, and yellowness over greenness, and assuming a parental generation of purebred round yellows and wrinkled greens, what proportions of each phenotype would you expect in the F_1 and F_2 generations?

Assume that each individual in the F_1 generation is produced by a cross between a round yellow and a wrinkled green, and that the F_2 generation is produced by random mating between members of the F_1 generation.

4.3. Show that a population is in Hardy–Weinberg equilibrium with respect to a locus with two alleles if and only if $y^2 = 4xz$, where x, y and z are the usual genotype frequencies.

4.4. The ABO blood groups in humans are determined by a system of three alleles, A, B and O. Genotypes AA and AO are group A, BB and BO are group B, AB is group AB, and OO is group O. The frequencies of the blood groups in England are 32.1% A, 22.4% B, 7.1% AB and 38.4% O. Are these proportions consistent with the assumption of random mating?

Hint: a Punnett square with three alleles might be helpful.

(We are straying into the area of statistics here, and to do this question properly we need to perform a χ^2 test with one degree of freedom. Non-statisticians will be reduced to guessing whether the difference between the data and the expected values on the basis of the random mating assumption is significant.)

4.5. In *positive assortative mating*, individuals mate more frequently with those of the same genotype. As an extreme case, let the frequency of the mating $AA \times AA$ be x, of $AB \times AB$ be y, and of $BB \times BB$ be z, where x, y and z are the usual genotype frequencies. This is of course non-random mating, a Punnett square will not help, and we must work with genotype frequencies rather than simply allele frequencies.

   a) What genotypes result from the mating of two heterozygotes, and in what proportions?

   b) Derive the difference equations to be satisfied by x, y, z, p and q, and analyse them.

   c) What conclusions can you draw from your analysis?

4.6. In humans, sex is determined genetically. (This is not necessarily true in other organisms, e.g. the sex of some reptiles is determined by the temperature of incubation of the eggs.) Females have two X-chromosomes, whereas males have only one, inherited from their mother, and one Y-chromosome inherited from their father. A gene that appears on the X-chromosome but not on the Y is said to be

X-linked, and is always expressed in males. Colour-blindness affects about 1 in 20 Caucasian males. It is caused by a recessive allele of an X-linked gene. Estimate the frequency of colour-blindness among Caucasian females. You may assume that allele frequencies are at equilibrium.

4.7. Does the Hardy–Weinberg law hold for X-linked genes? If not, derive the law that does hold and analyse the resulting equations.

4.3 Selection Pressure

In the absence of selection, then, there is no evolution. Things work out very prettily, but nothing of interest happens. How do we include selection?

The *absolute fitness* of a genotype is defined in terms of its reproductive success. For separate non-overlapping generations, it is the number of copies of each gene that an individual of that genotype expects to contribute to the gene pool of the next generation. The *relative fitness* of a genotype is the ratio of its absolute fitness to the absolute fitness of a reference genotype. It may be *density-dependent* or *frequency-dependent*, i.e. it may depend on the size and genetic makeup of the population concerned, and of course it may also depend on the environment.

The Hardy–Weinberg assumptions imply that each genotype is equally fit, i.e. that the expected number of genes an individual contributes to the gene pool of the next generation is independent of genotype. Now let us assume that there is some selective advantage or disadvantage to allele A. Let us assume that the probability of survival of the genotype AA from the zygotic phase to the breeding phase is different. There may also be an effect on the heterozygote AB. Other selective effects could occur at different points in the life cycle, or work in different ways, and this may lead to slight differences in the equations, but the principles are the same.

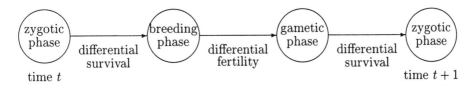

Figure 4.2   Diagram of a typical life cycle, showing the various points at which selection may act.

Now note that allele and genotype frequencies may be different at different points in the life cycle, and we must be careful and specify when we carry out our census. Usually we count the alleles and genotypes at the end of the gametic phase, so that we can consider first the sex and then the selection. We shall retain the assumption of random mating. This is important. Random mating equals Punnett square, which means that we can obtain equations solely in terms of allele frequencies. Otherwise we would have to work with genotype frequencies, and things would become much more complicated, as in Exercise 4.5. Let the allele frequencies at the end of the gametic phase of generation n be p_n and q_n. Then, at the beginning of the zygotic phase, by random mating, the genotype frequencies x_n, y_n and z_n are given by $x_n = p_n^2$, $y_n = 2p_nq_n$, and $z_n = q_n^2$. Let the probability of survival from zygotic phase to breeding phase for the various genotypes be in the ratio $w_x : w_y : w_z$. Usually we take (say) $w_z = 1$, so that w_x and w_y are the *relative selective values* of genotypes AA and AB, i.e. the probabilities of survival of genotypes AA and AB relative to the probability of survival of genotype BB. In this simple situation relative selective value and relative fitness are equal. Then at the breeding phase the ratios of the genotypes AA, AB and BB have been modified to

$$w_x p_n^2 : 2w_y p_n q_n : w_z q_n^2,$$

so that allele frequencies are now in the ratio

$$w_x p_n^2 + w_y p_n q_n : w_y p_n q_n + w_z q_n^2.$$

Because there is no differential fertility or differential survival of gametes, the alleles are in this same ratio at the end of the gametic phase of generation $n+1$. Converting to frequencies, we obtain

$$p_{n+1} = f(p_n) = \frac{(w_x p_n + w_y q_n)p_n}{w_x p_n^2 + 2w_y p_n q_n + w_z q_n^2}, \tag{4.3.1}$$

or

$$p_{n+1} = p_n + g(p_n) = p_n + p_n q_n \frac{(w_x - w_y)p_n + (w_y - w_z)q_n}{w_x p_n^2 + 2w_y p_n q_n + w_z q_n^2}. \tag{4.3.2}$$

This is the famous *Fisher–Haldane–Wright (FHW) equation* of mathematical population genetics. It is a complete description of the change in gene frequencies that occur as a result of this selection process, and we shall interpret and analyse it. One way of analysing it is to use the standard methods of Section A.1 of the appendix. There are steady states at $p = 0$ and $p = 1$, where either A or B is fixed in the population. The cobweb map may be sketched by analysing the sign of g in $(0, 1)$, and this determines the existence of any interior steady state p^*. The stability of the steady states may be determined by finding f'.

The equation may be interpreted in terms of mean fitnesses. We define the *mean fitness* w_p of A by taking a weighted mean over all the homozygotes and half the heterozygotes carrying allele A,

$$w_p = \frac{w_x p^2 + w_y pq}{p^2 + pq} = w_x p + w_y q, \tag{4.3.3}$$

the mean fitness w_q of B similarly,

$$w_q = \frac{w_y pq + w_z q^2}{pq + q^2} = w_y p + w_z q, \tag{4.3.4}$$

and the *overall mean fitness* $\bar{w}$ by

$$\bar{w} = w_x p^2 + 2w_y pq + w_z q^2 = pw_p + qw_q. \tag{4.3.5}$$

Equation (4.3.1) also looks simpler if we drop the suffix n from p and q, and use a prime to denote the generation $n + 1$. It becomes

$$p' = \frac{w_p p}{\bar{w}}, \tag{4.3.6}$$

and Equation (4.3.2) becomes

$$\delta p = p' - p = \frac{\alpha_p p}{\bar{w}} = \frac{(w_p - \bar{w})p}{\bar{w}}. \tag{4.3.7}$$

The quantity $\alpha_p = w_p - \bar{w}$ is called the *mean excess fitness* of allele A. This equation may be interpreted as saying that the relative amount $\frac{\delta p}{p}$ by which the genotype A increases in a generation is given by its excess fitness in the population relative to the fitness of the population itself. It may also be written

$$\delta p = p' - p = pq \frac{w_p - w_q}{\bar{w}} = pq \frac{(w_x - w_y)p + (w_y - w_z)q}{\bar{w}}. \tag{4.3.8}$$

An interior steady state $p^* = 1 - q^*$ exists if A and B are equally fit, $w_p^* = w_q^*$, $(w_x - w_y)p^* + (w_y - w_z)q^* = 0$, but this is only biologically realistic if $w_y - w_x$ and $w_y - w_z$ have the same sign. In other words, the heterozygote either has to be superior, fitter than both homozygotes, or inferior, less fit than both. This is unusual, but does sometimes occur, as we shall see in Exercise 4.9. Despite this, many *gene polymorphisms*, i.e. loci with two or more alleles co-existing in a population, do occur in real life. Some may be neutral with regard to fitness, and some may be moving towards fixation, but if they are steady states they are often maintained by a balance between selection and mutation, which we shall discuss in Section 4.6. The occurrence of polymorphisms in the population is reflected within individuals; of your own genes, between six and seven per cent are heterozygous.

The analysis of the equations is sometimes simpler if we consider the gene ratio $u = p/q$ or $v = q/p$. For u, we obtain

$$\delta u = u' - u = \frac{w_p - w_q}{w_q} u = \frac{(w_y - w_z) + (w_x - w_y)u}{w_z + w_y u} u, \qquad (4.3.9)$$

with a similar equation for v.

EXERCISES

4.8.   a) Using (4.3.1), or otherwise, show that $p = 0$ is stable if $w_y < w_z$, and interpret this condition biologically.

     b) Give a similar result for the other fixation steady state $p = 1$.

     c) Find the mathematical and biological conditions under which there is an interior steady state p^* which is stable.

4.9. Sickle-cell anaemia is a genetic disease of humans. If we denote the allele associated with the disease by B, and the normal allele of the gene by A, then the disease has the following properties:

    − heterozygotes AB do not have the disease (or have a very mild form of it) and are also protected from malaria;

    − homozygotes BB have the disease, which may be lethal, and which also predisposes the sufferer to malaria.

Let the relative fitnesses of the genotypes AA, AB and BB in a malarial area be 1, $1 + s$ and $1 - t$, where $0 < t \leq 1$, $0 < s < \infty$.

     a) Show that the difference equation for the frequency p of the allele A is given by

$$p' = p + pq \frac{s(1 - 2p) + t(1 - p)}{1 + 2spq - tq^2}$$

     where q is the frequency of the allele B.

     b) Analyse this equation by finding the steady states and using a cobweb map (see Section A.1.1 of the appendix). Comment on your results.

     c) Assume that sickle-cell anaemia reduces the fitness of BB homozygotes to 20% of that of AA homozygotes, but is so mild in heterozygotes as to make no difference to fitness, and that the gene offers heterozygotes complete protection from malaria.

Let the frequency q of the allele B be steady at 0.2, a frequency attained in some African areas. Neglecting deaths from causes other than malaria, estimate the chance of someone of genotype AA dying from malaria before reaching maturity.

4.4 Selection in Some Special Cases

4.4.1 Selection for a Dominant Allele

Suppose that A is dominant and advantageous. Then, if we take $w_z = 1$, we have $w_x = w_y = 1+s$, say. The parameter s is called the *selection coefficient* and measures the strength of selection for the allele A. The FHW Equation (4.3.2) becomes

$$p_{n+1} = p_n + sp_nq_n \frac{q_n}{1 + s(p_n^2 + 2p_nq_n)}. \qquad (4.4.10)$$

It is easy to see that (p_n) is a monotonic increasing sequence, bounded below by 0 and above by 1, so that $p = 0$ is an unstable and $p = 1$ a stable steady state. The departure from $p = 0$ is exponential and the approach to $p = 1$ algebraic. This may be seen analytically. For p close to 0, q is close to 1 and the equation is given to leading order in p by

$$p_{n+1} = (1 + s)p_n,$$

so that $p_n \approx p_0(1 + s)^n$. For the behaviour near $p = 1$, it is easier to use the gene ratio $u = p/q$. The equation becomes

$$\delta u = u' - u = \frac{su}{1 + (1 + s)u}.$$

To first order for q small, u large, this gives the approximate equation $u_{n+1} = u_n + s/(1 + s)$, so that for q_0 small, $u_n = u_0 + ns/(1 + s)$,

$$q_n = \frac{(1 + s)q_0}{1 + s + nsq_0}.$$

4.4.2 Selection for a Recessive Allele

Now suppose that A is recessive and advantageous. Again taking $w_z = 1$, we now have $w_y = 1$, $w_x = 1 + s$. The FHW Equation (4.3.2) becomes

$$p_{n+1} = p_n + sp_nq_n \frac{p_n}{1 - sp_n^2}. \qquad (4.4.11)$$

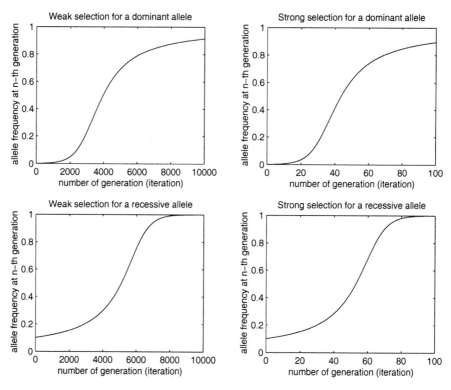

Figure 4.3   Numerical solutions for weak and strong selection for a dominant and a recessive allele. We have taken $s = 0.002$ for weak and 0.2 for strong selection; the only difference this makes that is visible in the results is the different time scale. Selection is much stronger at high frequencies and weaker at low frequencies when the allele is recessive.

Again (p_n) is an increasing sequence bounded below by 0 and above by 1, and so $p = 0$ is an unstable and $p = 1$ a stable steady state. However, in this case the departure from $p = 0$ is algebraic and the approach to $p = 1$ exponential. For a given value of selection coefficient s, selection is much weaker at low frequencies of A than it was in the dominant case (see Figure 4.3). This is because heterozygotes, which include most copies of the allele A when it is rare (see Figure 4.1), do not benefit from the advantageous effect of A if A is recessive.

As can be seen from Figure 4.3, even an exponential rate of increase means small changes in frequency if p_0 is small, and an algebraic rate of increase is much slower. In Sri Lanka, spraying with DDT had by 1963 reduced malaria cases to 17, but by 1968 resistant mosquitoes were prevalent, and over a million cases of malaria were reported. An initial slow increase followed by a sharp acceleration in the numbers of resistant mosquitoes was observed; this is typical of a selection process.

4.4.3 Selection against Dominant and Recessive Alleles

Selection against a dominant allele is the same as selection for a recessive, and vice versa. Think of the graphs in Figure 4.3 with the alternative allele frequency $q = 1 - p$ plotted on the y-axis, so that the graphs are upside down. A deleterious dominant allele will tend to disappear from a population exponentially, while a deleterious recessive will tend to disappear algebraically (in the absence of mutations).

4.4.4 The Additive Case

In general, if the trait considered is quantitative (like height) rather than qualitative (like polydactyly), the allele A may be neither completely dominant nor completely recessive, and the heterozygote may be intermediate in fitness between the homozygotes. In this case, $w_x < w_y < w_z$ if A is deleterious, $w_x > w_y > w_z$ if A is advantageous. Let us take A to be advantageous, and assume that each A allele adds the same amount to the selection coefficient, the so-called *additive* or *semi-dominant* case. This is the simplest case of all to deal with, as it means essentially that how the alleles are combined into genotypes is irrelevant. The term semi-dominant is rather confusing, as neither A nor B is dominant, both having equal effects on the selection coefficient of the heterozygote. In this sense, there are *no* dominance effects in the additive case. We write $w_x = 1 + 2s$, $w_y = 1 + s$, $w_z = 1$. Then the FHW Equation (4.3.2) becomes very simple in form,

$$p_{n+1} = p_n + \frac{sp_nq_n}{1 + sp_n}. \tag{4.4.12}$$

The advantageous gene spreads through the population. It can be shown analytically that both the departure from $p = 0$ and the approach to $p = 1$ are exponential.

EXERCISES

4.10. Confirm that if selection is for a recessive allele, departure from $p = 0$ is algebraic and approach to $p = 1$ exponential.

4.11. Confirm that if selection is for an additive allele, both departure from $p = 0$ and approach to $p = 1$ are exponential.

4.5 Analytical Approach for Weak Selection

The Fisher–Haldane–Wright equation is a nonlinear difference equation. It is easy to obtain numerical results, analyse its qualitative behaviour, and obtain analytic results near the steady states, but it has no analytic solution in general. However, it is amenable to more thorough analysis if selection is weak. The palaeontological record suggests that in most evolutionary changes, time scales are of the order of thousands of generations, so that s is about 0.001 or smaller, and we would therefore expect this to be a good approximation.

Let $w_x = 1 + O(s)$, $w_y = 1 + O(s)$, and $w_z = 1 + O(s)$, where the $O(s)$ terms here are simply constants multiplied by s. Then s is a measure of the strength of selection. We shall start with version (4.3.7) of the FHW equation. We have $w_p = 1 + O(s)$ and $\bar{w} = 1 + O(s)$, and $w_p - \bar{w} = O(s)$, and is proportional to s. Now assume s is small. As a consequence we can do two useful things. First, we can neglect terms of $O(s)$ compared to 1 in the denominator of (4.3.7). Second, we can approximate the difference equation by a differential equation. The equation becomes

$$\dot{p} = \alpha_p p = (w_p - \bar{w})p. \tag{4.5.13}$$

This will be the basis of the replicator equation that we shall see in Section 4.10. Among other things, it follows that for weak selection, the time for a given change in frequency is inversely proportional to s.

Now let us be specific about the $O(s)$ terms in the selection coefficients, and write $w_x = 1 + hs$, $w_y = 1 + ks$, $w_z = 1$. The equation becomes

$$\dot{p} = spq\left((h - k)p + kq\right). \tag{4.5.14}$$

This is an equation with separable variables, and the time for any change in allele frequency, from p_0 to p_1 say, may be found by integration,

$$t = \frac{1}{s} \int_{p_0}^{p_1} \frac{dp}{p(1 - p)((h - k)p + k(1 - p))}. \tag{4.5.15}$$

In the additive case with A advantageous, taking $k = 1$ so that $h = 2$, we obtain

$$\frac{dp}{dt} = sp(1 - p),$$

the logistic Equation (1.3.7) with $r = s$ and $K = 1$. Fisher looked at a spatial version of this,

$$\frac{\partial p}{\partial t} = sp(1 - p) + D\frac{\partial^2 p}{\partial x^2}, \tag{4.5.16}$$

where the diffusion term represents random motion of the alleles in space. He showed that the advantageous gene spreads as a travelling wave through the population. We shall investigate such travelling waves in Chapter 5.

EXERCISES

4.12. The industrial revolution in England resulted in industrial melanism, the spread of a gene for dark coloration in certain species of moth. The original grey moth was camouflaged on lichen-covered trees but easy to see on trees where the lichen had been killed by pollution, and the dark form was camouflaged on lichen-free trees. In one species, the peppered moth *Biston betularia*, a single dominant gene caused the colour difference, and the frequency of this gene increased from less than 0.01 to 0.90 in 35 years, or 35 generations. (In fact, although the gene is dominant now, its dominance has increased over time, which ties in with Section 4.8, but I suggest ignoring this awkward fact for the purposes of this question. There are also suggestions in the recent literature that this is not such a clear case of natural selection as it seems.)

a) Making the assumption that the selection coefficient s is small, estimate the value of s by integrating the weak selection equation.

b) *Computer exercise.* Solve the difference equation numerically for this value of s, and comment on the accuracy of the small s approximation.

4.6 The Balance Between Selection and Mutation

The central dogma of modern evolutionary biology dates back to Weismann at the end of the nineteenth century and is summarised in Figure 4.4.

It states that the *germ line*, i.e. the reproductive cells that will contribute to future generations, and the *soma*, i.e. the non-reproductive cells of the current generation, are separate, in the sense that no information may pass from the soma to the germ line. Thus if a blacksmith develops muscular arms, then his (or her) offspring will not inherit this trait. Similarly, a *somatic mutation*, a

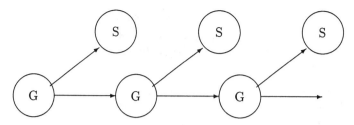

Figure 4.4   Central dogma of modern evolutionary biology. G stands for germ line, S for soma. In modern molecular terms, we would replace G by DNA and S by P (for protein).

change in the genetic information in a cell of the soma, is not inherited. Instead, evolution works on germ-line mutations. These may be caused by transcription errors or some mutagenic agent, and may for example change allele A to B. They arise without reference to the adaptive needs of the organism, and are much more likely to be deleterious than beneficial. However, they are fundamental to the process of evolution because they help to maintain a supply of genetic variation for selection to act on. We shall see that such variation is crucial.

We shall first include mutations in the simplest Hardy–Weinberg situation, before generalising. Assume that there is a probability u that a given allele A mutates to allele B in a generation, and a probability v of the reverse mutation, both u and v being constant. Then, with Hardy–Weinberg assumptions, of the fraction p of gametes expected to be A only $(1 - u)p$ are in fact A, and of the fraction q of gametes expected to be B vq are in fact A, so that after random mating,

$$p_{n+1} = (1 - u)p_n + vq_n.$$

This is a linear first-order difference equation, which we can write in the form

$$p_{n+1} - p^* = (1 - u - v)(p_n - p^*),$$

and which has the solution

$$p_n = p^* + (p_0 - p^*)(1 - u - v)^n.$$

Here $p^* = \frac{v}{u+v}$, and is the steady state of the equation, stable for typical values of u and v of about 10^{-6}. The time scale for changes in gene frequency by mutation can be seen to be of the order of $(u+v)^{-1}$, or millions of generations. This is much slower than the time scale for changes in gene frequency by even very weak selection, so this steady state is very likely to be disrupted by selective effects and should not be taken seriously.

If selection is now included, by differences in survivorship as before, we simply need to alter the ratio of gene frequencies at the breeding phase from $p : q$ to $w_x p^2 + w_y pq : w_y pq + w_z q^2 = w_p p : w_q q$, and then take account of mutation. The equation becomes

$$p' = (1 - u)\frac{w_p p}{\bar{w}} + v\frac{w_q q}{\bar{w}},$$

so that

$$\delta p = \frac{\alpha_p p}{\bar{w}} - u\frac{w_p p}{\bar{w}} + v\frac{w_q q}{\bar{w}}.$$

If selection is weak, i.e. $w_i = 1 + O(s)$ for each genotype i and s is small, and if mutation rates are low, i.e. u and v are also small, then to leading order in small quantities,

$$\delta p = \delta p_{sel} + \delta p_{mut}$$

where δp_{sel} and δp_{mut} are the changes that would arise from selection and mutation alone. A steady state arises when the effects of selection and mutation cancel each other out; see the exercise below.

EXERCISES

4.13. In a selection-mutation balance, where selection is against a deleterious gene, assume that the mutation parameters are much less than the selection coefficient, which is itself small, and neglect terms which are second order in small quantities.

a) If the gene selected against is recessive, show that its steady state value is about $p^* = \sqrt{v/s}$, in the notation of Section 4.6.

b) If the gene is dominant, show that its steady state value is about $p^* = v/s$.

c) In cystic fibrosis the deleterious allele is lethal and recessive, and the mutation rate to it is about 4×10^{-4}. Estimate the allele frequency.

4.7 Wright's Adaptive Topography

First we show that natural selection acts to increase mean fitness, given by $\bar{w}$ in Equation (4.3.5). A straightforward calculation, making use of Equation (4.3.8), shows us that

$$\delta \bar{w} = \bar{w}(p + \delta p) - \bar{w}(p) = \frac{(\delta p)^2}{pq}(pw_x + qw_z + \bar{w}). \qquad (4.7.17)$$

It follows that $\delta \bar{w} \geq 0$, with equality occurring if and only if $\delta p = 0$. Now, assuming that w_x, w_y and w_z are constants, and differentiating $\bar{w}$ with respect to p, it follows that

$$\frac{d\bar{w}}{dp} = 2((w_x - w_y)p + (w_y - w_z)q),$$

so that, making use of the FHW Equation (4.3.8) again,

$$\delta p = \frac{1}{2}\frac{pq}{\bar{w}}\frac{d\bar{w}}{dp}. \qquad (4.7.18)$$

Hence p increases when $\bar{w}$ has positive slope and decreases when $\bar{w}$ has negative slope. A graph of $\bar{w}$ as a function of p is known as *Wright's adaptive topography*. Under selection, the allele frequency p of the population moves uphill on the graph and the mean selective value $\bar{w}$ increases to a maximum.

EXERCISES

4.14. Starting with Equations (4.3.5) and (4.3.8), prove Equations (4.7.17) and (4.7.18).

4.15. Sketch Wright's adaptive topography for the model of sickle-cell anaemia in Exercise 4.9, and confirm the results of part (b) of that exercise.

4.8 Evolution of the Genetic System

The theory we have outlined above may be extended in many ways. We have essentially considered infinite populations, so that when an advantageous allele arises it always goes to fixation. In a finite population this is not necessarily true, as stochastic effects might then be important. The theory of *genetic drift* tells us how allele frequencies change in finite populations. Even if we stick to the deterministic theory we may also consider, for example, more than two alleles at a locus, or more than one locus, or populations structured spatially or otherwise. The two-locus theory is fundamentally different from the theory above if the fitness effects at the two loci interact. In such cases, we may ask

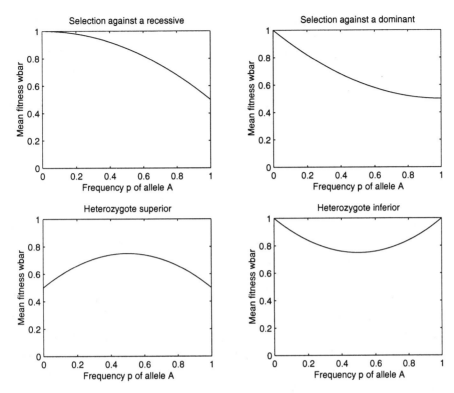

Figure 4.5 Wright's adaptive topography: the graph of $\bar{w}$ as a function of p shows us how p will change under selection. In the top two diagrams, $p \to 0$ as $t \to \infty$, in the third, $p \to \frac{1}{2}$, and in the fourth, $p \to 0$ or 1 depending on the initial conditions.

how the genetic system itself evolves, e.g. why do certain genes form complexes, sitting very close to each other on a chromosome and often being inherited together, how do mutation rates evolve, and why are there two sexes (and not one or three) in humans? This is much more the kind of question that is asked in modern population genetics. To give a flavour of the kind of problems that can arise, we shall consider here the question of why the most common allele at a locus, the so-called wild-type allele, is almost always dominant. Our theory so far says that an advantageous dominant mutant allele initially invades a wild-type allele more quickly than a recessive mutant does, but a recessive increases more quickly at higher frequencies, and the overall time to fixation does not depend on the dominance; thus dominant and recessive wild-type alleles should be equally common. One theory put forward to explain the preponderance of dominant wild-type alleles is the following *modifier theory*. Let A and B be two alleles at the *primary locus*, with A advantageous over B. Now let M and N

be two alleles at a second *modifier locus*, and let the dominance relationship between A and B be determined by the modifier genotype. Specifically, taking the simplest case, let A be dominant over B if the modifier genotype is MM, B be dominant over A if the modifier genotype is NN, and let there be no dominance if the modifier genotype is MN. Otherwise let the modifier genotype have no effect on fitness. We are interested in whether M goes to fixation. If it does, then the advantageous allele A becomes dominant, and so we would expect dominance to evolve. Let the fitness of the AA genotype be 1 and that of the BB genotype $1 - s$, whatever the modifier genotype is. The dominance assumptions imply that the fitness of the heterozygote AB is 1 if the modifier genotype is MM, $1 - s$ if it is NN, and $1 - \frac{1}{2}s$ if it is MN. Let the frequency of allele A be p, and that of allele M be m. Then, with random mating and weak selection, and if there is no *epistasis* (essentially, which allele is present at one locus is not affected by which is present at the other), the equations determining the changes in frequency of p and m are

$$\dot{p} = sp(1-p)\left((1 - p) + (2p - 1)(1 - m)\right) - up + v(1-p), \quad \dot{m} = sp(1-p)m(1-m).$$

We have included mutation from A to B and vice versa, and assume that $u, v \ll s \ll 1$. The steady states (p^*, m^*) of this system are at $S_1 = (p_1^*, 0)$ and $S_2 = (p_2^*, 1)$. A calculation similar to that for the standard selection-mutation balance of Section 4.6, Exercise 4.13, gives $p_1^* \approx 1 - u/s$, $p_2^* = 1 - \sqrt{u/s}$. The eigenvalues at S_1 are approximately u and $-s$, and those at S_2 approximately $-\sqrt{su}$ and $-2s$. S_1 is unstable, with growth rate $1+u$, and S_2 stable, with decay rate $1 - \sqrt{su}$. The modifier gene does go to fixation, so that dominance evolves. The rate of invasion $1 + u$ of the modifier gene depends on the mutation rate at the primary locus, which is often on the order of 10^{-6}, so that the invasion is extremely slow. There is still controversy over whether such a slow process could explain the preponderance of dominant wild-type genes.

4.9 Game Theory

In the next two sections we move from population genetics to game theory. Game theory was developed in the 1940s by the great mathematician John von Neumann and Oskar Morgenstern, to investigate strategies in human economic interactions, where one individual may win or lose money to another. The individuals are assumed to behave rationally according to some criterion of self-interest. In 1967 the evolutionary biologist Bill Hamilton published a paper on an unbeatable strategy for sex ratio allocation which used game-theoretic ideas. The criterion of rationality is replaced by that of population dynamics and stability, and that of self-interest by Darwinian fitness. The idea was extended in the 1970s by John Maynard Smith, who looked for an *evolutionarily stable strategy* or *ESS* for animal interactions where the pay-off obtained by following one strategy depends on strategies that others are following. This is an important feature of game theory that is often not easy to incorporate into population genetics models. Animals with the best strategy are assumed to produce more offspring, which adopt the same strategy as their parents, so that the best strategy will be favoured by evolution. In this sense it is a model for phenotypic evolution, and concentrates on *why* evolution has favoured certain phenotypes rather than *how* such phenotypes might have been incorporated into the population. It has been criticised for this, but it is often the case that the results obtained by game theory still hold when genetics are included, and game theory is often much simpler to deal with. Moreover, for many cases the genetic basis of traits of interest is unknown.

The classic application of game theory in biology is to animal contests. We may think for example of stags competing with each other for hinds, although the purpose of the model is to explore the logical possibilities inherent in contests rather than to analyse any particular case. Let us assume that two alternative strategies are available to an animal competing with others for some resource: "hawk", H, fights, continuing until it wins or loses, "dove", D, displays, but retreats if its opponent escalates. A mixed strategy, P, "play H with probability p and D with probability $1 - p$", may also be possible. This *strategy set* is very naïve, but it suits the purpose of the model. Let us assume that the pay-offs for the various strategies are as shown in the *pay-off matrix* below. The pay-offs are in terms of a change in Darwinian fitness.

Pay-off matrix		
	on encountering this strategy	
	H	D
Pay-off to H	$\frac{1}{2}(G - C)$	G
this strategy D	0	$\frac{1}{2}G$

The resource is assumed to be worth a gain G to fitness; a hawk-dove contest results in the resource going to the hawk, while two doves are assumed either to share the resource or to receive the whole resource with probability $\frac{1}{2}$. If two hawks meet, then one will suffer a cost C of losing the fight, whereas the other will gain the resource; the mean pay-off is $\frac{1}{2}(G - C)$.

To model evolution, we imagine a population of individuals adopting different pure (H or D) or mixed (P) strategies. Individuals pair off at random, and accumulate the appropriate pay-offs. They then produce offspring identical to themselves, in numbers equal to a constant initial fitness, plus the pay-off. A strategy is defined to be an *evolutionarily stable strategy*, or ESS, if it is proof against invading mutant strategies. In other words, a population of individuals all of whom adopt the same ESS will never switch to another strategy by natural selection.

Let us develop the theory in a more general case. We shall consider games with n pure strategies. We shall restrict ourselves to *normal form games*, i.e. games where the pay-off is determined by a matrix. Let $U = (u_{ij})$ be the $n \times n$ pay-off matrix for such a game, so that u_{ij} is the pay-off for an i-strategist against a j-strategist, for $i, j = 1, \cdots, n$. Mixed strategies are determined by the probability column vector $p = (p_i)$, where p_i is the probability of using strategy i. The ith pure strategy is represented by the ith unit vector e_i. The pay-off for an i-strategist against a q-strategist is given by $(Uq)_i = e_i^T U q$, where the superscript T denotes transpose. The pay-off for a p-strategist against a q-strategist is given by $p^T U q = \sum_{i,j=1}^{n} u_{ij} p_i q_j$. Let us write $W(p, q)$ for this pay-off, so that

$$W(p, q) = p^T U q. \qquad (4.9.19)$$

The vectors defining strategies are probability vectors, and therefore live in the *simplex S_{n-1}* defined by

$$S_{n-1} = \{p = (p_i) \in \mathbb{R}^n : p_i \geq 0, \sum_{i=1}^{n} p_i = 1\}. \qquad (4.9.20)$$

The vertices of the simplex correspond to the n pure strategies. The boundary ∂S_{n-1} consists of all vectors $p \in S_{n-1}$ with at least one component equal to zero, i.e. all strategies whose *support*, defined by $\text{supp}(p) = \{i : p_i > 0\}$, is a proper subset of $\{1, 2, \cdots, n\}$. The interior $S_{n-1} \setminus \partial S_{n-1}$ consists of all vectors $p \in S_{n-1}$ with all components greater than zero.

The strategy p is defined to be an *evolutionarily stable strategy* if, whenever a small number of mutant q-strategists is introduced into a population of p-strategists, the mutant strategy *cannot invade*, i.e. the old type p fares better than the newcomers q. In other words,

$$W(q, \epsilon q + (1 - \epsilon)p) < W(p, \epsilon q + (1 - \epsilon)p) \qquad (4.9.21)$$

for all q and all ϵ sufficiently small. This is the crucial definition, but it is not an easy condition to check, especially if p is not known to begin with. We shall derive some consequences and an alternative form for it. Since W is continuous in the second variable, it follows by taking the limit as $\epsilon \to 0$ that

$$W(q, p) \leq W(p, p) \qquad (4.9.22)$$

for all $q \neq p$. This is the definition of a *Nash equilibrium*, one of the most important concepts in non-evolutionary game theory. It was introduced in the economic context by John Nash, who received a Nobel prize for his work on game theory. It can be interpreted as saying that there is no better reply to strategy p than p itself. However, there may be an alternative reply that is just as good. We can say more about the condition (4.9.21) for an ESS. Since W is linear in the second variable, it gives

$$\epsilon W(q,q) + (1 - \epsilon)W(q,p) < \epsilon W(p,q) + (1 - \epsilon)W(p,p)$$

for all $q \neq p$ and all ϵ sufficiently small. For all $q \neq p$, either

$$W(q,p) < W(p,p), \tag{4.9.23}$$

or

$$W(q,p) = W(p,p) \text{ and } W(q,q) < W(p,q). \tag{4.9.24}$$

These alternatives, rather than the original definition (4.9.21), are what we usually check in order to verify that a given strategy p is an ESS. If the first of them (4.9.23) is true for all $q \neq p$, this is the definition of a *strict Nash equilibrium*; such a strategy p is the *unique* best reply to itself. It is clear that strict Nash implies ESS, and ESS implies Nash, but that the reverse implications are not generally true. The second condition (4.9.24) states that in the case that there is an alternative best reply q to the strategy p, then p fares better against q than q does against itself.

If there are only two pure strategies A and B, a compete analysis is possible. Let us define

$$V(A) = W(A,A) - W(B,A), \quad V(B) = W(B,B) - W(A,B); \tag{4.9.25}$$

then $V(A)$ and $V(B)$ are the advantages that the resident pure strategy has over the alternative pure strategy in environments A and B respectively, and it turns out that it is these quantities that determine the outcome of the game. Now let P, play A with probability p, be the resident mixed strategy, and Q, play A with probability q, a mutant mixed strategy. The crucial quantity in deciding whether Q can invade P is $W(Q,P) - W(P,P)$. It is easy to show that

$$W(Q,P) - W(P,P) = (q-p)(W(A,P) - W(B,P))$$
$$= (q-p)(pV(A) - (1-p)V(B)) = (p-q)V(B)(1-p/p^*), \tag{4.9.26}$$

where

$$p^* = V(B)/(V(A) + V(B)). \tag{4.9.27}$$

It is clear that $p^* \in (0,1)$ if and only if $V(A)$ and $V(B)$ are either both positive or both negative, and otherwise (unless $p^* = 0$), $1 - p/p^* > 0$ for all $p \in (0,1)$.

Let us apply this to the hawk-dove game, with hawk as A and dove as B. The advantage that the hawk has as resident is $V(A) = V(H) = W(H,H) - W(D,H) = \frac{1}{2}(G - C)$, while the advantage that the dove has as resident is $V(B) = V(D) = W(D,D) - W(H,D) = -\frac{1}{2}G$. If $G > C$, $V(A) = V(H) > 0$ and $V(B) = V(D) < 0$ have opposite signs, $W(Q,P) - W(P,P) > 0$ as long as $q > p$, so any mutant strategy with a greater probability of playing hawk can invade a given resident strategy, and the only ESS is to play hawk with probability 1. On the other hand, if $G < C$, $V(A) = V(H)$ and $V(B) = V(D)$ are both negative, and

$$W(Q,P) - W(P,P) = -\frac{1}{2}C(q - p)(p - p^*), \qquad (4.9.28)$$

where $p^* = V(D)/(V(H) + V(D)) = G/C$. The easiest way to find the ESSs of the system is to plot the sign of this quantity in the (p,q)-plane, as in Figure 4.6(a). Clearly, if $p < p^*$ then any strategy with $q > p$ can invade, since the diagram shows that $W(Q,P) - W(P,P) > 0$, while if $p > p^*$ then any strategy with $q < p$ can invade. The only ESS is $p = p^* = G/C$, which has $W(Q,P^*) - W(P^*,P^*) = 0$, but $W(Q,Q) < W(P^*,Q)$, as in Equation (4.9.24). In particular, if the cost C is large relative to the benefit G, the strategy "dove" should be used in a large fraction of encounters. It seems to be true that animals that could do a lot of damage to each other, so that C is large, tend to employ ritual encounters rather than all-out fights.

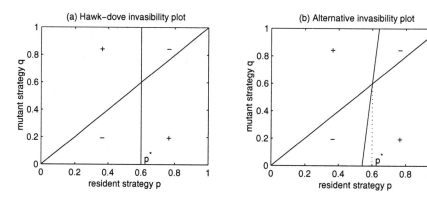

Figure 4.6   Invasibility plots. In (a), the hawk-dove invasibility plot. $+$ denotes points in the (p,q)-plane where $W(Q,P) - W(P,P) > 0$, so that the mutant Q can invade the resident P, and $-$ denotes points where $W(Q,P) - W(P,P) < 0$. In (b), a hypothetical alternative invasibility plot is shown.

A different question may be asked. Assume that a game-playing population is subject to continual small mutations in its strategy; after each mutation the resident strategy persists if the mutant cannot invade, and the mutant invades if it can. Does the population tend toward an ESS from any other strategy? If so, the ESS is called *convergence-stable*. This is a question in *adaptive dynamics*. It can be partially answered by looking at the invasibility plots in Figure 4.6. In (a), start at a point on the main diagonal with $p < p^*$, and introduce a mutant strategy q. If $q > p$, it can invade, but if $q < p$, it cannot. After sufficiently many small mutations, the system arrives at the ESS, which is therefore convergence-stable. But now look at (b), and do the same thing. Again the system arrives at P^*, but when it arrives there *any other strategy can invade*, and so P^* is not an ESS. Convergence stability and evolutionary stability are not equivalent concepts. Other sign patterns can be found where a strategy is evolutionarily stable but not convergence-stable. These differences can lead among other things to populations splitting and pursuing two different strategies, and hence possibly to speciation.

EXERCISES

4.16. There are games with no ESS, as we can see from Figure 4.6, and games with several ESSs. However, an ESS in the interior of S_{n-1} must be unique. Prove that if there are two distinct ESSs p and q in S_{n-1}, then both must be in ∂S_{n-1}, as follows.

a) Assume for contradiction that p is an interior point. We know $W(r,p) \leq W(p,p)$ for all $r \in S_{n-1}$. Choose $r = (1+\epsilon)p - \epsilon q \in S_{n-1}$, $\epsilon > 0$.

b) Deduce that $W(r,p) = W(p,p)$ and $W(r,r) \geq W(p,r)$, contradicting the alternatives (4.9.23) or (4.9.24) required for p to be an ESS.

4.17. The alternatives (4.9.23) and (4.9.24) tell us how to check that a given strategy p is an ESS, but they do not tell us how to find p. A result that is often helpful is the *Bishop–Cannings theorem*: if p is a mixed ESS and i is in the support of p, then

$$W(e_i, p) = W(p, p). \tag{4.9.29}$$

Prove this theorem.

4.18. Consider the pay-off matrix below.

Pay-off matrix			
		on encountering this strategy	
		D	C
Pay-off to	D	2	4
this strategy	C	1	3

This is called the Prisoner's Dilemma game. Originally D stood for defector, who shops their partner in crime, C for co-operator, who does not, but the model has been used in more general investigations of the evolution of co-operation.

a) Do you play D or C? (Assume no conferring between partners.) What is the ESS?

Now assume the same game is played between the same partners ten times. One strategy is still D, but the other is TFT, tit for tat, who plays C first but subsequently plays what their partner played last time.

b) Find the pay-off matrix and the ESS(s).

c) Can this explain the evolution of co-operation?

4.10 Replicator Dynamics

The last section and this one ask quite different questions, respectively as follows.

- Given that all individuals employ the same strategy, which strategy will they employ?

- Given the strategies that can be employed, how will the numbers employing each strategy change under natural selection?

The reason for the change is that it is often unrealistic to assume that each individual in the population employs the same strategy. This might be because different strategies require different resource allocation decisions, a hawk preferring to invest in size and a dove in speed, for example. The prediction that each individual in the population should play hawk with probability $\frac{G}{C}$ must be discarded. If only two phenotypes are possible, hawk and dove, should it be replaced by the prediction that the population evolves until a fraction $\frac{G}{C}$ of them are hawks? (The phenotypes we consider need not employ pure strategies, but this is conceptually the easiest case and we shall confine ourselves to it.)

The first question above is about steady states, but the second requires a model of evolutionary dynamics. Let us consider a population, each one of which employs one of n pure strategies, and let the frequency of the ith phenotype be x_i. Then $x = (x_i)$ is a point in the simplex S_{n-1}, and represents the *state* or phenotypic composition of the population. (It is *not* a strategy.) Let the state of the system evolve according to

$$\dot{x}_i = (w_i(x) - \bar{w}(x))x_i \qquad (4.10.30)$$

for each $i = 1, \cdots, n$, where $w_i(x)$ is the fitness of the ith phenotype, and $\bar{w}(x) = \sum_{j=1}^{n} x_j w_j(x)$ the mean fitness of the population, when the state of the population is x. This is based on Equation (4.5.13) extended from S_1 to S_{n-1}.

Now recall that the pay-offs of the game were interpreted as a change in fitness from some base level, and note that this base level will cancel out when we subtract $\bar{w}$ from w_i. We can write

$$\dot{x}_i = (W(e_i, x) - W(x, x)) x_i = ((Ux)_i - x^T U x) x_i. \qquad (4.10.31)$$

This is the system of equations that we take to represent the evolutionary process. Note that the simplex S_{n-1} is invariant under Equation (4.10.31); if $x(0) \in S_{n-1}$ then $x(t) \in S_{n-1}$ for all $t > 0$. Note also that if p is an ESS then $x = p$ is a steady state of the replicator equations. (For if $i \in \text{supp}(p)$, use Bishop–Cannings, and if $i \notin \text{supp}(p)$, then $x_i = p_i = 0$.) It can be shown that this steady state is stable, but this is beyond the scope of this book.

The male side-blotched lizard, *Uta stansburniana*, employs one of three mating strategies. Orange-throated O males maintain territories large enough to contain several females, blue-throated B males maintain territories large enough to contain a single female, while yellow-throated Y males do not maintain a territory at all. If the population is predominantly orange, the yellow strategy is best, because while the oranges are occupied in defending their territories, the yellows sneak in and copulate with the females. If yellow predominates, the blue strategy is best, because their territories are small enough to defend against the yellows. If blue predominates, the yellow sneakers are rare, it pays to defend a larger territory, and orange is best. In a field study, the frequencies of the three colour morphs cycled with a period of about six years. A possible pay-off matrix is given below.

Pay-off matrix				
		on encountering this strategy		
		Y	O	B
Pay-off	Y	0	1	-1
to this	O	-1	0	1
strategy	B	1	-1	0

This is equivalent to the children's rock-scissors-paper game. Let p be the symmetric strategy $(\frac{1}{3}, \frac{1}{3}, \frac{1}{3})$. Then p is a Nash equilibrium, but it is easy to show that for any strategy q, $W(q,q) = W(p,q) = W(q,p) = W(p,p) = 0$, so that there is no ESS. The replicator equations are

$$\dot{x} = x(y-z), \quad \dot{y} = y(z-x), \quad \dot{z} = z(x-y).$$

The evolutionary trajectories on the simplex S_2, which is the part of the plane $x+y+z=1$ in the positive quadrant and is an equilateral triangle, are shown below. The function xyz is a constant of motion, and the closed curves are the curves $xyz = c$, for various constants c. The cyclic behaviour of the system is clear.

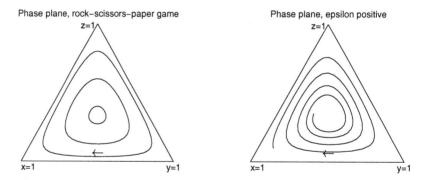

Figure 4.7   The evolutionary trajectories for the rock-scissors-paper game. The phase plane is the simplex S_2, a triangle whose corners are the points $(1,0,0)$, $(0,1,0)$ and $(0,0,1)$ in (x,y,z)-space. On the left, $\epsilon = 0$, and the trajectories are closed; on the right $\epsilon = 0.1$, representing a penalty for like encounters, and they spiral in towards the mid-point.

Now assume that there is a small penalty when an individual meets another of the same phenotype, so that the zeroes on the diagonal of the pay-off matrix are each replaced by $-\epsilon$. The symmetric strategy becomes an ESS, and the symmetric steady state of the replicator equations becomes asymptotically rather than neutrally stable. (The result of a reward for like encounters is discussed in Exercise 4.20.) If there were a phenotype that played the symmetric strategy, a population consisting of that phenotype could not be invaded, and if the only possible phenotypes were the pure strategies, the population would tend to the symmetric state.

EXERCISES

4.19. Write down the replicator equations for the hawk-dove game, and find their asymptotic behaviour as $t \to \infty$. Compare with the results obtained in Section 4.9.

4.20. Modify the lizard rock-scissors-paper game by assuming that there is a small penalty to be paid if a lizard encounters another of the same colour.

a) Show that the symmetric strategy $p = (\frac{1}{3}, \frac{1}{3}, \frac{1}{3})$ is now an ESS.

b) How does this modification change the dynamics of the replicator system?

c) What happens if the penalty is replaced by a reward?

Hints.

i) What happens to the function xyz under the replicator equations?

ii) It can be shown using Lagrange multipliers or otherwise that xyz attains its maximum on S_2 and $x^2 + y^2 + z^2$ its minimum on S_2 at the symmetric point $(\frac{1}{3}, \frac{1}{3}, \frac{1}{3})$.

4.21. *Computer exercise.* Write a program for plotting evolutionary trajectories for normal form games on S_2, given the pay-off matrix and the initial frequencies of the three pure strategies.

4.11 Conclusions

– Hardy–Weinberg law: with no stochastic effects, mutations, migration or selective pressure, allele frequencies remain constant, irrespective of whether the allele is dominant or recessive.

– Random mating (or more strictly random union of gametes) means Punnett square, and simplifies the equations. At the beginning of the zygotic phase, $x = p^2$, $y = 2pq$, $z = q^2$.

– With selective pressure, advantageous alleles spread through a population. The rate of spread is given by the Fisher–Haldane–Wright (FHW) equation. It is proportional to the amount by which their mean fitness exceeds the mean fitness of the population.

– The rate of spread of an advantageous allele by natural selection is extremely slow to begin with, and is slower for recessive than for dominant genes.

– Deleterious genes may be maintained in a population through mutation, and will occur at higher frequency if they are recessive.

– Let the mean fitness of a population depend only on the frequency p of an allele. A sketch of mean fitness $\bar{w}$ as a function of p is known as Wright's adaptive topography. As the population evolves under natural selection, p moves up the gradient of $\bar{w}$ to a maximum, so that $\bar{w}$ increases.

– An evolutionarily stable strategy (ESS) is proof against invading mutant strategies. For all $q \neq p$, such a strategy p satisfies either condition (4.9.23) or condition (4.9.24).

– There are games without an ESS, and games with several ESSs, which must necessarily lie on the boundary of the simplex S_{n-1}. An ESS in the interior of S_{n-1} is necessarily unique.

– For normal-form games, ESSs are convergence-stable, so that a population subject to continual small mutations will end up at an ESS. This is not necessarily so in general.

– The replicator equation is used to model the dynamics of game theory. If p is an ESS, then $x = p$ is a stable steady state of the replicator equation.

5
Biological Motion

- The mobility of cells or organisms, both random and in response to environmental influences, plays a crucial role in many biological phenomena. Well-established models for random motion from the physical sciences have been adapted to biological situations, and extensions have been made to investigate novel phenomena without a physical counterpart.

- The interaction between kinetics and motion leads to phenomena that would be difficult to predict from a consideration of each in isolation, and has led to a large body of work since the 1930s but particularly over the last thirty years.

5.1 Introduction

In many sciences, including biology, we need to model motion that has some random element to it. In physics, for example, we might be concerned with *molecular diffusion*, the random motion of molecules in a fluid. In addition to undergoing diffusion, molecules within a fluid may also be carried along by a current in the fluid, a process known as *advection*. A microscopic theory of random motion is a description of the statistical properties of this motion for a molecule or an ensemble of molecules, a problem investigated by Einstein. We might also hope to gain from a consideration of these properties an idea of how the bulk properties of the fluid, such as its pressure, temperature, density and velocity field, vary with space and time. This is a macroscopic theory of random motion, which may be derived using a random walk approach or a continuum approach. In this book we shall restrict ourselves to a continuum approach, although this precludes us from deriving detailed models of individual-based behaviour and stochastic effects. There is some material on the website about the alternatives.

Motion in the biological sciences, at least at the level of cells or organisms, is of course very different from molecular diffusion and advection, but there are also some striking similarities. Thus inorganic models for motion provide a point of departure for the mathematics of spatial variation in populations, as was recognised almost immediately that the problem was addressed. One important difference is that the interaction between the motion and the biological kinetics is often crucial, as was recognised in papers by Fisher (1937) on the spread of an advantageous gene, and Kolmogorov, Piscounov and Petrovsky (1937) on the invasion of virgin territory by a population. A second difference is that the motion of biological cells or organisms is often influenced by individuals or substances in the environment, including other members of the same species. These influences lead to *chemotaxis*, where the cells or organisms climb a chemical gradient, and many similar phenomena. The equations are often far more complicated than those of molecular diffusion, and in many cases modelling questions are still open.

The questions raised in this chapter have led to feedback between biology and mathematics, the biology leading to new mathematical equations and hence novel mathematical techniques and results, and the mathematics leading to a deeper understanding of the biological phenomena, and in turn to new questions. This is the essence of mathematical biology.

5.2 Macroscopic Theory of Motion; A Continuum Approach

5.2.1 General Derivation

Our theory of motion is firmly based on the principle of conservation of matter. Using the continuum approach, we consider conservation of particles in an arbitrary volume V fixed in space and enclosed by a surface S. Then

$$
\begin{array}{ccccccc}
\text{Particles in } V & = & \text{Particles} & + & \text{Net particles} & + & \text{Net creation} \\
\text{at } t + \delta t & & \text{in } V \text{ at } t & & \text{entering } V & & \text{of particles in } V.
\end{array}
$$

The equations will include sources and sinks (births and deaths, growth and decay), in the term "net creation of particles". A description of the particles entering V requires the concept of a *flux vector*. This describes the rate and direction in which particles move under the influence of advection, diffusion, and other motive forces.

The particle flux at $(\mathbf{x}, t)$, denoted by $\mathbf{J}(\mathbf{x}, t)$, is defined as follows. First, let $\mathbf{J}$ have magnitude $J = |\mathbf{J}|$ and direction $\mathbf{m} = \mathbf{J}/J$, so $\mathbf{J} = J\mathbf{m}$. Define

– $\mathbf{m}$ to be the direction of net flow,

– J by placing a small test surface of area δS and normal $\mathbf{m}$ "at" $(\mathbf{x}, t)$; then $J\delta S$ is the net number of particles crossing the test surface in the (positive) $\mathbf{m}$-direction per unit time, the so-called *current* across the surface.

The particle flux is related to the mean velocity $\overline{\mathbf{v}}$ of the particles by $\mathbf{J} = u\overline{\mathbf{v}}$ (by definition of $\overline{\mathbf{v}}$), where u is the concentration, in particles per unit volume.

Now, since we need the rate at which particles cross the surface of an arbitrary volume V, i.e. the current across this surface, we need to know the rate at which particles cross surfaces with other orientations. Given some small test surface of area δS and normal $\mathbf{n}$, the rate at which particles cross it depends on $\cos\theta$, where θ is the angle between the flux vector $\mathbf{m}$ and the normal $\mathbf{n}$ to the test surface. Since $\mathbf{m} \cdot \mathbf{n} = \cos\theta$, this current is given by

$$I = J\cos\theta\delta S = J\mathbf{m} \cdot \mathbf{n}\delta S = \mathbf{J} \cdot \delta\mathbf{S},$$

where $\delta\mathbf{S} = \mathbf{n}\delta S$ is the usual oriented surface element.

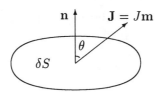

Figure 5.1   Flux in a general direction.

Now we can state the principle of conservation of matter in mathematical terms. We have, to leading order in δt,

$$\int_V u(\mathbf{x}, t + \delta t)dV = \int_V u(\mathbf{x}, t)dV - \int_S \mathbf{J}(\mathbf{x}, t) \cdot \mathbf{dS}\delta t + \int_V f(\mathbf{x}, t)dV\delta t,$$

where we have defined $f(\mathbf{x}, t)$ to be the sink/source density (net number of particles created per unit time and per unit volume) at $(\mathbf{x}, t)$.

Subtracting $\int_V u(\mathbf{x}, t)dV$ from both sides, dividing by δt, taking the limit as $\delta t \to 0$ and using the divergence theorem, which states that

$$\int_S \mathbf{J} \cdot d\mathbf{S} = \int_V \nabla \cdot \mathbf{J}dV,$$

we have

$$\int_V \left(\frac{\partial u}{\partial t} + \nabla \cdot \mathbf{J} - f \right) dV = 0,$$

as long as the functions involved are sufficiently smooth. But this is true for arbitrary volumes V, so the integrand is zero,

$$\frac{\partial u}{\partial t} + \nabla \cdot \mathbf{J} - f = 0. \tag{5.2.1}$$

This is the equation of conservation of matter with a source term, and is crucial to the rest of the book.

We need to model the flux $\mathbf{J}$. First, consider the flux due to advection with velocity $\mathbf{v}$. The particles are moving in the direction of $\mathbf{v}$, and the rate at which they cross a test surface of area δS placed perpendicular to the flow is $|\mathbf{v}|u\delta S$, where u is the concentration (particles per unit volume). Thus $\mathbf{J}_{\text{adv}} = \mathbf{v}u$. The advection equation with a source term is

$$\frac{\partial u}{\partial t} = -\nabla \cdot \mathbf{J}_{\text{adv}} + f = -\nabla \cdot (\mathbf{v}u) + f. \tag{5.2.2}$$

Second, consider the flux due to diffusion. Empirically, the net flow of particles is down the concentration gradient and proportional to its magnitude, so

$$\mathbf{J}_{\text{diff}} = -D\nabla u. \tag{5.2.3}$$

This mathematical model of diffusive flux is known as *Fick's law*. The diffusion equation with a source term is

$$\frac{\partial u}{\partial t} = -\nabla \cdot \mathbf{J}_{\text{diff}} + f = \nabla \cdot (D\nabla u) + f. \tag{5.2.4}$$

If we have both advection and diffusion, $\mathbf{J} = \mathbf{J}_{\text{adv}} + \mathbf{J}_{\text{diff}}$, and the advection-diffusion equation with a source term is

$$\frac{\partial u}{\partial t} = -\nabla \cdot (\mathbf{v}u) + \nabla \cdot (D\nabla u) + f. \tag{5.2.5}$$

EXERCISES

5.1. Consider the conservation equation with no source term, Equation (5.2.1) with $f = 0$, to be solved in a region V with zero-flux boundary conditions $\mathbf{J} \cdot \mathbf{n} = 0$ on S. Show that conservation holds, in the sense that $\int_V udV$ is constant.

5.2.2 Some Particular Cases

In one-dimensional flow in a three-dimensional domain, working in Cartesian coordinates x, y, z, all motion is in the x-direction and all dependent variables depend on x (and t) only. With constant advective velocity v and diffusion coefficient D, the advection-diffusion equation with no source term becomes

$$\frac{\partial u}{\partial t} = -v\frac{\partial u}{\partial x} + D\frac{\partial^2 u}{\partial x^2}. \tag{5.2.6}$$

This may also be considered as flow in a one-dimensional domain.

In cylindrically symmetric flow in a three-dimensional domain, working in cylindrical polar coordinates R, ϕ, z, all motion is in the radial R direction and all dependent variables depend on radius R (and t) only. With constant advective velocity v and diffusion coefficient D, the advection-diffusion equation with no source term becomes

$$\frac{\partial u}{\partial t} = -v\frac{\partial u}{\partial R} + D\frac{1}{R}\frac{\partial}{\partial R}\left(R\frac{\partial u}{\partial R}\right). \tag{5.2.7}$$

This may also be considered as radially symmetric flow in two dimensions, working in plane polar coordinates R, ϕ.

In spherically symmetric flow, working in spherical polars r, θ, ϕ, all motion is in the radial r direction and all dependent variables depend on radius r (and t) only. With constant advective velocity v and diffusion coefficient D, the advection-diffusion equation with no source term becomes

$$\frac{\partial u}{\partial t} = -v\frac{\partial u}{\partial r} + D\frac{1}{r^2}\frac{\partial}{\partial r}\left(r^2\frac{\partial u}{\partial r}\right). \tag{5.2.8}$$

All these equations may be derived directly from the general advection-diffusion Equation (5.2.5) with $f = 0$, using the formulae for the vector differential operators in the appropriate coordinate system given in Section C.2.3 of the appendix, or they may be derived directly, as below.

Example 5.1

Derive the equation of conservation of matter for one-dimensional flow by adapting the general argument in the previous subsection.

Let I be the interval $I = (x, x + \delta x)$, and let $t' = t + \delta t$. The equation of conservation of particles may be written in words as follows.

Particles in I at t'	=	Particles in I at t	+	Particles entering	−	Particles leaving
			+	Particles created	−	Particles destroyed.

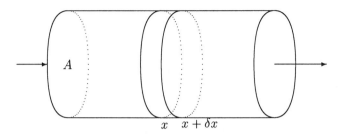

Figure 5.2   Conservation in one dimension.

Define

- $u(x, t)$, concentration (particles per unit volume) at (x, t),

- $J(x, t)$, particle flux (net number of particles crossing a surface perpendicular to the x-axis in the positive x-direction, per unit time and per unit area), at (x, t),

- $f(x, t)$, sink/source density (net number of particles created per unit time and per unit volume) at (x, t).

Let A be the cross-sectional area of the tube. Then the equation becomes

$$u(x, t + \delta t)A\delta x = u(x, t)A\delta x + J(x, t)A\delta t - J(x + \delta x, t)A\delta t$$
$$+ f(x, t)A\delta x\delta t + h.o.t. \quad (5.2.9)$$

Dividing through by $A\delta x\delta t$ and taking limits as $\delta t \to 0$, $\delta x \to 0$, we obtain

$$\frac{\partial u}{\partial t} = -\frac{\partial J}{\partial x} + f. \quad (5.2.10)$$

This is the equation of conservation of matter in one dimension.

This is essentially a three-dimensional derivation of a one-dimensional result. A purely one-dimensional derivation could have been obtained at the cost of a slight change in interpretation of u, f and J, with u the amount of matter per unit *length*, f the net source strength per unit length, and $J = I$, the particle current, or the number of particles passing x per unit time. The derivation is then exactly the same except that A never appears.

How is the principle of conservation of matter manifested in solutions of Equation (5.2.10)?

Example 5.2

Consider the solution of Equation (5.2.10) with $f = 0$ on the whole of $\mathbb{R}$, subject to some initial condition $u(x, 0) = u_0(x)$. The number of particles at $t = 0$ is given by $N_0 = \int_{-\infty}^{\infty} u_0(x)dx$, and the number of particles at a general time t by $N(t) = \int_{-\infty}^{\infty} u(x,t)dx$. Show that conservation of particles holds, $N(t) = N_0$ for all t. You may assume that $J(x,t) \to 0$ as $|x| \to \infty$.

We have

$$\frac{dN}{dt}(t) = \int_{-\infty}^{\infty} \frac{\partial u}{\partial t}(x,t)dx$$

$$= -\int_{-\infty}^{\infty} \frac{\partial J}{\partial x}(x,t)dx = -[J(x,t)]_{-\infty}^{\infty} = 0, \quad (5.2.11)$$

and the result follows.

EXERCISES

5.2. Derive the equation of conservation of matter as in Example 5.1 above for spherically symmetric flow in spherical polars r, θ, ϕ.

[Consider conservation of matter over the shell V between two spheres both centred at the origin and with radii r and $r + \delta r$.]

5.3. Derive the equation of conservation of matter as in Example 5.1 above for cylindrically symmetric flow in cylindrical polars R, ϕ, z.

[The cylindrical shell between two circular cylinders with axis the z-axis and radii R and $R + \delta R$ is infinite, so you should consider conservation of matter over V, a finite part of it which lies between two arbitrarily chosen z-planes, e.g. $z = 0$ and $z = h$.]

5.4. Derive the equation of conservation of matter as in Example 5.1 above for two-dimensional radially symmetric flow in plane polars R, ϕ.

[You will need to re-interpret flux and concentration slightly (because, for example, concentration can no longer be the amount of matter per unit *volume*). You should then consider conservation of matter over the annulus (or ring) S between two circles both centred at the origin and with radii R and $R + \delta R$.]

5.5. Show that the law of conservation of matter holds for the diffusion equation $u_t = Du_{xx}$ on a finite interval (a,b) with homogeneous Neumann boundary conditions $u_x(a,t) = u_x(b,t) = 0$, where D is a constant. That is, define $N(t) = \int_a^b u(x,t)dx$, and show that $\frac{dN}{dt} = 0$.

5.6. What boundary conditions would be appropriate for conservation for the advection-diffusion equation $u_t = -vu_x + Du_{xx}$ on a finite interval (a, b), where v and D are constants? Interpret these boundary conditions.

5.7. Let v and D be constant, and let u satisfy the advection-diffusion equation

$$u_t = -vu_x + Du_{xx}.$$

Let $z = x - vt$, $s = t$, and $U(z, s) = u(x, t)$. Use the *chain rule* applied to the equation $U(z, s) = u(x, t)$ to show that U satisfies the diffusion equation

$$U_s = DU_{zz}.$$

(In other words, if you choose coordinates moving with the advective flow, or drift, then the advective term disappears.)

5.3 Directed Motion, or Taxis

Biological organisms and even cells not only move at random, but sense their environment and respond to it. The response often involves a *taxis*, a movement towards or away from an external stimulus. Some examples of these *taxes* are *chemotaxis*, a response to a chemical gradient, *phototaxis*, a response to a light source, *geotaxis*, a response to a gravitational field, *galvanotaxis*, a response to an electric field, and *haptotaxis*, a response to an adhesive gradient. Tactic responses to conspecifics or to predators and prey are also observed. In fact, it has been said that nothing is certain except death and taxes.

At the cellular level and even the organismal level one of the most important of these taxes is chemotaxis. The chemical may be produced by conspecifics, as when ants follow pheromone trails or bacteria or slime moulds aggregate. At the cellular level, tumours produce a variety of chemicals that influence the motion as well as the growth of nearby cells, as we shall see in Chapter 8. A chemotactic response may be positive, leading to motion up the chemical gradient, or negative.

To fix ideas, let us consider a positive chemotactic response of unicellular organisms such as bacteria. Let us first assume that a chemical gradient has been produced externally and the bacteria are merely responding to it. Bacterial motion involves a sequence of runs and tumbles. At the run stage, the bacterium moves a certain distance in a combination of motion in a certain direction with random diffusion. At the tumble stage, the bacterium reorients itself in preparation for a new run. Rather than actively changing their direction of motion in response to the stimulus, bacteria seem to turn more frequently when the chemical concentration is low. We shall model the flux directly by appealing to Occam's razor, i.e. by looking for the simplest possible model. Clearly a positive chemotactic response should be in the direction of the chemical gradient, and should increase with the magnitude of that gradient. The simplest model is that the average velocity with which the bacteria respond to the gradient is proportional to the gradient, and the flux therefore should be proportional to the product of the gradient and the density of bacteria. Denoting the concentration of the chemical by c and the density of the bacteria by u, we have

$$\mathbf{J}_{\text{chemo}} = \chi u \nabla c, \tag{5.3.12}$$

where $\chi > 0$ is a constant of proportionality, known as the *chemotactic co-efficient* or *chemotactic sensitivity*. This is the most widely used model for chemotactic flux. For negative chemotaxis, χ is negative.

Things get more complicated if the bacteria themselves are producing the chemical. A typical model in this case is

$$\frac{\partial u}{\partial t} = f(u) - \nabla \cdot (\chi u \nabla c) + \nabla \cdot (D_u \nabla u),$$

$$\frac{\partial c}{\partial t} = g(u, c) + \nabla \cdot (D_c \nabla c), \tag{5.3.13}$$

where f is the net rate at which bacteria are produced and g is the net rate at which the chemical is produced, i.e. the rate at which bacteria produce it minus the rate at which it breaks down. A typical choice for f is $f(u) = 0$, on the assumption that birth and death rates for bacteria are much slower than chemical production and loss rates, and for g is $g(u, c) = \alpha u - \beta c$. This model will be looked at in more detail in Chapter 7.

Let us consider another example of taxis, the response of unicellular organisms to a force field. Since they are at the microscopic end of the spectrum, such organisms are dominated by viscous forces rather than by inertia. In the terminology of fluid mechanics, they live life at low Reynolds number. When subjected to a force field they will very quickly accelerate to a terminal velocity and remain there, and when subjected to no forces they will come to rest almost immediately. Consider a bacterium of length 10^{-6} metres moving at a typical speed of 10^{-5} metres per second. If no forces act, it will come to rest in a time of about 10^{-6} seconds, and in that time will travel about 10^{-12} metres, a distance small compared with the diameter of a hydrogen atom! Of course, it will not in fact stop, since it is still subject to diffusion, but its drift velocity will become zero. Consider now the response of unicellular plankton, plants and animals that drift near the surface of lakes and seas, to a gravitational field. The resultant force will generally be upward, since most plankton are buoyant. In the absence of other forces, a single planktonic organism will almost immediately reach a terminal velocity (upward), which will be proportional to the gravitational force g per unit mass. A model for the planktonic gravitational flux is therefore

$$\mathbf{J}_{\text{geo}} = \alpha u g \mathbf{k},$$

where $\mathbf{k}$ is the unit upward-pointing vector, α is a measure of buoyancy, and u is the planktonic density. More generally, if a force field with potential Φ acting on a particle leads very quickly to a terminal velocity $\mathbf{v} = \alpha \nabla \Phi$, for some constant α, the relevant equation for the flux is

$$\mathbf{J}_{\text{force}} = \alpha u \nabla \Phi.$$

The equation for particle concentration is

$$\frac{\partial u}{\partial t} = -\nabla \cdot (\alpha u \nabla \Phi) + \nabla \cdot (D \nabla u).$$

5.4 Steady State Equations and Transit Times

The diffusion equation and its relatives are partial differential equations, and hence can be difficult to solve. However there are many biological situations where a steady state is reached, with all variables independent of time. If the problem is one-dimensional or radially or spherically symmetric the partial differential equation then becomes an ordinary differential equation. We shall consider various examples where this is the case. It is important to realise that the steady state assumption does not mean that fluxes are necessarily zero, only that they do not change through time. We shall calculate fluxes, and the times that molecules or other particles might take to move from one point to another under the influence of diffusion. We consider the implications for molecular diffusion as a transport mechanism, and conclude that by itself it is inefficient except over short distances.

5.4.1 Steady State Equations in One Spatial Variable

The conservation Equation (5.2.1) in the steady state gives

$$-\nabla \cdot \mathbf{J} + f = 0, \tag{5.4.14}$$

with $\mathbf{J}$ independent of t. If there are no sources or sinks ($f = 0$) and the problem is one-dimensional, cylindrically symmetric (or radially symmetric in two dimensions), or spherically symmetric, we obtain

$$\frac{dJ}{dx} = 0, \quad \frac{1}{R}\frac{d}{dR}(RJ) = 0, \quad \text{or} \quad \frac{1}{r^2}\frac{d}{dr}(r^2 J) = 0 \qquad (5.4.15)$$

respectively, so that $J = C$, $J = C/R$, or $J = C/r^2$ respectively, where C in each case is a constant of integration. If the flux is purely diffusional, then $J = -D\frac{du}{dx}$, $J = -D\frac{du}{dR}$, or $J = -D\frac{du}{dr}$ respectively (using Fick's law, Equation (5.2.3), in the appropriate coordinate system, see Section C.2.3 of the appendix), and if D is constant, then $u = Ax + B$, $u = A\log R + B$, or $u = -A/r + B$ respectively, where $A = -C/D$ and B is a second constant of integration. The constants of integration are determined by the boundary conditions.

EXERCISES

5.8. a) Derive an equation for plankton density as a function of depth. You may assume that buoyancy effects lead to a terminal velocity $\mathbf{w} = \alpha g\mathbf{k}$ in the upward direction, but that the plankton also diffuse.

b) What boundary conditions would you apply to this equation in this situation?

c) Find the steady state density profile for plankton with depth.

5.4.2 Transit Times

The general idea of transit times, or residence times, is as follows. Assume that particles enter a region V through a subset S_1 and leave it through a subset S_2 of its surface. Let the entry current be I_1 and the exit current I_2. We shall consider the steady state situation, so $I_1 = I_2 = I$, say. Let N be the (constant) number of particles in V. Then the probability that a particle chosen at random from those in V leaves V in the next interval of time δt is $I\delta t/N$, and it follows that the *transit time* τ, defined to be the average time taken for a particle to cross V from S_1 to S_2, is given by $\tau = N/I$.

Example 5.3

A substance diffuses in one dimension (with constant diffusion coefficient) from $x = 0$, where its concentration is u_0, to $x = L$, where its concentration is 0.

(Particles arriving at L are assumed to be removed immediately.) Find the steady state flux and the transit time from 0 to L.

The problem is one-dimensional, and we interpret u as a *line concentration*, amount of matter per unit length, and J as the rate at which matter passes a point, equivalent to current. In terms of u, the boundary value problem is given by

$$0 = D\frac{d^2 u}{dx^2} \text{ in } (0, L), \quad u(0) = u_0, \quad u(L) = 0,$$

which has solution $u(x) = u_0(1 - \frac{x}{L})$. The flux is therefore $J = -D\frac{du}{dx} = D\frac{u_0}{L}$ (a constant as expected), and hence the diffusional current is $I = J = D\frac{u_0}{L}$. The number of particles in the tube is $N = \int_0^L u(x)dx = \frac{1}{2}Lu_0$, so that the transit time from 0 to L is

$$\tau = \frac{N}{I} = \frac{\frac{1}{2}Lu_0}{D\frac{u_0}{L}} = \frac{1}{2}\frac{L^2}{D}.$$

Hence the average time to diffuse a distance L is $\tau = \frac{1}{2}\frac{L^2}{D}$, or the average distance through which diffusion works in a time τ is $L = \sqrt{2D\tau}$. A typical magnitude for the diffusion coefficient of a small molecule like oxygen in a medium like water is $D = 10^{-5}\,\text{cm}^2\,\text{s}^{-1}$. Such molecules can diffuse across a cell 10 microns across in 0.1 seconds, but would take 27 years to diffuse along a nerve axon 1 metre long. This is a bit too long to wait. Similarly, the diffusion coefficient of a small molecule in air is about $D = 10^{-1}\,\text{cm}^2\,\text{s}^{-1}$. If one relied on diffusion to carry molecules of perfume across a crowded room, it would take of the order of one month, and perfume manufacturers would be out of business. On the other hand, insect sex pheromones can attract a mate over a distance of the order of one kilometre. Clearly other processes such as advection are important in such cases.

EXERCISES

5.9. Consider the situation described in Example 5.3 above if the substance advects (with constant advective velocity v) as well as diffusing.

a) Find the steady state line concentration profile u and the flux (equivalent to the diffusion current) $J = I$.

b) Find the amount of matter N in $(0, L)$.

c) Show that the transit time is given by

$$\tau = \frac{L}{v} - \frac{D}{v^2}\left(1 - \exp\left(-\frac{vL}{D}\right)\right).$$

d) Show that u, J and τ tend to the corresponding expressions for the case $v = 0$, found in Example 5.3, as $v \to 0$.

5.10. *Steady spherically symmetric flow.* In spherically symmetric flow, the diffusion equation is given by

$$\frac{\partial u}{\partial t} = \frac{1}{r^2} \frac{\partial}{\partial r} \left(r^2 D \frac{\partial u}{\partial r} \right),$$

with flux $J = -D\frac{\partial u}{\partial r}$ in the radial direction.

a) Solve the equation with D constant in the steady state in a spherical shell $a < r < b$, with boundary conditions $u(a) = 0$, $u(b) = u_0$, and find the diffusive flux.

b) Find the rate at which material flows through the outside (or the inside) curved surface of the spherical shell (the diffusive current), and the amount of material within the shell.

c) Deduce the transit time τ from a point on $r = b$ to a point on $r = a$.

d) Find an approximation to τ if $a \ll b$.

[The element of volume in spherical polars is $r^2 \sin\theta\, dr d\theta d\phi$.]

5.11. *Steady two-dimensional radially symmetric flow.* In two-dimensional radially symmetric flow, the diffusion equation is given by

$$\frac{\partial u}{\partial t} = \frac{1}{R} \frac{\partial}{\partial R} \left(RD \frac{\partial u}{\partial R} \right),$$

with flux $J = -D\frac{\partial u}{\partial R}$ in the radial direction.

a) Solve the two-dimensional steady state diffusion equation with D constant in an annulus $a < R < b$, with boundary conditions $u(a) = 0$, $u(b) = u_0$, and find the diffusive flux.

b) Find the rate at which material flows through the outside (or the inside) curved surface of the annulus (the diffusive current), and the amount of material within the annulus.

c) Deduce the transit time τ from a point on $R = b$ to a point on $R = a$.

d) If $a \ll b$, show that $\tau \approx (b^2/2D) \log(b/a)$.

[The element of area in plane polars is $RdRd\phi$.]

5.4.3 Macrophages vs Bacteria

Macrophages are part of the body's immune system, which provides a defence against foreign materials; in particular, alveolar macrophages clear the lung surface of inhaled bacteria by engulfing and digesting them. In this section we shall consider whether diffusion is sufficient for macrophages to control bacteria in the lung. The relevant diffusion coefficient may be estimated from data.

Assume that the alveolus may be approximated by a disc of radius b. Assume that there is a single macrophage that starts on the circle $R = b$ and is confined to the disc $R \leq b$, and one bacterium with radius $R = a$ inside $R = b$, for simplicity centred at the origin. How long does it take the macrophage to find the bacterium? By Exercise 5.11, the transit time τ is given approximately by

$$\tau \approx \frac{b^2}{2D} \log\left(\frac{b}{a}\right).$$

The parameters $a \approx 20$ microns and $b \approx 300$ microns are known. How do we estimate D? A macrophage moves at a characteristic speed u. The direction of motion is fairly constant for time δt, then re-orientation occurs. It may be shown by a random walk argument that

$$D \approx \frac{1}{4} \frac{\delta x^2}{\delta t},$$

where $\delta x = u \delta t$. With $u \approx 3$ microns per minute and $\delta t \approx 5$ minutes this gives $D \approx 11$ square microns per minute. This gives a transit time of about 160 hours, far greater than the bacteria doubling time of around 5 hours, so we need around 30 macrophages per alveolus to find bacteria solely by random motion.

The alternative hypothesis, that macrophages head directly towards bacteria, gives $\tau = \frac{b}{u} \approx 1\frac{1}{2}$ hours. It may be shown that even a small amount of directed motion can lead to a transit time that is much closer to $1\frac{1}{2}$ hours than to 160 hours, drastically reducing the number of macrophages required.

5.5 Biological Invasions: A Model for Muskrat Dispersal

Even with spatial symmetries, time-dependent problems are generally more difficult than steady-state problems, as they involve solving a partial rather than an ordinary differential equation. However there are various situations where they are tractable, some of which are discussed in Section C.2 of the appendix. In this section we give an application of a time-dependent problem, where the mathematical problem is to solve the diffusion equation with a linear growth term on the whole of two-dimensional space.

Muskrats (*Ondatra zibethica*) were introduced inadvertently to Europe in 1905 by a landowner in Bohemia, and spread as an invasion wave through Europe, as shown in Figure 5.3. The speed of the wave seems to tend to a constant. Can these data be explained by diffusion alone, or by diffusion and growth? We shall make a mathematical model to try to answer this question.

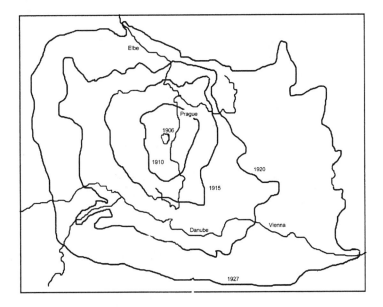

Figure 5.3   Data for the invasion of muskrats in Europe, after Elton (*The Ecology of Invasion by Animals and Plants*, Methuen, 1958). A plot of the square root of area against time is very close to a straight line.

A possible model for the spread of the muskrats after their introduction is through random motion in two dimensions combined with exponential growth,

$$\frac{\partial u}{\partial t} = \alpha u + \nabla \cdot (D \nabla u). \qquad (5.5.16)$$

Assuming that M individuals were released at the origin, the appropriate initial condition is

$$u(\mathbf{x}, 0) = M \delta(\mathbf{x}). \qquad (5.5.17)$$

Similar problems are discussed in Section C.2.1 of the appendix, and the solution for D constant is given by combining Equation (C.2.10) with the method leading to Equation (C.2.24), namely

$$u(R, t) = \frac{M}{4\pi Dt} \exp\left(\alpha t - \frac{R^2}{4Dt}\right). \tag{5.5.18}$$

This is a wave of invasion, as can be seen in Figure 5.4. The wave extends all the way to infinity if $t > 0$, so we have to decide on a definition of the wave front. (Note that this means that some muskrats invade at speeds far faster than any real muskrat could attain. The model may be modified to avoid this problem, but we shall accept it as it stands if it gives good agreement with data.)

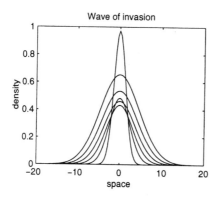

Figure 5.4   Invasion wave for Equation (5.5.18). The initial peak is first damped by diffusion and then begins to grow and spread.

Let the wave front $R = R_1(t)$ be the circle where $u = u_1$ (some predetermined small value of the density). Then

$$u_1 = \frac{M}{4\pi Dt} \exp\left(\alpha t - \frac{R_1^2(t)}{4Dt}\right),$$

$$\log\left(\frac{4\pi D u_1 t}{M}\right) = \alpha t - \frac{R_1^2(t)}{4Dt}.$$

For t large the only term that can balance αt is $\frac{R_1^2(t)}{4Dt}$, which is therefore $O(t)$. Neglecting $\log t$ and constants compared to terms of $O(t)$, we obtain

$$R_1(t) \approx \sqrt{4\alpha D}\, t,$$

for t large, a linear equation, qualitatively different behaviour from that of the diffusion equation with $\alpha = 0$, where the wave front grows like $\sqrt{t}$. The speed of the wave of invasion tends to a constant $\sqrt{4\alpha D}$.

There are some problems with the model, including the unbounded invasion speeds of some muskrats that we have already mentioned. In addition the model predicts unlimited densities, e.g. $u(0, t) \to \infty$ as $t \to \infty$. There are problems for small values of the density, i.e. near the wave front, because of the breakdown of the law of large numbers used in the derivation of the diffusion equation. An individual-based model is strongly suggested here. Moreover, the speed of the invasion is determined by the behaviour exactly where these problems occur. Although microorganisms follow random walks to a good approximation, it is difficult to justify taking muskrat motion as diffusive. *However*, the model gives good results, both qualitatively ($R_1(t) \sim t$), and quantitatively (the correct speed is predicted).

EXERCISES

5.12. a) Verify that Equation (5.5.18) satisfies the partial differential Equation (5.5.16) and the initial condition (5.5.17). [Refer to Section C.2.1 of the appendix if necessary.]

   b) Show that the integral of Equation (5.5.18) over the whole of $\mathbb{R}^2$ is $e^{\alpha t}$.

5.13. In our model for muskrat dispersal, we chose a definition $R_1(t)$ of "wave front". We would like to know whether our choice affected our conclusions, so in this exercise we investigate an alternative definition.

   a) Integrate Equation (5.5.18) over an appropriate domain to show that the number of muskrats that are further than a distance $R_2(t)$ from the origin at time t is given by

$$N(t) = M \exp\left(\alpha t - \frac{R_2^2(t)}{4Dt}\right).$$

   [The element of area in plane polar coordinates is $RdRd\phi$.]

   b) If the wave front is defined to be the circle $R = R_2(t)$ outside which there are no more than m muskrats, for some predetermined small number m, show that $R_2(t) \approx \sqrt{4\alpha D}\,t$, so that the wave front still propagates with constant speed $\sqrt{4\alpha D}$.

   c) In the case $\alpha = 0$, show that the choice of definition does affect the conclusions.

5.14. Skellam's (1951) investigation of the spread of oak trees in Britain since the last Ice Age makes the following assumptions.

   (i) Oaks are produced at a per capita rate α and diffuse with diffusion coefficient D.

(ii) The generation time for oaks is at least 60 years.

(iii) Even in a virgin environment, an oak can produce no more than 9 million mature offspring.

(iv) Acorns are dispersed by falling from the tree, with RMS (root-mean-square) dispersal distance no more than 50 metres. [The RMS dispersal distance for a particle diffusing in n dimensions over a time t is $\sqrt{2nDt}$.]

  a) Write down the equation for oak density u, using assumption (i).

  b) Use the other assumptions to estimate upper bounds for D and α. [You may assume that there is initially a single oak at the origin.]

  c) Hence test the hypothesis that oaks invaded Britain by dispersal of the acorns, in the way suggested by assumption (i). The total distance covered by the wave of invasion is about 1000 km, in less that 20000 years.

  d) Comment on the estimates for the parameters.

5.6 Travelling Wave Solutions of General Reaction-diffusion Equations

In this section we shall generalise and improve on the analysis of the last section. The general scalar reaction-diffusion equation is given by

$$\frac{\partial \tilde{u}}{\partial t} = f(\tilde{u}) + \nabla \cdot (D\nabla \tilde{u}). \tag{5.6.19}$$

(The reason for calling the dependent variable $\tilde{u}$ rather than u will become clear later.) Henceforth we shall assume that the diffusion coefficient is constant, and re-scale the spatial variables so that $D = 1$. One typical example of such equations is the Fisher Equation (4.5.16) for the spread of an advantageous gene, where $\tilde{u}$ is the frequency of the gene, $f(\tilde{u}) = s\tilde{u}(1-\tilde{u})$, and s is a selection coefficient. The Fisher equation is called *monostable*, because one of its spatially uniform solutions, $\tilde{u} = 1$, is stable as a solution of the corresponding kinetic equation $\frac{d\tilde{u}}{dt} = f(\tilde{u})$, while the other, $\tilde{u} = 0$, is unstable. The solutions $\tilde{u} = 1$ and $\tilde{u} = 0$ represent fixation of the advantageous and the disadvantageous gene respectively. A second example is the Nagumo equation which arises in nerve conduction, where $\tilde{u}$ is a normalised membrane potential, $f(\tilde{u}) = k\tilde{u}(\tilde{u}-a)(1-$

$\tilde{u}$), and $0 < a < 1$, $k > 0$. This equation is called *bistable*, because two of its spatially uniform solutions, $\tilde{u} = 0$ and $\tilde{u} = 1$, are stable as a solution of $\frac{d\tilde{u}}{dt} = f(\tilde{u})$, while $\tilde{u} = a$ is unstable. The solutions $\tilde{u} = 0$ and $\tilde{u} = 1$ represent the depolarised and polarised states of the nerve, both of which are stable if longer-term effects are neglected, but polarisation waves can travel down the nerve, changing its state from $\tilde{u} = 0$ to $\tilde{u} = 1$.

Section 5.5 suggests that such equations have solutions whose speed tends to a constant as $t \to \infty$. In two dimensions the wave front is circular, and the curvature of the wave front tends to zero as $t \to \infty$. This in turn suggests that they have planar travelling wave solutions, i.e. solutions that travel without change of shape at a constant velocity, and which depend on space only in the direction of travel. We shall look for such solutions in this section. It is an improvement on the approach of the last section because

– it deals with nonlinear terms, essential in realistic models, and

– it copes with cases where linearisation does not work, e.g. the bistable equation.

Let f be a continuously differentiable function on $[0,1]$ satisfying $f(0) = f(1) = 0$. We are interested in solutions satisfying $0 \le \tilde{u} \le 1$, and which represent an invasion or a polarisation or a similar effect, depending on the context, so that $\tilde{u}$ increases from 0 to 1 as the wave passes. By definition, a planar travelling wave solution is a solution of the form $\tilde{u}(\mathbf{x},t) = u(x + ct)$, where x is distance measured in the positive (if $c < 0$) or negative (if $c > 0$) direction of travel, and c is the constant wave speed. We shall take $c > 0$ so that the wave travels from right to left. If $\tilde{u}$ increases from 0 to 1 as the wave passes we obtain

$$cu' = f(u) + u'', \tag{5.6.20}$$

where

$$u(s) \to 0 \text{ as } s \to -\infty, \quad u(s) \to 1 \text{ as } s \to \infty, \tag{5.6.21}$$

$s = x + ct$ and prime denotes differentiation with respect to s.

Let us define $v = u'$. Equation (5.6.20) becomes

$$u' = v, \quad v' = -f(u) + cv \tag{5.6.22}$$

with

$$(u(s), v(s)) \to (0,0) \text{ as } s \to -\infty, \quad (u(s), v(s)) \to (1,0) \text{ as } s \to \infty. \tag{5.6.23}$$

There are critical points at $(0,0)$ and $(1,0)$ and we look for a trajectory connecting them (a so-called heteroclinic orbit). The existence of such a trajectory depends on the function f. The characters of the critical points are determined by the Jacobian of the system (5.6.22),

$$J(u,v) = \begin{pmatrix} 0 & 1 \\ -f'(u) & c \end{pmatrix} \tag{5.6.24}$$

which has $\operatorname{tr} J(u,v) = c$, $\det J(u,v) = f'(u)$, $\operatorname{disc} J(u,v) = c^2 - 4f'(u)$.

5.6.1 Node-saddle Orbits (the Monostable Equation)

Let f satisfy $f'(0) > 0$, $f'(1) < 0$, $f(u) > 0$ for $u \in (0,1)$. The prototype is the Fisher equation. Then $(0,0)$ is an unstable focus ($0 < c < 2\sqrt{f'(0)}$), or an unstable node ($c \geq 2\sqrt{f'(0)}$); $(1,0)$ is a saddle point. For a non-negative solution we require $c \geq 2\sqrt{f'(0)}$. In this case the phase plane is as shown in Figure 5.5.

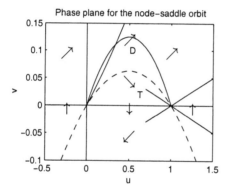

Figure 5.5   Phase plane for the node-saddle orbit. The idea is to show that (under the condition $c \geq 2\sqrt{f'(0)}$) the trajectory labelled T tends to the origin as s decreases.

If we can prove the existence of a set D that has the property that no trajectory can leave it as s *decreases* (a negatively invariant set) and that contains T for large values of its argument (near $(1,0)$), where T is the trajectory entering the saddle point at $(1,0)$ as shown, then as $s \to -\infty$ the only possibility for the behaviour of T is that it tends to $(0,0)$. It can be shown that such a set exists as long as c is sufficiently large. For define D to be the region

$$D = \{(u,v)|0 < u < 1, 0 < v < mf(u)\}$$

where m is a constant to be determined. Since f is continuously differentiable, it follows that there exists a constant K such that $|f'(u)| \leq K$ for all $u \in [0,1]$. It is easy to see that no trajectory leaves D as s decreases by crossing the

u-axis, since $v' < 0$ there. The only other possibility is by crossing the curve $\Phi(u,v) = 0$, where $\Phi(u,v) = v - mf(u)$. But on this curve

$$\Phi'(u,v) = v' - mf'(u)u' = -f(u) + cv - mf'(u)v \geq (-\frac{1}{m} + c - mK)v > 0$$

as long as $c > 1/m + mK$. In particular, if $K = f'(0)$, as in the Fisher case $f(u) = su(1-u)$, and we choose $m = 1/\sqrt{f'(0)}$, then $\Phi'(u,v) > 0$ on the curve of $\Phi(u,v) = 0$ as long as $c > 2\sqrt{f'(0)}$. But $\Phi'(u,v) > 0$ means that trajectories are leaving D, and D is negatively invariant as required. In fact there is also such a solution for $c = 2\sqrt{f'(0)}$, and this is the only solution that is stable as a solution of the original partial differential equation. In the Fisher equation this represents the spread of an advantageous gene through the population; in ecological terms it represents the spread of a species with a purely compensatory growth function.

5.6.2 Saddle-saddle Orbits (the Bistable Equation)

Let f satisfy $f'(0) < 0$, $f'(1) < 0$, $f(u) < 0$ for $u \in (0,a)$, $f(u) > 0$ for $u \in (a,1)$. Then $(0,0)$ and $(1,0)$ are both saddle points, and $(a,0)$ is an unstable node or focus. The prototype here is the Nagumo equation, but it may also be thought of as representing the spread of a species with a critically depensatory growth function. We obtain some information by multiplying (5.6.19) by u' and integrating from $s = -\infty$ to $s = \infty$:

$$c \int_{-\infty}^{\infty} u'^2(s)ds = \int_{-\infty}^{\infty} f(u(s))u'(s)ds + \left[\frac{1}{2}u'^2\right]_{-\infty}^{\infty}$$

or, since $u'(s) \to 0$ as $s \to \pm\infty$,

$$c \int_{-\infty}^{\infty} u'^2(s)ds = \int_{0}^{1} f(u)du.$$

A solution of the type we are seeking is therefore only possible if $\int_0^1 f(u)du > 0$. Assuming that this holds, it may then be shown that there is a *unique c* for which a travelling wave solution exists. The method of proof is a *shooting method*. It is shown that the trajectory leaving one of the saddle points misses the other saddle point by passing to one side if c is small, and misses it by passing to the other side if c is large. The result then follows since the solution of the equations is continuously and monotonically dependent on c.

EXERCISES

5.15. Show that the trajectory T is indeed in D for large values of the argument.

5.7 Travelling Wave Solutions of Systems of Reaction-diffusion Equations: Spatial Spread of Epidemics

In Section 3.3 we looked at a model for an SIR epidemic, given by

$$\frac{dS}{d\tau} = -\beta IS, \quad \frac{dI}{d\tau} = \beta IS - \gamma I, \quad \frac{dR}{d\tau} = \gamma I.$$

In such an epidemic in a closed population, where birth and non-disease-related death are neglected, individuals move on infection from a susceptible to an infected class and then to a removed class, where they take no further part in the epidemic because they are immune, or removed by an isolation policy, or dead. There we took no account of spatial structure, assuming that the population as a whole was well mixed. However data on real epidemics often shows that they sweep across continents as waves, and the well-mixed assumption is clearly unrealistic.

In this section we shall use the ideas of this chapter to extend the non-spatial SIR model to include spatial effects. S, I and R must then be thought of as the population *densities* of susceptibles, infectives, and removed individuals, depending on position as well as time, and β also has a slightly different interpretation, as the rate of infection per susceptible *per unit density of infectives*. We shall consider the case of rabies, which is spread mainly by foxes. Rabies is invariably lethal, so that the removed class in this case is dead. Healthy foxes tend to stay in their own territory, but when rabid will travel large distances more or less at random and attack other foxes. For this reason a model for rabies has been proposed which includes diffusion of infective but not of susceptible (nor of course removed) foxes, namely

$$\frac{\partial S}{\partial \tau} = -\beta IS, \quad \frac{\partial I}{\partial \tau} = \beta IS - \gamma I + D\frac{\partial^2 I}{\partial \xi^2}, \quad \frac{\partial R}{\partial \tau} = \gamma I,$$

where D is the diffusion coefficient of the infectious foxes. An epidemic is a wave of infectives propagating into a population of susceptibles, and we shall eventually look for solutions as functions of a travelling wave variable.

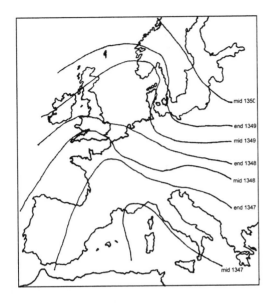

Figure 5.6 The Black Death in Europe, 1347–1350, after Langer (*Scientific American*, 114-121, February 1964). This epidemic was one of the reasons that the human population of the world fell in the fourteenth century, as shown in Figure 1.1.

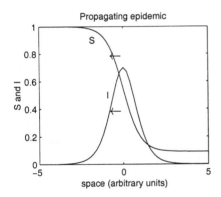

Figure 5.7 Diagrammatic representation of a propagating epidemic.

Let $S \to N$, $I \to 0$, $R \to 0$ as $\xi \to -\infty$. Non-dimensionalise S, I and R by

$$\tilde{u} = \frac{S}{N}, \quad \tilde{v} = \frac{I}{N}, \quad \tilde{w} = \frac{R}{N}.$$

Re-scaling the time and space variables by $t = \gamma\tau$, $x = \xi\sqrt{\frac{\gamma}{D}}$, we obtain

$$\frac{\partial \tilde{u}}{\partial t} = -R_0 \tilde{u}\tilde{v}, \quad \frac{\partial \tilde{v}}{\partial t} = R_0 \tilde{u}\tilde{v} - \tilde{v} + \frac{\partial^2 \tilde{v}}{\partial x^2}, \quad \frac{\partial \tilde{w}}{\partial t} = \tilde{v},$$

where $R_0 = \frac{\beta N}{\gamma}$, the basic reproductive ratio. This has essentially the same interpretation as in Chapter 3, as the expected number of infectious contacts made by a single infective introduced into a population with $(S, I, R) =$

$(N, 0, 0)$. We saw there that if $R_0 < 1$ the disease dies out in the spatially uniform case, whereas if $R_0 > 1$ an epidemic occurs.

We look for travelling wave solutions of the form $\tilde{u}(x, t) = u(s) = u(x + ct)$, $\tilde{v}(x, t) = v(s) = v(x + ct)$, where c is some constant wave speed. The equations become

$$cu' = -R_0 uv, \quad cv' = R_0 uv - v + v'', \quad cw' = v. \tag{5.7.25}$$

These are to be solved subject to

$$u(s) \to 1 \text{ as } s \to -\infty, \quad u(s) \to u_1 \text{ as } s \to \infty,$$

$$v(s) \to 0 \text{ as } s \to \pm\infty, \tag{5.7.26}$$

$$w(s) \to 0 \text{ as } s \to -\infty, \quad w(s) \to w_1 \text{ as } s \to \infty.$$

In Chapter 3 we used the fact that the equations without the v'' term were separable to find a first integral of them, which led us to some useful results. Amazingly, the equations are still separable *with* the v'' term, once we notice that in the same way as $\dfrac{v'}{u'} = \dfrac{dv}{du}$, $\dfrac{v''}{u'} = \dfrac{d(v')}{du}$. For, dividing the second and third equations of system (5.7.25) by the first,

$$\frac{dv}{du} = -1 + \frac{1}{R_0 u} + \frac{1}{c}\frac{d(v')}{du}, \quad \frac{dw}{du} = -\frac{1}{R_0 u}.$$

Integrating,

$$v = -u + \frac{1}{R_0}\log u + \frac{1}{c}v' + A, \quad w = -\frac{1}{R_0}\log u + B,$$

where A and B are constants of integration. Since u and v tend to limits as $s \to \pm\infty$, then the first of these shows that v' does too, so that

$$v'(s) \to 0 \text{ as } s \to \pm\infty. \tag{5.7.27}$$

Applying the conditions as $s \to \pm\infty$, $A = 1$, $B = 0$. The equations become $w = w(u)$,

$$u' = -\frac{1}{c}uv, \quad v' = c(u + w(u) - 1 + v), \tag{5.7.28}$$

where

$$w(u) = -\frac{1}{R_0}\log u.$$

Applying the conditions as $s \to \infty$, $u_1 + w_1 = 1$, where

$$1 - w_1 = \exp(-R_0 w_1). \tag{5.7.29}$$

This is Equation (3.3.8), and always has the solution $w_1 = 0$, representing no epidemic. We showed in Chapter 3 (see Figure 3.3 and Exercise 3.5) that if

$R_0 < 1$ then there is no other positive solution, whereas if $R_0 > 1$ then there is another solution w_1 satisfying $0 < w_1 < 1$ and $R_0(1 - w_1) = R_0 u_1 < 1$.

It follows that in this spatially inhomogeneous case too there can be no epidemic if $R_0 < 1$. Let us take $R_0 > 1$, so that $0 < u_1 < 1/R_0 < 1$. The only critical points of the system $(5.7.28)$ are at $(u_1, 0)$ and $(1, 0)$. For $(u_1, 0)$, the Jacobian matrix $J(u_1, 0)$ has trace $\operatorname{tr} J(u_1, 0) = c$, determinant $\det J(u_1, 0) = (R_0 u_1 - 1)/R_0 < 0$, and $(u_1, 0)$ is a saddle point. For $(1, 0)$, $J(1, 0)$ has trace $\operatorname{tr} J(1, 0) = c$, determinant $\det J(1, 0) = (R_0 - 1)/R_0 > 0$, and $(1, 0)$ is an unstable node if $c^2 \geq 4(R_0 - 1)/R_0$, unstable focus if $c^2 < 4(R_0 - 1)/R_0$. We require a trajectory from $(1, 0)$ to $(u_1, 0)$, marked T in the phase plane below, and hence must have $c^2 \geq 4(R_0 - 1)/R_0$, since v must always be non-negative.

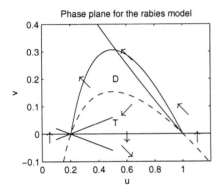

Phase plane for the rabies model

Figure 5.8 Proof of a travelling wave trajectory in the rabies model. The idea is to show that (for $c \geq 2\sqrt{1 - 1/R_0}$) the trajectory marked T tends to the disease-free steady state $(1, 0)$ as s decreases.

Such a solution exists if the trajectory which enters the saddle point at $(u_1, 0)$ from the positive quadrant as $s \to \infty$ tends to $(1, 0)$ as $s \to -\infty$. If we can find a region D such that no trajectory leaves D as s *decreases* (a negatively invariant set), then T must come from $(1, 0)$. It may be shown that

$$D = \{(u, v) | u_1 < u < 1, 0 < v < 2(1 - u - w(u))\} \qquad (5.7.30)$$

is such a set.

Hence we have a travelling wave for $R_0 > 1$. It may be shown that the wave with minimal speed $c = 2\sqrt{(R_0 - 1)/R_0}$ is the only one which can be stable as a solution of the original system of partial differential equations, and in dimensional variables this is about 40 km per year, depending on the initial susceptible fox density. This agrees well with observed values.

Can the fact that no epidemic can propagate if $R_0 < 1$ be used to control rabies? Since $R_0 = \frac{\beta N}{\gamma}$, and β and γ are parameters of the disease that are not under our control, the only possibility is to reduce N, the initial density of susceptible foxes, below $\frac{\gamma}{\beta}$. This must be done by vaccination rather than culling, since vacant territory is quickly re-colonised by foxes. Further analysis shows that if this can be done in a strip approximately 14 km wide, then this may stop the epidemic. Such a barrier has been set up to prevent rabies spreading into Denmark, and has so far been effective (although the related bat rabies virus is endemic there).

The modelling presented here seems to give some useful insights, but there are some problems with the whole underlying rationale if some of the population densities become very low. In particular, the modelling may be extended to provide an analysis of secondary outbreaks following on behind the primary one, but the population density of infectives between outbreaks falls into the "attofox" range. It is clear that stochastic effects will be crucial in determining such secondary outbreaks.

EXERCISES

5.16. Show that the region D defined by Equation (5.7.30) is negatively invariant.

5.8 Conclusions

- Theories of motion are based on the principle of conservation of matter.

- The continuum approach requires the concept of flux $\mathbf{J} = J\mathbf{m}$, where $\mathbf{m}$ is a unit vector in the direction of net flow and J is the rate at which particles cross a test surface placed perpendicular to the flow, per unit area.

- The flux is related to the mean velocity of the particles by $\mathbf{J} = \bar{\mathbf{v}}u$.

- Advective flux is $\mathbf{J}_{\text{adv}} = \mathbf{v}u$, diffusive flux is $\mathbf{J}_{\text{diff}} = -D\nabla u$.

- The continuum approach leads to the equation $u_t = -\nabla \cdot (\mathbf{v}u) + \nabla \cdot (D\nabla u) + f$ in n dimensions in an inhomogeneous medium with sources and/or sinks.

- Chemotaxis is motion up (or down) a chemical gradient. Chemotactic flux is usually modelled by $\mathbf{J}_{\text{chemo}} = \chi u \nabla c$, where c is the concentration of the chemical and χ is the chemotactic sensitivity, positive for positive chemotaxis (up a gradient) and negative for negative chemotaxis.

- Unicellular organisms lead life at low Reynolds number, dominated by viscous forces, and very quickly accelerate to terminal velocity under the influence of a force. If a force field with potential Φ leads to a terminal velocity $\alpha\nabla\Phi$, the flux due to the force field is $\alpha u \nabla \Phi$.

– Steady state diffusion (or dimensional analysis) shows that the average time to diffuse a distance L is proportional to L^2/D, or the average distance through which diffusion works in a time τ is proportional to $\sqrt{D\tau}$. Diffusion is a good transport mechanism over cellular distances of about 10 microns, but not over much larger distances.

– A small amount of directed motion drastically increases the efficiency with which macrophages find bacteria.

– Biological invasions can be propagated by a combination of kinetics and diffusion. They then expand with a speed proportional to t, whereas an invasion via diffusion alone would expand with a speed proportional to $\sqrt{t}$. For purely compensatory kinetics, the speed is determined by conditions near the wave front.

– The scalar reaction-diffusion equation $u_t = f(u) + \nabla \cdot (D\nabla u)$, with $f(0) = f(1) = 0$, can admit travelling wave solutions from 0 to 1.

– If the growth rate $f(u)$ in the scalar reaction-diffusion equation is positive between 0 and 1, the equation is called monostable. The travelling wave solutions are node-saddle orbits in the phase plane, and exist for all speeds $c \geq c^*$, some critical speed. Only the wave with speed c^* is stable as a solution of the original partial differential equation.

– If the growth rate $f(u)$ in the scalar reaction-diffusion equation is negative between 0 and a and positive between a and 1, the equation is called bistable. The travelling wave solutions are saddle-saddle orbits in the phase plane, and exist for a unique speed $c = c^*$ only. They travel from 0 to 1 if $\int_0^1 f(u)du > 0$, 1 to 0 if this integral is negative, and are stationary if the integral is zero. The wave is stable as a solution of the original partial differential equation.

6
Molecular and Cellular Biology

6.1 Introduction

Molecular biology is one of the most important and rapidly developing areas in the life sciences, and now forms the basis of subjects such as physiology, immunology and genetics. Cellular biology is the study of cells, which make up all living creatures, and which occupy an intermediate level of biological complexity between molecules and multicellular organisms. Mathematics has been applied in many areas of molecular and cellular biology; this chapter is concerned mainly with the kinetics of chemical processes in cells. These are involved in

– metabolism and its control,

– information gathering, interpretation, transmission, and replication,

– defence against invading organisms,

– transport of essential substances and destruction or removal of noxious ones, and

– mechanical work,

although we shall not consider the last two of these.

Biological systems differ from purely chemical systems in the sheer complexity of the reaction schemes and the numbers of chemicals involved, and in their repertoire of kinetic behaviour. But their complex kinetic behaviour is not merely a result of complex reaction schemes. For many years the conventional

175

wisdom was that thermodynamic considerations precluded chemical reactions from doing anything other than run down monotonically to equilibrium, despite theoretical and experimental work showing, for example, that oscillations were possible. As Prigogine pointed out in the 1960s, complex behaviour is permissible thermodynamically if the system is far from equilibrium. One of the functions of metabolism is to maintain biological systems far from equilibrium, allowing complex behaviour to take place. Sustained oscillations as seen in many metabolic processes and excitability as seen in cardiac and nerve cells would not be possible in other circumstances; oscillations can occur in closed chemical systems but only last for a limited time.

Control and communication are themes that will recur throughout this chapter. Metabolic processes are exquisitely controlled, by feedback and feedforward processes and by the use of ultra-efficient catalysts known as enzymes. These are proteins whose three-dimensional structure is essential to the precision of their operation, as has become clear since the work of Monod and Jacob in the early 1960s. Turning to the theme of communication, we shall consider the mechanisms of nerve cells which allow them to transmit information through nerve impulses, mechanisms which were uncovered in the early 1950s by Hodgkin and Huxley in a stunning blend of experimental and theoretical work. Finally we shall look at modern models stimulated by the HIV epidemic that investigate the immune system and its response to invading organisms.

6.2 Biochemical Kinetics

Biochemical kinetics concerns the concentrations of chemical substances in biological systems as functions of time. Biochemical processes are often controlled by enzyme catalysts that are present in very low concentrations, but nevertheless have a large effect on the rate of the process. As a consequence, the various chemical reactions in even the simplest processes may take place on very different time scales. A numerical analyst would view this with dismay, knowing that it leads to difficulties with stiff systems of equations, but it can also be exploited to give good approximations to the solution by the method of *matched asymptotic expansions*. In this section we shall look at the simplest example of such a system, known as Michaelis–Menten kinetics after its discoverers.

The law of mass action states that if chemical A reacts with chemical B to produce chemical C by the reaction

$$A + B \overset{k}{\rightarrow} C,$$

then the rate of the reaction is given by kAB, where A and B now denote *concentrations* of the chemicals. We have

$$\frac{dC}{d\tau} = -\frac{dA}{d\tau} = -\frac{dB}{d\tau} = kAB. \tag{6.2.1}$$

In fact, the law of mass action is only an approximation, but it is a very good one for dilute solutions of the chemicals concerned. The constant k is called the *rate constant* of the reaction. Here we have neglected the back reaction $C \to A + B$, but is a consequence of thermodynamic principles that reactions can take place in either direction, and we write

$$A + B \underset{k_-}{\overset{k_+}{\rightleftharpoons}} C,$$

and

$$\frac{dC}{d\tau} = -\frac{dA}{d\tau} = -\frac{dB}{d\tau} = k_+ AB - k_- C. \tag{6.2.2}$$

Now consider a reaction that is catalysed by an enzyme. Enzymes are proteins, ubiquitous and crucial in biochemistry, that catalyse a biochemical reaction by lowering the *activation energy* required for the reaction to proceed. They are generally specific to some substrate and catalyse its conversion to a product, remaining themselves unchanged by the reaction. In the simplest and archetypal situation, Michaelis–Menten kinetics, they accomplish this in two steps, first forming a complex with the substrate, which then breaks down to the product and the enzyme. Diagrammatically,

$$S + E \underset{k_{-1}}{\overset{k_1}{\rightleftharpoons}} C \overset{k_2}{\to} P + E. \tag{6.2.3}$$

(Here the back reaction $P + E \to C$ is considered so slow as to be negligible.) The equations for the chemical concentrations are now

$$\frac{dS}{d\tau} = k_{-1}C - k_1 SE, \quad \frac{dE}{d\tau} = (k_{-1} + k_2)C - k_1 SE,$$

$$\frac{dC}{d\tau} = k_1 SE - (k_{-1} + k_2)C, \quad \frac{dP}{d\tau} = k_2 C. \tag{6.2.4}$$

Since $\frac{d}{d\tau}(E + C) = 0$, then $E + C = E_0$, a constant. E_0 is the total amount of enzyme, free and bound, and is conserved. This makes intuitive sense, as the enzyme is only a catalyst of the overall reaction. Note also that $\frac{d}{d\tau}(S+C+P) = 0$, so that there is another conservation equation, and $S + C + P = S_0$, another constant. Substrate occurs in its original form, or bound to the enzyme, or converted to its product. Our problem becomes

$$\frac{dS}{d\tau} = k_{-1}C - k_1 S(E_0 - C), \quad \frac{dC}{d\tau} = k_1 S(E_0 - C) - (k_{-1} + k_2)C,$$

$$E = E_0 - C, \quad P = S_0 - S - C. \tag{6.2.5}$$

Initial conditions are usually taken to be

$$S(0) = S_0, \quad E(0) = E_0, \quad C(0) = 0, \quad P(0) = 0, \quad (6.2.6)$$

corresponding to the situation where free enzyme is added to its substrate.

A typical solution of the equations is shown in Figure 6.1. There are two

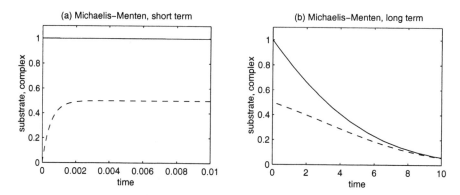

Figure 6.1   The substrate and complex concentrations in a Michaelis–Menten reaction as functions of time. The extremely fast rise of the complex concentration from its initial value of zero is invisible in the right-hand panel, but is seen in the left-hand panel on a very short time scale. The parameter values are taken from data for the hydrolysis of benzoyl-L-arginine ethyl ester by trypsin, $k_1 = 4 \times 10^6 \, \text{M}^{-1} \, \text{s}^{-1}$, $k_{-1} = 25 \, \text{s}^{-1}$, $k_2 = 15 \, \text{s}^{-1}$, with $S_0 = K_m = (k_{-1} + k_2)/k_1 = 10^{-5} \, \text{M}$, $E_0 = 10^{-3} K_m = 10^{-8} \, \text{M}$. Note that $E_0 \ll S_0$. This is typical; enzymes are extremely efficient catalysts and are usually present in very small concentrations.

distinct parts to it; first the concentration of the complex rises very quickly while the substrate concentration remains substantially unchanged, and then both concentrations change on a much slower time scale as the substrate is converted to the product by the enzyme. The second part of the solution is where all the action is (in terms of converting substrate to product), and it is intuitively clear (to biochemists) that it may be analysed by making the approximation $\frac{dC}{d\tau} \approx 0$. This is called the *quasi-steady-state hypothesis*. With this approximation we have $k_1 S(E_0 - C) = (k_{-1} + k_2)C$, so

$$C = \frac{k_1 S E_0}{k_{-1} + k_2 + k_1 S} = \frac{S E_0}{K_m + S},$$

and

$$\frac{dS}{d\tau} = -k_2 C = -\frac{V_m S}{K_m + S},$$

where $V_m = k_2 E_0$ and $K_m = (k_{-1} + k_2)/k_1$, the *Michaelis constant*. This equation may be integrated by separating the variables. The *saturation function*, the fraction of binding sites on the enzyme that are occupied, is given by

$$Y(S) = \frac{C}{E + C} = \frac{S}{K_m + S}. \qquad (6.2.7)$$

Half the binding sites are occupied when $S = K_m$. Biologically, it is important to know the overall velocity V of the reaction, i.e. the rate at which product is formed, which in this approximation is the same as the rate at which substrate is consumed, and which is given by

$$V = \frac{dP}{d\tau} = V_m Y(S) = \frac{V_m S}{K_m + S}. \qquad (6.2.8)$$

This is known as the *Michaelis–Menten rate equation*. Equation (6.2.8) highlights the importance of the saturation function. Similar equations hold more generally if conditions are such that the quasi-steady-state hypothesis holds in a general enzyme reaction scheme. The saturation function is then found by putting the right hand sides of all the enzyme equations equal to zero, including the enzyme conservation equation, and solving the resulting system of algebraic equations.

How do we proceed if we are not blessed with the necessary biochemical intuition to arrive at the quasi-steady-state hypothesis, or wish to confirm that our intuition makes sense and to decide under what circumstances it is valid? The intuition is based on the observation that there is a fast and a slow time scale, which suggests the use of the method of *matched asymptotic expansions*. This involves obtaining expressions for what is happening on each time scale, and then matching them smoothly together. (The method is also useful when there are different *spatial* scales in a problem, such as in *boundary layer* problems in fluid mechanics where the flow is very different near to and away from a boundary.)

First we need to be precise about what is meant by a fast and a slow time scale, and to do this we must non-dimensionalise the equations. Let us define

$$s = \frac{S}{S_0}, \quad c = \frac{C}{E_0}, \quad e = \frac{E}{E_0}, \quad p = \frac{P}{S_0}, \quad t = k_1 E_0 \tau. \qquad (6.2.9)$$

The choice of non-dimensionalisation for the chemical concentrations is straightforward, but the non-dimensionalisation for τ depends on a careful examination of the equations to determine possible time scales. Looking at the S equation in the system (6.2.4), we see that the maximal specific rate at which S may be taken up, when $C = 0$, is $k_1 E_0$. On the other hand, looking at the E equation in system (6.2.4), we see that the maximal specific rate at which E may be taken up, again when $C = 0$, is $k_1 S_0$. Both these time scales will be important, but we start with the first, called the *outer* time scale for reasons that will become apparent. The equations become

$$\frac{ds}{dt} = \kappa_d c - s(1 - c), \quad \epsilon\frac{dc}{dt} = s(1 - c) - \kappa_m c, \qquad (6.2.10)$$

where

$$\epsilon = \frac{E_0}{S_0}, \quad \kappa_e = \frac{k_{-1}}{k_1 S_0} = \frac{K_e}{S_0}, \quad \kappa_m = \frac{k-1 + k_2}{k_1 S_0} = \frac{K_m}{S_0}.$$

The constant $K_e = k_{-1}/k_1$ is the *equilibrium constant* of the reaction between S and E, and κ_e is a non-dimensional version of this; κ_m is a non-dimensional version of the Michaelis constant K_m. The equations are to be solved with initial conditions

$$s(0) = 1, \quad c(0) = 0. \qquad (6.2.11)$$

If $\epsilon \ll 1$, which is often the case, we try to solve this problem by looking for s and c as power series in ϵ,

$$s(t) = \sum_{n=0}^{\infty} \epsilon^n s_n(t), \quad c(t) = \sum_{n=0}^{\infty} \epsilon^n c_n(t).$$

Substituting these into Equations (6.2.10), and equating powers of ϵ, we obtain, to leading order,

$$\frac{ds_0}{dt} = \kappa_d c_0 - s_0(1 - c_0), \quad 0 = s_0(1 - c_0) - \kappa_m c_0, \qquad (6.2.12)$$

so that

$$\frac{ds_0}{dt} = -\frac{\kappa s_0}{\kappa_m + s_0}, \quad c_0 = \frac{s_0}{\kappa_m + s_0}, \qquad (6.2.13)$$

to be solved with $s_0(0) = 1$, $c_0(0) = 0$, where we have defined $\kappa = \kappa_m - \kappa_d = k_2/(k_1 S_0)$. Integrating the first of these,

$$\kappa_m \log s_0 + s_0 = A - \kappa t, \qquad (6.2.14)$$

where A is a constant of integration. Higher order corrections may easily be found. But there is a problem with this solution. If we choose $A = 1$ to satisfy the initial condition for s, then the initial value of c_0 is $1/(\kappa_m + 1)$, and the initial condition for c cannot be satisfied.

The problem stems from our implicit assumption that s and c are analytic functions of ϵ (and hence may be expanded as we proposed as power series in ϵ). Unfortunately this assumption is not true. Looking at this another way, we have assumed that the solution of the problem P_ϵ, consisting of equations (6.2.10) with initial conditions (6.2.11), tends in the limit as $\epsilon \to 0$ to the solution of the problem P_0 with $\epsilon = 0$. But the problem P_0 is quite different from the problem P_ϵ, in that it consists of a single differential equation and an algebraic equation rather than a system of two differential equations. We cannot expect to satisfy two initial conditions with only one differential equation, and so P_0 does not in general have a solution. A problem P_ϵ whose solution does not tend to the solution of P_0 in the limit as $\epsilon \to 0$ is called a *singular perturbation* problem.

So how do we deal with this difficulty? We use the method of *matched asymptotic expansions*. The idea is as follows. Throughout, we shall only describe how to find a leading order approximation to the solution, but better approximations may be found by going to higher order in ϵ. To leading order, Equation (6.2.14) with the second of (6.2.13) defines a solution of the problem P_ϵ which is fine *except* (i) it does not satisfy the initial conditions and so is not valid near $t = 0$, and (ii) it contains an unknown constant of integration A. Let us call it the *outer solution*. Near $t = 0$ we need to find another solution, called the *inner solution*, that satisfies the initial conditions. We shall then require that these solutions match together smoothly, i.e. satisfy some *matching conditions*, which will determine the constant of integration.

We examine the region near $t = 0$ by defining a new independent (time) variable T and dependent variables S and C (not to be confused with the original dimensional variables) by

$$T = \frac{t}{\epsilon}, \quad S(T) = s(t), \quad C(T) = c(t).$$

In terms of the original dimensional variables $T = k_1 S_0 \tau$, so our second time scale is coming into play here. The equations become

$$\frac{dS}{dT} = \epsilon \left(\kappa_d C - S(1 - C) \right), \quad \frac{dC}{dt} = S(1 - C) - \kappa_m C, \tag{6.2.15}$$

with $S(0) = 1$, $C(0) = 0$. Expanding S and C as power series in ϵ, substituting into the equations and equating powers of ϵ, we obtain the leading order approximation

$$S_0(T) = 1, \quad C_0(T) = \frac{1}{1 + \kappa_m} \left(1 - e^{-(1 + \kappa_m)T} \right).$$

As $T \to \infty$, $(S_0(T), C_0(T)) \to (1, \frac{1}{1 + \kappa_m})$. For matching, we have to choose A in the outer solution so that the *common part* of the solution is equal,

$$\lim_{t \to 0} (s_0(t), c_0(t)) = \lim_{T \to \infty} (S_0(T), C_0(T)).$$

These are the *matching conditions*. They are satisfied by taking $A = 1$ in Equation (6.2.14), and matching is complete.

The condition $\epsilon \ll 1$ is crucial to the method. It is equivalent to the quasi-steady-state hypothesis, the requirement that after an initial short time period the right-hand side of the C equation may be neglected. It is very often satisfied because enzymes are so efficient that they need to be present in very small concentrations, so that $E_0 \ll S_0$.

EXERCISES

6.1. How would you estimate the parameters V_m and K_m in the Michaelis–Menten Equation (6.2.8) by a *Lineweaver–Burke plot*, a plot of $1/V$ against $1/S$?

6.2. We have obtained leading order solutions of the Michaelis–Menten equations that are valid in the inner and outer regions, but no solutions uniformly valid in time. These are found by adding together the inner and outer solutions and subtracting the common part. Show that the uniformly valid solution for the substrate is given simply by $s_{0,\text{unif}}(t) = s_0(t)$, the outer solution, but that the uniformly valid solution for the complex is

$$c_{0,\text{unif}}(t) = \frac{s_0(t)}{\kappa_m + s_0(t)} - \frac{1}{\kappa_m + 1} \exp\left(-\frac{(\kappa_m + 1)t}{\epsilon}\right).$$

6.3. Occasionally in enzyme kinetics E_0 and S_0 are the same order of magnitude, but $E_0 \ll K_m$. Hence E_0/S_0 is no longer a small parameter, and the asymptotic analysis in this section fails for the Michaelis–Menten kinetic Equations (6.2.4) with the usual initial conditions (6.2.6). However, matched asymptotic expansions may still be found for this problem in terms of the new small parameter $\epsilon = E_0/K_m$.

a) In the outer region we non-dimensionalise E, S and τ as before, but define $c = \frac{K_m C}{S_0 E_0}$, motivated by the C equation in (6.2.4). Show that $e = 1 - \alpha\epsilon c$, and the substrate and complex equations in this region are given by

$$\frac{ds}{dt} = \frac{K_d}{K_m}c - s + \epsilon\alpha sc, \quad \epsilon\frac{dc}{dt} = s - \epsilon\alpha sc - c,$$

where $\alpha = S_0/E_0$.

b) Show that leading order solutions in the outer region are given by

$$s_0(t) = c_0(t) = Ae^{-Kt},$$

where $K = \frac{K_m - K_d}{K_m} = \frac{k_2}{k_{-1}+k_2}$, and A is a constant of integration.

c) Using the time variable $T = t/\epsilon = (k_{-1} + k_2)\tau$ in the inner region, show that leading order solutions there are given by

$$S_0(T) = 1, \quad C_0(T) = 1 - e^{-T}.$$

d) Use matching conditions to determine A, and obtain leading order uniformly valid solutions for the substrate and the complex in this case.

6.4. If the back reaction $C \overset{k_{-2}}{\underset{}{\leftarrow}} P + E$ is not negligibly slow, the analysis of this section needs modification.

a) Write down the modified version of Equations (6.2.4).

b) Making the quasi-steady-state hypothesis, show that

$$C = \frac{k_1 E_0 S + k_{-2} E_0 P}{k_1 S + k_{-2} P + k_{-1} + k_2}.$$

c) Show that the overall velocity of the reaction is given by

$$V = \frac{dP}{d\tau} = \frac{k_1 k_2 S - k_{-1} k_{-2} P}{k_1 S + k_{-2} P + k_{-1} + k_2}.$$

d) Deduce that after a very long time, when equilibrium is established, the concentrations S^* and P^* of the substrate and product satisfy *Haldane's relation*

$$\frac{P^*}{S^*} = \frac{k_1 k_2}{k_{-1} k_{-2}}. \tag{6.2.16}$$

6.3 Metabolic Pathways

Metabolism consists of chemical processes which either store energy in molecules (anabolism) or release it from them (catabolism). A metabolic pathway is a sequence of chemical reactions in such a process, whose products are known as metabolites. Each reaction in the pathway is catalysed by a specific enzyme whose structure is specified by a specific gene. In aerobic cellular respiration in humans and many other organisms, energy is extracted from glucose and put into short-term storage by the process of phosphorylation (the addition of a phosphate group to a molecule), which here involves converting molecules of ADP (adenosine diphosphate) to ATP (adenosine triphosphate). Two of the early steps in the pathway consume a molecule of ATP, but eventually 40 molecules of ATP are produced for each molecule of glucose, a net gain of 38. An active cell requires more than two million molecules of ATP per second to drive its biochemical machinery.

It is crucial that complex pathways such as this should have multiple control mechanisms to ensure that metabolite concentrations are kept at the correct level. High concentrations of a metabolite at one point of a pathway are often able to inhibit a step earlier on in the pathway, providing negative feedback control, or activate a step later on, providing feedforward control. Metabolites with such properties are known as modifiers or effectors, or more specifically as inhibitors or activators. Obvious targets for modification (either inhibition or activation) are the enzymes that catalyse these reactions. One way to inhibit an enzyme is to bind to its active site, where its substrate would normally bind, but to do this the inhibitory molecule has to be *isosteric* to ("the same shape as") the substrate itself. This kind of inhibition is known as *competitive inhibition*. Much less restrictive and more common is to bind to a different site, which may nevertheless have an effect on the chemistry of the enzyme. It may change it from a more active to a less active form or vice versa, and may therefore lead to its inhibition or its activation. Such a site (and the enzyme that possesses it) is called *allosteric* ("another shape"). The change in chemistry is probably brought about by a conformational change in the folding of the enzyme. The importance of allosteric enzymes is such that they have been called the "second secret of life", DNA being the first, and in this section we shall investigate their effects.

6.3.1 Activation and Inhibition

Consider a hypothetical enzyme that catalyses the production of a product P from a substrate S in the normal way,

$$E + S \underset{k_{-1}}{\overset{k_1}{\rightleftharpoons}} C_s \overset{k_2}{\rightarrow} E + P, \tag{6.3.17}$$

but also reacts with a modifier M,

$$E + M \underset{k_{-3}}{\overset{k_3}{\rightleftharpoons}} C_m. \tag{6.3.18}$$

Let us also assume that both S and M may be bound on the same enzyme molecule, possibly changing the rate constants,

$$C_m + S \underset{k'_{-1}}{\overset{k'_1}{\rightleftharpoons}} C_{sm} \overset{k'_2}{\rightarrow} C_m + P, \tag{6.3.19}$$

$$C_s + M \underset{k'_{-3}}{\overset{k'_3}{\rightleftharpoons}} C_{sm}. \tag{6.3.20}$$

The law of mass action then leads to a system of seven differential equations for S, M, P, E and the three complexes C_s, C_m and C_{sm}. There is a conservation law $E + C_s + C_m + C_{sm} = E_0$ for the enzyme, one for the modifier and one for the substrate-product. If $E_0 \ll S_0$, the quasi-steady-state hypothesis may be applied to the equations for the enzyme and its three complexes, leading to four algebraic equations, three of which are linearly independent. An expression for the velocity of the overall reaction, i.e. the rate of production of the product, may be found, but unless some simplifying assumptions are made the algebra required is hairy.

Example 6.1

Consider the extreme case where binding of one *ligand* ("binding molecule") makes it impossible to bind the other, so that $k_1' = 0$, $k_3' = 0$. (This is equivalent to competitive inhibition.) Find the effect of the modifier on the velocity of the reaction.

The equations are

$$\frac{dS}{d\tau} = -k_1 SE + k_{-1}C_s, \quad \frac{dM}{d\tau} = -k_3 ME + k_{-3}C_m,$$

$$\frac{dC_s}{d\tau} = k_1 SE - (k_{-1} + k_2)C_s, \quad \frac{dC_m}{d\tau} = k_3 ME - k_{-3}C_m,$$

with conservation equations $E + C_s + C_m = E_0$, $S + C_s + P = S_0$. Using the quasi-steady-state hypothesis, we put $k_1 SE = (k_{-1} + k_2)C_s$, $k_3 ME = k_{-3}C_m$, leading to

$$C_s = \frac{K_m E_0 S}{K_m M + K_e S + K_m K_e}, \quad C_m = \frac{K_m E_0 M}{K_m M + K_e S + K_m K_e}, \quad (6.3.21)$$

where $K_m = (k_{-1} + k_2)/k_1$ is the Michaelis constant for the substrate, $K_e = k_{-3}/k_3$ is the equilibrium constant for the modifier. Thus the saturation function is

$$Y(S) = \frac{C_s}{E_0} = \frac{S}{K_m(1 + M/K_e) + S}, \quad (6.3.22)$$

and the velocity of the reaction is

$$V = V_m Y(s), \quad (6.3.23)$$

where $V_m = k_2 E_0$, the maximum velocity of the reaction. The effect of the modifier is inhibitory, increasing the effective Michaelis constant of the enzyme from K_m to $K_m(1 + M/K_d)$, thus decreasing the velocity of the reaction for given S while leaving the maximum velocity unchanged.

6.3.2 Cooperative Phenomena

For many enzymes and other proteins such as haemoglobin, the curve of reaction velocity against substrate concentration is sigmoid (S-shaped, with a single change from positive to negative curvature), rather than hyperbolic as in the Michaelis–Menten reaction scheme (see Figure 6.2). This possibility of having

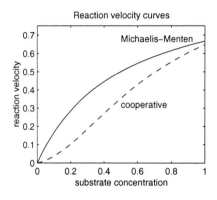

Figure 6.2   Possible curves of reaction velocity vs substrate concentration for reactions catalysed by an enzyme. The Michaelis–Menten curve is hyperbolic, while the cooperative reaction curve has a sigmoid shape.

sharper changes of reaction velocity may allow for much more precise control of metabolic pathways. The sigmoid shape often arises from *cooperative* enzyme binding effects, where a ligand may bind at several sites on an enzyme, and binding at one site changes the rate constants of the binding reactions at other sites. The standard model for such effects makes the following assumptions.

– Cooperative enzymes are composed of several identical reacting units, called *protomers*, that occupy equivalent positions within the enzyme.

– Each protomer contains one binding site for each ligand.

– The binding sites within each enzyme are equivalent.

– If the binding of a ligand to one protomer induces a conformational change in the protomer, a conformational change may be induced in the other protomers, changing the rate constants of the binding reactions.

A dimer (an enzyme with two protomers) may therefore bind two molecules of substrate consecutively, as shown below:

$$E + S \underset{k_{-1}}{\overset{2k_1}{\rightleftharpoons}} C_1 \overset{k_2}{\rightarrow} E + P, \quad C_1 + S \underset{2k'_{-1}}{\overset{k'_1}{\rightleftharpoons}} C_2 \overset{2k'_2}{\rightarrow} C_1 + P. \qquad (6.3.24)$$

There are factors of 2 in the scheme because there are two sites to bind to on E, and two sites to unbind from on C_2. In a non-cooperative enzyme $k'_1 = k_1$, $k'_{-1} = k_{-1}$, $k'_2 = k_2$; whether a substrate is bound at one site makes no difference to the rate constants at the other. The equations to be satisfied by the various chemical concentrations are as follows.

$$\frac{dS}{d\tau} = -2k_1 SE + k_{-1}C_1 - k_1' SC_1 + 2k_{-1}' C_2,$$

$$\frac{dE}{d\tau} = -2k_1 SE + k_{-1}C_1 + k_2 C_1,$$

$$\frac{dC_1}{d\tau} = 2k_1 SE - k_{-1}C_1 - k_2 C_1 - k_1' SC_1 + 2k_{-1}' C_2 + 2k_2' C_2, \qquad (6.3.25)$$

$$\frac{dC_2}{d\tau} = k_1' SC_1 - 2k_{-1}' C_2 - 2k_2' C_2,$$

$$\frac{dP}{d\tau} = k_2 C_1 + 2k_2' C_2,$$

with the conservation equations $E + C_1 + C_2 = E_0$, $S + C_1 + 2C_2 + P = 0$. The quasi-steady-state approximation gives

$$SE = \frac{1}{2}K_m C_1, \quad SC_1 = 2K_m' C_2, \qquad (6.3.26)$$

where $K_m = \frac{k_{-1}+k_2}{k_1}$ and $K_m' = \frac{k_{-1}'+k_2'}{k_1'}$ are the Michaelis constants for the first and the second Michaelis–Menten schemes in Equation (6.3.24). After some algebra, we obtain expressions for the quantities of interest. The saturation function Y is given by

$$Y(S) = \frac{C_1 + 2C_2}{2(E + C_1 + C_2)} = \frac{S(K_m' + S)}{K_m K_m' + 2K_m' S + S^2}.$$

If $K_m' = K_m$, the non-cooperative case, we have $Y(S) = \frac{S}{K_m+S}$, as in the Michaelis–Menten reaction. The velocity of the reaction is given by

$$V = \frac{dP}{d\tau} = \frac{2E_0 S(k_2 K_m' + k_2' S)}{K_m K_m' + 2K_m' S + S^2}. \qquad (6.3.27)$$

If $k_2' = k_2$, this reduces to

$$V = V_m Y(S)$$

with $V_m = 2k_2 E_0$, which should be compared with the corresponding formula (6.2.8) for the Michaelis–Menten reaction.

A highly cooperative limiting case occurs when the binding of the second substrate molecule is greatly facilitated by the presence of the first, so that very little C_1 is present, while E and C_2 are both present in non-negligible quantities. Thus, using Equation (6.3.26), $K_m' \ll S \ll K_m$, while $K_m K_m'$ and S^2 are the same order of magnitude. Defining K by $K^2 = K_m K_m'$, the limiting overall reaction velocity is given by

$$V = V_m Y(S) = \frac{V_m S^2}{K^2 + S^2}, \qquad (6.3.28)$$

where $V_m = 2k'_2 E_0$. A generalisation of this often used to model cooperative enzymes is the *Hill equation*,

$$V = \frac{V_m S^n}{K^n + S^n}. \tag{6.3.29}$$

The exponent n is usually determined empirically as the slope of a *Hill plot*, a plot of $\log(V/(V_m - V))$ against $\log(S)$, and despite the interpretation of the Hill equation as the reaction velocity in a limiting case of an n-stage binding process, the best fit to data often gives a non-integer value of n.

EXERCISES

6.5. The model for cooperative enzymes given above assumes that when the substrate itself binds to one of the protomers in an enzyme it changes the conformational states of the other protomers. An alternative hypothesis is put forward in the Monod–Wyman–Changeux model for allostery, that the conformational states of the protomers in an enzyme change when a modifier binds at an allosteric site, or even spontaneously. Only two conformational states are possible for each protomer, and every protomer must be in the same state, so that only two conformational states are possible for the enzyme as a whole. When a conformational change occurs it is like an umbrella being blown inside out by the wind, with every spoke flipping simultaneously. In other respects, the sites act independently of each other. The conformational states are usually denoted by R and T, and for a dimer the possible reactions are as shown:

$$R_2 \underset{k_1 S}{\overset{2k_{-1}}{\rightleftharpoons}} R_1 \underset{2k_1 S}{\overset{k_{-1}}{\rightleftharpoons}} R_0 \underset{k_{-0}}{\overset{k_0}{\rightleftharpoons}} T_0 \underset{k'_{-1}}{\overset{2k'_1 S}{\rightleftharpoons}} T_1 \underset{2k'_{-1}}{\overset{k'_1 S}{\rightleftharpoons}} T_2. \tag{6.3.30}$$

Here we have assumed that the conformational change can only occur to the bare enzyme, and not to any of its complexes, but in general transitions between R_1 and T_1 and between R_2 and T_2 are also possible.

a) Making the quasi-steady-state hypothesis, show that $2k_{-1}R_2 = k_1 S R_1$, $k_{-1}R_1 = 2k_1 S R_0$, with similar equations for the T conformational states, and $k_0 R_0 = k_{-0} T_0$.

b) Show that the saturation function is given by

$$Y(S) = \frac{SK_1^{-1}(1 + SK_1^{-1}) + K_0^{-1}(SK_1'^{-1}(1 + SK_1'^{-1}))}{(1 + SK_1^{-1})^2 + K_0^{-1}(1 + SK_1'^{-1})^2},$$

where $K_1 = k_{-1}/k_1$, $K_1' = k_{-1}'/k_1'$.

c) Show that if the equilibrium constant $K_0 = k_{-0}/k_0 = \infty$, $Y(S) = S/(S + K_1)$, and interpret this result.

6.6. Aerobic cellular respiration occurs in three stages, (i) glycolysis (sugar-splitting), (ii) the Krebs cycle, and (iii) the electron transport (or cytochrome) system. Each of these stages is a metabolic pathway of many elementary steps, eleven in the case of glycolysis. Although the ultimate aim is to produce ATP for energy by phosphorylating ADP, two of the first three steps of glycolysis work in the opposite direction and require the addition of a molecule of ATP, which is dephosphorylated to ADP.

In the 1950s it was noted that oscillations could occur in the glycolytic pathway. They were first observed in yeast cells but have since been seen in many cells of the body. The function of the oscillations, if any, is unclear, although it has recently been proposed that such oscillations in pancreatic β-cells could lead to pulsatile insulin secretion by these cells. However, the oscillations are well studied and are a prototype for periodic phenomena in biochemistry. It seems that the oscillations occur in the third step in the pathway, where the enzyme phosphofructokinase (PFK) catalyses the phosphorylation of fructose 6-phosphate to fructose 1,6-bis-phosphate, and the consequent dephosphorylation of ATP to ADP. PFK is an allosteric enzyme which has two substrates ATP and ADP and which occurs in an active and an inactive form. It is activated by binding with n molecules of ADP, denoted by S_2, in the reaction

$$nS_2 + E \underset{k_{-3}}{\overset{k_3}{\rightleftharpoons}} ES_2^n.$$

(This is known as the concerted transition theory of Monod, Wyman, Changeux and Jacob.) The activated form of the enzyme binds with one molecule of ATP (denoted by S_1) and dephosphorylates it to ADP, in a Michaelis–Menten-type reaction,

$$S_1 + ES_2^n \underset{k_{-1}}{\overset{k_1}{\rightleftharpoons}} S_1 ES_2^n \overset{k_2}{\rightarrow} ES_2^n + S_2.$$

In addition, S_1 is constantly added and S_2 removed. Diagrammatically,

$$\overset{a}{\rightarrow} S_1, \quad S_2 \overset{b}{\rightarrow} .$$

a) Apply the law of mass action to this scheme to obtain the equations

$$\frac{dS_1}{d\tau} = a - k_1 S_1 X_1 + k_{-1} X_2,$$

$$\frac{dS_2}{d\tau} = k_2 X_2 - k_3 S_2^n E + k_{-3} X_1 - b S_2,$$

$$\frac{dE}{d\tau} = -k_3 S_2^n E + k_{-3} X_1,$$

$$\frac{dX_2}{d\tau} = k_1 S_1 X_1 - (k_{-1} + k_2) X_2.$$

b) Assuming that the quasi-steady-state hypothesis holds, use these equations and the enzyme conservation equation $E + X_1 + X_2 = E_0$ to derive the substrate equations in the form

$$\frac{dS_1}{d\tau} = a - k_2 E_0 F(S_1, S_2), \quad \frac{dS_2}{d\tau} = k_{-1} E_0 F(S_1, S_2) - b S_2,$$

where

$$F(S_1, S_2) = \frac{K_m^{-1} S_1}{K_e S_2^{-n} + 1 + K_m^{-1} S_1},$$

and $K_m = (k_{-1} + k_2)/k_1$, $K_e = k_{-3}/k_3$.

c) Non-dimensionalise the equations by defining $s_1 = K_m^{-1} S_1$, $s_2 = K_e^{1/n} S_2$, $t = (k_2 E_0/K_m)\tau$, to obtain

$$\frac{ds_1}{dt} = \alpha - f(s_1, s_2), \quad \frac{ds_2}{dt} = \gamma f(s_1, s_2) - \beta s_2,$$

where $\alpha = a/(k_2 E_0)$, $\beta = b K_m/(k_2 E_0)$, $\gamma = \frac{k_{-1}}{k_2} K_e^{1/n} K_m^{-1}$, and

$$f(s_1, s_2) = \frac{s_1}{s_2^{-n} + 1 + s_1}.$$

It may be shown that this system has oscillatory solutions, but the algebra is horrible.

6.4 Neural Modelling

The nervous system is the main means by which humans and other animals coordinate short-term responses to stimuli. It consists of *receptors*, such as eyes, which receive signals from the outside or inside world, *effectors*, such as muscles, which respond to these signals by producing an effect, and *nerve cells*, also known as *neurons* or *neurones*, which communicate between the receptors and the effectors. Neurons usually consist of a cell body (the *soma*) and cytoplasmic extensions (the *axon* and many *dendrites*) through which they connect (via *synapses*) to a network of other neurons, and to receptors and effectors. An axon may be a metre long or more. Within neurons, signals are transmitted by *action potentials* or *nerve impulses*. Cells that have the ability to transmit action potentials, such as cardiac cells, smooth and skeletal muscle cells, secretory cells and neurons, are called *excitable cells*. The action potentials are initiated by inputs from the dendrites arriving at the *axon hillock*, where the axon meets the soma. They then travel down the axon to terminal branches which have synapses to the next cells in the network. The action potential is electrical, but it would be erroneous to think of the axon as like an electric cable. For one thing, its resistance is far higher. As Hodgkin pointed out, the resistance of a nerve fibre $1\,\mathrm{m}$ long with a diameter of $1\,\mu\mathrm{m}$ is the same as that of a copper wire with a diameter of $0.6\,\mathrm{mm}$ and a length ten times the distance between here and the planet Saturn. For another, the current is a flow predominantly of ions rather than electrons, and its direction is predominantly not longitudinal but transverse, passing into and out of the cell through ion channels in the membrane. These channels open and close in response to voltage changes, and each is specific to a particular ion. In addition, some channels may be temporarily inactivated, after which they need time to recover to the active state before they can open again.

The modern understanding of nerve conduction is based on the work of Hodgkin and Huxley, who shared the Nobel Prize for Physiology and Medicine with Eccles in 1963. They worked on a nerve cell with the largest axon known, the squid giant axon. They manipulated ionic concentrations outside the axon and discovered that sodium and potassium currents were controlled separately, and not a result of general changes in membrane permeability to all ions, as had previously been believed. They used a technique called a voltage clamp to control the membrane potential and deduce how ion conductances would change with time at fixed voltages, hence finding models for the functions $\tau_m(V)$, $m_\infty(V)$, etc in Equations (6.4.31). They used a technique called a space clamp to remove the spatial variation inherent in the travelling action potential. This involved threading a silver wire down the inside of the axon. Moreover they made a mathematical model of all these processes with functional forms fitted to data, published in 1952, in a *tour de force* of mathematical modelling. This is now seen as one of the major successes of mathematical biology, and is the basis of study of many excitable systems in neurobiology and elsewhere. The result was a system of three ordinary differential equations and one partial differential equation, which Huxley solved numerically on a hand-cranked calculator. It took him three weeks to compute a single action potential.

As Hodgkin and Huxley recognised, to understand the transmission of action potentials along axons it helps to understand first what would happen if there were no spatial effects. Consider a space-clamped squid giant axon at its resting potential; it is negatively charged or *polarised*, with a potential difference of about −70 mV between the inside and the outside of the cell. There is a high concentration of potassium and a low concentration of sodium inside the cell, compared to external concentrations. There are metabolic costs involved in maintaining the resting neuron away from equilibrium, electrically polarised and with different sodium and potassium ion concentrations inside and outside the cell. This is essential among other things for the propagation of the nerve impulse, and is maintained by a sodium-potassium pump powered by the short-term energy stored in the cellular ATP. In other words, the cell pumps ions.

Now partially *depolarise* the cell. If the depolarisation is small enough, the cell returns almost immediately to its steady state. But a larger depolarisation, to a potential difference of about −25 mV, causes the following changes to occur. Time scales are about 1 ms, or 10 ms for changes described as "slow".

− Sodium channels open, triggered by the partial depolarisation, and positively-charged sodium ions flood into the cell, further increasing the depolarisation. This is a positive feedback effect that eventually leaves the cell positively charged or *reverse polarised*. This is the *upstroke* phase.

− Over a slow time scale, potassium channels open, and potassium ions flood out of the cell. For a time, sodium ions are still flooding in and just about keep pace, and the potential difference falls slowly. This is the *excited* phase.

− Eventually the outward potassium current overwhelms the inward sodium current, making the cell more negatively charged, which closes even more sodium channels in a second positive feedback process. The cell overshoots its equilibrium polarisation, and becomes *hyperpolarised*. This is the *downstroke* phase.

− Most of the sodium channels are now inactive, and need time to recover before they can open again. This is the *refractory* period, when the cell has more than recovered its polarisation but is nevertheless not responsive to any further stimuli.

− Slowly most of the potassium channels close, and the sodium channels become active, although still mostly closed, and the cell returns to its original
state. This is the *recovery* phase.

When spatial effects are included, the initial depolarisation is due to excitatory inputs from other cells arriving at the axon hillock. From there the nerve
impulse propagates down the axon, since when sodium ions flood into the cell
they move down the axon, following the potential gradient, depolarising the
cell and triggering the same changes there.

The Hodgkin–Huxley equations involve four variables, the potential difference V, the sodium activation variable m, the sodium inactivation variable h,
and the potassium activation variable n. They are given by

$$C_m \frac{dV}{dt} = -\bar{g}_{Na}m^3 h(V - V_{Na}) - \bar{g}_K n^4 (V - V_K) - g_L(V - V_L),$$

$$\tau_m(V)\frac{dm}{dt} = m_\infty(V) - m, \quad \tau_h(V)\frac{dh}{dt} = h_\infty(V) - h,$$

$$\tau_n(V)\frac{dn}{dt} = n_\infty(V) - n. \quad (6.4.31)$$

Here C_m is the membrane capacitance, $g_{Na} = \bar{g}_{Na}m^3 h$ is the sodium conductance, $g_K = \bar{g}_K n^4$ the potassium conductance, and g_L the so-called leakage
conductance for other ionic currents, predominantly chloride. The m equation
shows that if V is kept constant, then m tends exponentially to $m_\infty(V)$ with
time constant $\tau_m(V)$, and similar interpretations hold for the h and n equations. The functions m_∞ and n_∞ increase with V, since these are activation
variables and represent channel opening, whereas h_∞, an inactivation variable,
decreases with V and represents channel closing.

The variables V and m are about an order of magnitude faster than h and n,
with time constants of about $1\,\mathrm{ms}$ compared to about $10\,\mathrm{ms}$. Relatively simple
models of nerve impulses involve two variables v and w. The excitation variable
v represents the fast variables and may be thought of as potential difference;
and the recovery variable w represents the slow variables and may be thought
of as potassium conductance. The space-clamped equations are of the form

$$\epsilon \frac{dv}{dt} = f(v, w), \quad \frac{dw}{dt} = g(v, w), \quad (6.4.32)$$

and are known as the *generalised FitzHugh–Nagumo equations*, after FitzHugh,
who first reduced the Hodgkin–Huxley models to two variables, and Nagumo,
who built an electrical circuit that mimics the behaviour of one of FitzHugh's
models. A typical phase plane for the equations is shown in Figure 6.3. The
traditional form for g is a straight line $g(v, w) = v - c - bw$, and for f a
cubic $f(v, w) = v(v - a)(1 - v) - w$, or a piecewise linear function $f(v, w) =$

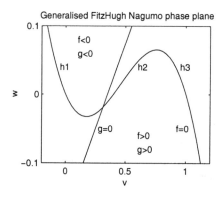

Figure 6.3   The phase plane for the generalised FitzHugh–Nagumo Equations (6.4.32). The essential feature is that the nullcline $f(v, w) = 0$ is of cubic shape. There are three solutions of $f(v, w) = 0$ for $w_* \leq w \leq w^*$ given by $v = h_1(w)$, $v = h_2(w)$ and $v = h_3(w)$, with $h_1(w) \leq h_2(w) \leq h_3(w)$. The nullcline $g(v, w) = 0$ is assumed to have positive slope and exactly one intersection with the curve $f(v, w) = 0$. The signs of f and g above and below the nullclines are as shown.

$H(v - a) - v - w$, where H is the Heaviside function. A numerical solution of the equations when f is a cubic is shown in Figure 6.4. Let us consider the problem

$$\epsilon \frac{dv}{dt} = f(v, w) = v(v - a)(1 - v) - w, \quad \frac{dw}{dt} = g(v, w) = v - bw, \quad (6.4.33)$$

with $v(0) = v_0$, $w(0) = w_0$. This is a singular perturbation problem, and it may be analysed using the method of matched asymptotic expansions, as in the analysis of the Michaelis–Menten equations of Section 6.2.

There are four phases of the solution, as shown in Figure 6.4. In phase 1, the upstroke phase, the excitation variable v is changing very quickly to attain $f = 0$. This corresponds to the inner solution of the Michaelis–Menten analysis, and we need to rescale time in order to analyse it. Defining a new short time scale by $T = t/\epsilon$, and defining $V(T) = v(t)$, $W(T) = w(t)$, we obtain

$$\frac{dV}{dT} = f(V, W) = V(V - a)(1 - V) - W, \quad \frac{dW}{dT} = \epsilon g(V, W) = \epsilon(V - bW).$$
$$(6.4.34)$$

To first order, $W = w_0$, and V satisfies $\frac{dV}{dT} = f(V, w_0)$ and $V(0) = v_0$. An exact form for the solution may be obtained by separation of variables, but we are only interested in the matching condition, $V(T) \to h_3(w_0)$ as $T \to \infty$. The time taken for this phase is negligible in terms of the t variable.

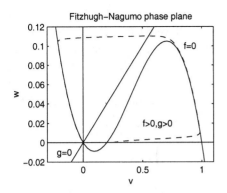 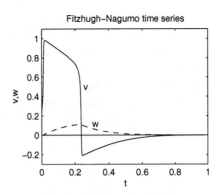

Figure 6.4   A numerical solution of the excitatory FitzHugh–Nagumo Equations (6.4.32) with $f(v,w) = v(v-a)(1-v) - w$, $g(v,w) = v - bw$, with $\epsilon = 0.01$, $a = 0.1$, $b = 0.5$. The equations have a unique globally stable steady state at the origin, but there is a threshold phenomenon. If v is perturbed slightly from the steady state, the system returns there immediately, but if it is perturbed beyond $v = h_2(0) = 0.1$, then there is a large excursion before an eventual return.

In phase 2, the excited phase, we move back to the long t time scale and consider Equations (6.4.33). To leading order we have $f(v,w) = 0$, so by continuity $v = h_3(w)$, and w satisfies $\frac{dw}{dt} = g(h_3(w), w) = G_3(w)$, say. Hence w increases until it reaches w^*, beyond which $h_3(w)$ ceases to exist. The time taken for this phase is

$$t_2 = \int_{w_0}^{w^*} \frac{1}{G_3(w)} dw.$$

In phase 3, the downstroke phase, v changes very rapidly as the solution jumps from the right-hand to the left-hand branch of the nullcline $f = 0$. We need a short time scale again, and define the time variable $\hat{T} = (t - t_0)/\epsilon$, with $\hat{V}(\hat{T}) = v(t)$, $\hat{W}(\hat{T}) = w(t)$. Here t_0 is a time within the phase, and to leading order we may take $t_0 = t_2$. We are back to the inner Equations (6.4.34), but with hats. Hence $\hat{W} = w^*$, and $\hat{V}$ satisfies $\frac{d\hat{V}}{d\hat{T}} = f(\hat{V}, w^*)$. This is to be solved with matching conditions to phase 2, $\lim_{\hat{T} \to -\infty} \hat{V}(\hat{T}) = v^*$. Again this can be done explicitly by separating variables, but again we are only interested in the matching condition to phase 4, $\lim_{\hat{T} \to \infty} \hat{V}(\hat{T}) = h_1(w^*)$. The time taken for this phase is negligible in terms of the t variable.

Finally, in phase 4, the recovery phase, we return to the long t time scale and Equations (6.4.33). To first order we have $f(v,w) = 0$, so now $v = h_1(w)$, using the matching condition to phase 3, and w satisfies $\frac{dw}{dt} = g(h_1(w), w) = G_1(w)$, say, with $w(t_2) = w^*$. Hence w decreases back to the steady state, as does v.

This describes the behaviour if the steady state is on the left hand branch of $f = 0$, as in Figure 6.3. But if the nullcline $g = 0$ were shifted to the left so that the steady state were on the middle branch of $f = 0$, as in Figure 6.5, we would have different behaviour. In phase 4, w would drop until it reached w_*, and we would then have a jump to the right-hand branch of $f = 0$. This behaviour would be repeated indefinitely, and we would have a periodic oscillation. The period of the oscillation is given to first order by

$$t_p = \int_{w_*}^{w^*} \left(\frac{1}{G_3(w)} - \frac{1}{G_1(w)} \right) dw.$$

This kind of oscillation, with alternate fast and slow phases, is known as a *relaxation oscillation*. When spatial effects are included it becomes a periodic wave train, corresponding to regular firing of the neuron.

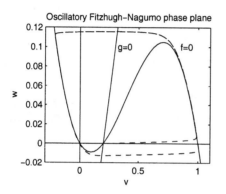
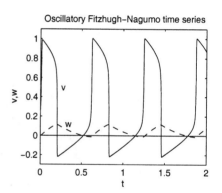

Figure 6.5   A numerical solution of the oscillatory FitzHugh–Nagumo Equations (6.4.32) with $f(v, w) = v(v - a)(1 - v) - w$, $g(v, w) = v - c - bw$, with $\epsilon = 0.01$, $a = 0.1$, $b = 0.5$, $c = 0.1$. The equations have a unique unstable steady state at $(0.1, 0)$, surrounded by a stable periodic relaxation oscillation.

EXERCISES

6.7. FitzHugh's original model was as follows:

$$\frac{dx}{dt} = c\left(y + x - \frac{1}{3}x^3 - I \right), \quad c\frac{dy}{dt} = a - x - by.$$

Here x is an excitability variable such as membrane potential, y is a recovery variable such as potassium channel activity, I is an input such as applied current (assumed constant in this exercise), and a, b and c are positive parameters, and $b < c$, $b < 1$, $b < c^2$.

a) Verify that the Jacobian at a steady state (x^*, y^*) is given by

$$J^* = \begin{pmatrix} c(1 - x^{*2}) & c \\ -1/c & -b/c \end{pmatrix},$$

and that the steady state is stable if

$$\frac{b}{c} - c(1 - x^{*2}) > 0, \quad 1 - b(1 - x^{*2}) > 0.$$

b) Show that the steady state of the model is unstable if x^* falls in the range $-\gamma < x^* < \gamma$, where $\gamma = \sqrt{1 - b/c^2}$.

c) Show that this constraint places x^* on the portion of the cubic nullcline between the two humps.

d) Show that if $I = I_c$, where $I_c = (a - \gamma)/b + \gamma - \gamma^3/3$, there is a steady state at $(\gamma, (a - \gamma)/b$, and that the Jacobian there has purely imaginary eigenvalues.

e) Show that if I increases from I_c, the eigenvalues at the steady state move into the right half plane, destabilising the steady state. (These are the conditions for a Hopf bifurcation to oscillatory solutions; see Section B.4.2 of the appendix.)

6.5 Immunology and AIDS

The body's main defence against the threat of invasion by pathogenic organisms such as bacteria and viruses is the immune system. Impairment of the immune system may be lethal, as the HIV/AIDS epidemic has shown. When a pathogenic organism invades, the immune system has three tasks: (a) to recognise the organism as foreign, (b) to produce weapons specifically fashioned to destroy the organism, and (c) to keep a record so that any further invasion by the same or a sufficiently similar organism can easily be repelled. This "memory" may fade or endure indefinitely.

Recognition depends on the shape of macromolecules (usually proteins) called *antigens* on the surface of the invading organism. In most cases the invader is destroyed in two ways: first by soluble proteins known as *antibodies* in the bodily fluids and second by *cytotoxic* (killer) cells. The first process takes care of free pathogens, and the second of pathogen-infected cells. The antibodies are produced by *B cells*, and the killer cells are *T cells*. Both B and T cells are *lymphocytes*, white blood cells produced in the bone marrow but matured in either the bone marrow (B cells) or the thymus gland (T cells). Both antibodies and killer cells are specific to the particular pathogen encountered, but the specificity is loose in that closely related pathogens are also recognised. The *clonal selection theory* states that enough different mature but inactive B and T cells always exist that essentially any pathogen can be recognised, and that a challenge by any pathogen results in a positive feedback effect that stimulates proliferation of the small fraction of cells that recognise that particular pathogen. This leads to large clones of cells descended from these precursors. An important feature of the system is extra-cellular signalling to produce more lymphocytes specific to the pathogen by positive feedback.

The recognition problem is enormous, given the variety of possible invading organisms and the necessity of not destroying one's own cells. It has been estimated that any animal must be capable of distinguishing as foreign at least 10^{16} antigens, but the human *immune repertoire*, the number of possible antibodies that a human immune system can make, is only about 10^7. However, each antibody is able to mount a defence against about 10^{11} different (but similarly shaped) antigens, so the chance that a given antibody responds to a given challenge is about 10^{-5}. With 10^7 different antibodies, if we assume that they are distributed uniformly across antigen space, the chance that a challenge is unanswered is about $(1 - 10^{-5})^{10^7} \approx \exp(-100) \approx 10^{-44}$. The total number of lymphocytes in the body is about 10^{12}, so that the expected number responding to a given challenge is about 10^7.

The ability of an antibody to recognise closely related antigens can be important in developing vaccines. The first vaccine, discovered by Edward Jenner in 1796, was an injection with the innocuous cowpox virus, which protects against the closely related but potentially lethal smallpox virus. This is the basis of the name "vaccine", which derives from the Latin *vaccinia* for the cowpox virus. Other vaccines are made from attenuated or inactivated microorganisms. The looseness of the specificity also helps the immune system to track mutating viruses. However, viruses have high mutation rates resulting from poor error-correcting systems in the replication of their genetic material, and a pathogen such as the influenza virus may mutate so quickly that it is unrecognisable on a second encounter. A new immune defence has to be mounted. Moreover, two strains of a virus may combine to produce a quite different and novel strain. The 1918 influenza pandemic, in which 20 million people died, is thought to have been produced by a recombination of a human and an avian strain of the virus. The human immunodeficiency virus mutates so quickly during a single infection that it is continually providing new challenges to the immune system.

We shall model the interaction inside the body between a virus and the immune system. A virus particle (or *virion*) on its own does absolutely nothing. It is a small amount of genetic material surrounded by one or more protective shells. If it gains entry to a host cell, it hijacks the cell's machinery for its own replication. It then leaves the cell, and the process is repeated. Different viruses target different host cell types for this purpose. Let V be the number of virions in an organism, X the number of uninfected target cells, and Y the number of infected cells. Our first model will neglect the immune response. The equations are given by

$$\frac{dV}{d\tau} = aY - bV, \quad \frac{dX}{d\tau} = c - dX - \beta XV, \quad \frac{dY}{d\tau} = \beta XV - fY. \quad (6.5.35)$$

The virus is replicated by the infected cells, so its rate of production is taken to be proportional to Y. Virions die at a specific rate b. The uninfected cells are constantly being produced by the organism at a rate c. They die at a specific rate d, and become infected by virus at a specific rate βv, entering the Y class. Infected cells die at a specific rate $f = e + d$, where d is the natural death rate and e the additional death rate owing to the infection. The relationship between the virus and the uninfected cells is analogous to the relationship between predator and prey discussed in Chapter 2, and with this analogy βX is the functional response of the virus to the uninfected cells. As usual, a saturating function might be a better model, but we use this form for simplicity.

The criterion for spread of the virus inside the body is similar to the criterion discussed in Chapter 3 for the spread of a disease in a population. Let R_0 be the basic reproductive ratio for the virus, defined to be the expected number of virions that one virion gives rise to in an uninfected cell population. A virion gives rise to infected cells at a rate βX for a time $1/b$, and each infected cell gives rise to virions at a rate a for a time $1/f$. Since $X = c/d$ for an uninfected population,

$$R_0 = \frac{\beta ca}{dbf}. \quad (6.5.36)$$

The criterion for the spread of the virus is $R_0 > 1$.

We non-dimensionalise the system by defining

$$x = \frac{d}{c}X, \quad y = \frac{d}{c}Y, \quad v = \frac{bf}{ac}V, \quad t = d\tau. \quad (6.5.37)$$

The non-dimensionalisation for X arises from its steady state in the absence of infection, that for Y is chosen to be the same, and that for V arises from its steady state value, as we shall see. There are at least three time scales in the problem, associated with d, f and b; we arbitrarily choose to non-dimensionalise τ with d. The equations become

$$\epsilon\frac{dv}{dt} = \alpha y - v, \quad \frac{dx}{dt} = 1 - x - R_0 xv, \quad \frac{dy}{dt} = R_0 xv - \alpha y, \qquad (6.5.38)$$

where

$$\epsilon = \frac{d}{b}, \quad \alpha = \frac{f}{d}. \qquad (6.5.39)$$

For typical parameter values $\epsilon \ll 1$.

The steady states of the non-dimensionalised system (6.5.38) are $S_0 = (0, 1, 0)$, the uninfected steady state, and $S^* = (v^*, x^*, y^*)$, where

$$v^* = 1 - \frac{1}{R_0}, \quad x^* = \frac{1}{R_0}, \quad y^* = \frac{1}{\alpha}\left(1 - \frac{1}{R_0}\right). \qquad (6.5.40)$$

For $R_0 > 1$, the normal situation, $(v(t), x(t), y(t)) \to (v^*, x^*, y^*)$ as $t \to \infty$. As in Chapter 3, the susceptible population (X in this case) is reduced by the disease until each virion is expected to give rise to exactly one new virion, $R_0 x^* = 1$.

This may be thought of as the primary phase of an infection, before the immune system has had time to kick in. Now we move on to the secondary phase, and include the effect of the immune system. Let us assume that the response is via killer cells Z, which are produced at a constant rate g and die at specific rate h. These cells kill infected cells at a rate $\gamma Y Z$. There is an analogy here of Z cells as predators and Y cells as prey, and we have again taken a linear functional response of Z to Y. The equations become

$$\frac{dV}{d\tau} = aY - bV, \quad \frac{dX}{d\tau} = c - dX - \beta XV,$$

$$\frac{dY}{d\tau} = \beta XV - fY - \gamma Y Z, \quad \frac{dZ}{d\tau} = g - hZ. \qquad (6.5.41)$$

If we now non-dimensionalise as before, with $z = hZ/g$ in addition, we obtain

$$\epsilon\frac{dv}{dt} = \alpha y - v, \quad \frac{dx}{dt} = 1 - x - R_0 xv,$$

$$\frac{dy}{dt} = R_0 xv - \alpha y - \kappa yz, \quad \frac{dz}{dt} = \lambda(1 - z), \qquad (6.5.42)$$

where

$$\lambda = \frac{h}{d}, \quad \kappa = \frac{\gamma g}{dh}. \qquad (6.5.43)$$

The steady states of the system are at $S_0 = (0, 1, 0, 1)$, the uninfected steady state, and $S^* = (v^*, x^*, y^*, z^*)$, where now

$$v^* = \frac{\alpha}{\alpha + \kappa}\left(1 - \frac{1}{R_0'}\right), \quad x^* = \frac{1}{R_0'}, \quad y^* = \frac{1}{\alpha + \kappa}\left(1 - \frac{1}{R_0'}\right), \quad z^* = 1,$$

$$\qquad (6.5.44)$$

and we have defined R_0' by

$$R_0' = \frac{\alpha}{\alpha + \kappa} R_0. \tag{6.5.45}$$

We see (i) that R_0' is the basic reproductive ratio in the presence of the immune response, (ii) if the infection persists then $R_0' x^* = 1$ as expected, and (iii) the infection persists as long as $R_0' > 1$. In order for the immune response to clear the infection we need the immune response parameter κ to satisfy

$$\kappa > \alpha(R_0 - 1). \tag{6.5.46}$$

One of the problems with HIV/AIDS is that the virus targets the killer cells themselves. The number of Z cells in the blood decreases from about 1000 per millilitre in the early stages of the disease to about 200 per millilitre in full-blown AIDS, a steady decrease that may take ten years, even in the absence of drug treatment. This is equivalent to a five-fold decrease in f/g and hence in κ, so that the inequality (6.5.46) is likely to fail after a time. But why doesn't the immune system clear the virus before this occurs? The answer seems to be that the virus can hide for long periods in so-called latent cells, Y cells that are not seen as infected by the immune system. The virus load increases very slowly over the course of the infection until full-blown AIDS occurs, when the virus breaks free of immune system control. This could be when κ is finally reduced below $\alpha(R_0 - 1)$.

EXERCISES

6.8. In HIV infection, reverse transcriptase inhibitors prevent infection of new cells.

   a) Assuming that such an inhibitor is available as a drug and is 100% efficient, write down the equations for the subsequent dynamics of the infected cells and free virus.

   b) Show that the number of infected cells falls exponentially, and assuming that the half-life of the virus is much less than that of the virus-producing cells, the amount of free virus falls exponentially after a shoulder phase.

6.9. Protease inhibitors of HIV prevent infected cells from producing infectious virus particles.

   a) Show that the equations after therapy with such a drug has begun are given by

$$\frac{dV}{d\tau} = -bV, \quad \frac{dW}{d\tau} = aY - bW, \tag{6.5.47}$$

with the X and Y equations as before.

b) Assuming that the uninfected cell population X remains roughly constant for the time-scale under consideration, and that $f \ll b$, show that the total amount $V + W$ of free virus falls exponentially after a shoulder phase.

6.10. In HIV, some cells enter a latent class on their infection. While in this class they do not produce new virions, but may later be reactivated to do so.

a) Explain the following model for this situation.

$$\frac{dV}{d\tau} = aY_1 - bV, \quad \frac{dX}{d\tau} = c - dX - \beta XV,$$

$$\frac{dY_1}{d\tau} = q_1\beta XV - f_1Y_1 + \delta Y_2, \quad \frac{dY_2}{d\tau} = q_2\beta XV - f_2Y_2 - \delta Y_2.$$

$$\text{(6.5.48)}$$

b) What is the basic reproductive ratio R_0 for this model?

6.6 Conclusions

- Law of mass action: the rate of a chemical reaction is proportional to the concentrations of the reactants. The constant of proportionality is called the rate constant.

- The Michaelis–Menten reaction is the simplest enzyme-catalysed reaction $E + S \rightleftharpoons C \rightarrow E + P$. The quasi-steady-state hypothesis $\frac{dC}{dt} \approx 0$ leads to the conclusion that the product is formed at a rate $V = V_mY(S) = V_mS/(K_m + S)$, where V_m is the maximum rate of production, K_m is the Michaelis constant, the substrate concentration at which the reaction proceeds at half its maximum speed, and $Y(S)$ is the saturation function, the fraction of enzyme binding sites that are occupied by the substrate.

- The quasi-steady-state hypothesis for the Michaelis–Menten reaction is valid after an initial short time period as long as the enzyme concentration is much less than the substrate concentration. The mathematical problem is a singular perturbation problem, and may be solved by the method of matched asymptotic expansions.

- Allosteric enzymes, whose efficiency as catalysts can be inhibited or activated, are very important in the control of metabolic pathways.

– Competitive inhibition by a modifier M with equilibrium constant K_e reduces the rate of formation of product from $V_m S/(K_m + S)$ to $V_m S/(K_m(1 + M/K_e) + S)$, effectively increasing the Michaelis constant from K_m to $K_m(1 + M/K_e)$.

– Cooperative enzymes (and other proteins) are composed of several identical protomers. The binding of a ligand to one of these may induce a conformational change that changes the rate constants of subsequent binding reactions. This may lead to sharper changes of reaction velocity with substrate concentration, and hence to more precise control of metabolic pathways.

– The Hill equation is often used as a phenomenological model of a cooperative protein, giving a reaction velocity $V = V_m S^n/(K^n + S^n)$.

– Potential differences and ion concentration differences are maintained between excitable cells and the extra-cellular medium. Partial depolarisation of such a cell results in a fast upstroke to an excited phase, followed by a fast downstroke to a refractory phase, followed by recovery to the initial state. The original four-variable Hodgkin–Huxley model of this process is usually reduced to a two-variable caricature, the generalised FitzHugh–Nagumo model, which is much easier to analyse.

– Small perturbations of models of excitable cells, which may be thought of as current input, lead to periodic oscillations in such cells. These are important in many physiological processes such as the heartbeat.

– The progress of a viral disease within an individual organism has parallels to the progress of a microparasitic disease in a population, the population now being the population of cells initially susceptible to infection. The basic reproductive ratio R_0 of the disease may be defined, and the disease reduces the number of susceptible cells x until this ratio becomes 1, $R_0 x^* = 1$.

– The immune system defends the body against bacteria and viruses. It reduces the basic reproductive ratio R_0 of a disease, and may clear it if R_0 becomes less than 1. The AIDS virus attacks the immune system itself, eventually destroying its ability to reduce R_0 below 1, and persists in the body before this happens by hiding in cells that are not recognised as infected by the immune system.

7
Pattern Formation

- Spontaneous generation of pattern is ubiquitous in biological systems, and is often of crucial adaptive significance. It often occurs through a loss of stability of a symmetric solution as a parameter of the system (such as the size of an embryo) changes, which may often be studied by a linear stability analysis.

- In a wide variety of biological systems, the underlying mechanism for pattern formation is an interaction between short-range activation and long-range inhibition.

7.1 Introduction

One of the major puzzles in biology is how a complex organism is formed from a simple egg. The process involves many instances where symmetry is broken, and a more symmetric situation leads to a less symmetric one. Many of these breaks in symmetry seem to have no external trigger, but are generated internally. How can this be done? And, crucially, how can it be done *robustly*, so that it is not disrupted by perturbations in the system or in the environment? Similar questions arise throughout biology, and the study of pattern formation is an important part of mathematical biology.

A seminal paper on this question was published by Alan Turing in 1952, and a substantial part of this chapter will be devoted to his idea. This empha-

Figure 7.1   The early stages of development of a vertebrate egg. The egg divides twice in a vertical plane, and then once in a horizontal plane.

sis should not be taken to imply that Turing's is the only or even the most important mechanism of pattern formation. We concentrate on it

– because it is one of the best-studied examples of the phenomenon, and

– because the concepts and methods that arise generalise to many other pattern formation mechanisms.

   Some of the main ideas are as follows.

– There is a symmetric steady state stable to symmetric perturbations, but not to asymmetric ones. These grow, resulting in pattern formation. (Usually, by symmetric we mean spatially uniform; this is the most tractable case.)

– Because this stability-instability requirement will generally only be satisfied for parameter values in a specific region of parameter space, qualitative changes in solution behaviour occur as parameters change. The study of such changes is *bifurcation theory*.

– In particular, we shall consider bifurcation with domain size as the bifurcation parameter, because this is a parameter that changes in biological development, an important area of application for pattern formation.

– An interaction between short-range activation and long-range inhibition is required for pattern formation in the simplest version of Turing's idea, and is a much more general mechanism for pattern formation.

7.2 Turing Instability

The Turing who published a paper on morphogenesis was the same Turing who worked on probability theory, decidability and the theory of computing, artificial intelligence, the Riemann hypothesis, group theory, quantum theory, relativity, cryptography and cracking the code used by the German Enigma machine in the Second World War, and computer design. He committed suicide in 1954 at the age of 41 after being arrested for homosexuality. His interests in biology included neurology and physiology, which he worked on for a year in Cambridge in 1947/48. His work on morphogenesis began in 1951. Morphogenesis is the development of the form or structure of an organism during the life history of the individual. It is often considered as taking place in two stages, the chemical and the mechanical, although it is more usual for the two processes to occur simultaneously. At the chemical (pre-pattern) stage, spatially inhomogeneous concentration profiles of various chemicals, called *morphogens* by Turing, are set up, ultimately through the action of genes. At the mechanical stage, these morphogens stimulate growth and differentiation, which lead to the development of the form itself. Turing considered a chemical theory, although he recognised the part that mechanics would have to play. His revolutionary idea was that passive diffusion could interact with chemical reaction in such a way that even if the reaction by itself has no symmetry-breaking capabilities, diffusion can de-stabilise the symmetric solutions so that the system with diffusion added can have them. This is counter-intuitive, and indeed diffusion in a single equation has a stabilising effect. In this chapter we shall look at cases where the interaction between diffusion and kinetics, sometimes with relevant boundary conditions, leads to pattern formation.

Can diffusion destabilise a spatially homogeneous steady state? If so, this is known as *diffusion-driven* or *Turing instability*. We consider the system of m reaction-diffusion equations given by

$$\frac{\partial \tilde{\mathbf{u}}}{\partial \tilde{t}} = \mathbf{f}(\tilde{\mathbf{u}}) + D\tilde{\nabla}^2 \tilde{\mathbf{u}} \qquad (7.2.1)$$

in a domain $\tilde{\Omega} \subset \mathbb{R}^N$, where $\tilde{\nabla}$ involves differentiation with respect to spatial variables $\tilde{\mathbf{x}}$. It is often useful to rescale the space variables in order to work with a problem on a standard domain Ω, and it turns out that rescaling the time variable as well simplifies the result. So we define $\mathbf{x} = \gamma\tilde{\mathbf{x}}$, $t = \gamma^2\tilde{t}$, and $\mathbf{u}(\mathbf{x}, t) = \tilde{\mathbf{u}}(\tilde{\mathbf{x}}, \tilde{t})$, to obtain

$$\frac{\partial \mathbf{u}}{\partial t} = \gamma^2\mathbf{f}(\mathbf{u}) + D\nabla^2\mathbf{u} = \alpha\mathbf{f}(\mathbf{u}) + D\nabla^2\mathbf{u}, \qquad (7.2.2)$$

say, on $\Omega \in \mathbb{R}^N$.

In this equation γ is a measure of the linear dimensions of the original domain, so that increasing γ is equivalent to increasing domain size. Let us assume that this has a spatially uniform steady state solution $\mathbf{u}^*$, so that $\mathbf{f}(\mathbf{u}^*) = \mathbf{0}$, and take homogeneous Neumann (zero-flux) boundary conditions on $\partial\Omega$. We choose these partly because they are natural in many cases, and partly because they are passive, unlikely in themselves to cause any pattern formation. Then $\mathbf{u} = \mathbf{u}^*$ is a solution of the differential equations and the boundary conditions. (Dirichlet boundary conditions $\mathbf{u} = \mathbf{u}^*$, or other boundary conditions that do not conflict with this solution, may also be considered.)

Now let $\hat{\mathbf{u}}$ be the perturbation from the steady state, $\hat{\mathbf{u}} = \mathbf{u} - \mathbf{u}^*$. The linearisation of Equation (7.2.2) about $\mathbf{u}^*$ is given by

$$\mathbf{v}_t = \alpha J^* \mathbf{v} + D\nabla^2 \mathbf{v} \qquad (7.2.3)$$

in Ω with homogeneous Neumann boundary conditions on $\partial\Omega$, where $\mathbf{v}$ is the linearised approximation to $\hat{\mathbf{u}}$ and J^* is the linearisation of f at the steady state, i.e. the Jacobian matrix

$$J^* = \left(\frac{\partial f_i}{\partial u_j}(\mathbf{u}^*) \right).$$

System (7.2.3) is linear with constant coefficients, and there is a standard method for finding its general solution, the method of *separation of variables*. First, let us assume that we know a function $F(\mathbf{x})$ that satisfies $-\nabla^2 F = \lambda F$ in Ω and homogeneous Neumann boundary conditions on $\partial\Omega$. (F is an eigen-function of $-\nabla^2$ on Ω with the boundary conditions, and λ its eigenvalue.) Now consider a function $\mathbf{v}$ of the form $\mathbf{v}(\mathbf{x}, t) = \mathbf{c} F(\mathbf{x}) \exp(\sigma t)$. It satisfies Equation (7.2.3) and the boundary conditions if

$$\sigma \mathbf{c} = \alpha J^* \mathbf{c} - \lambda D\mathbf{c} = A\mathbf{c}, \qquad (7.2.4)$$

say, so that σ and $\mathbf{c}$ are an eigenvalue and the corresponding eigenvector of the matrix $A = \alpha J^* - \lambda D$. Except for a couple of technicalities, finding the general solution of the equation and boundary conditions is just a matter of taking a general linear combination of such solutions. To be more precise, let us define the *spatial modes* to be the eigenfunctions $F_n(\mathbf{x})$ of $-\nabla^2$ on Ω with the appropriate boundary conditions, and the *spatial eigenvalues* λ_n to be the corresponding eigenvalues. An important fact is that the λ_n are all positive for Dirichlet boundary conditions, and all positive except for $\lambda_0 = 0$ for Neumann boundary conditions. We know from Fourier analysis (Section C.3 of the appendix) that any function on Ω may be written as a linear combination of the spatial modes, so that $\mathbf{v}$ may be written as $\mathbf{v}(\mathbf{x}, t) = \Sigma_{n=0}^{\infty} F_n(\mathbf{x}) \mathbf{G}_n(t)$. It follows after separation of variables and some linear algebra (Sections C.4 and C.5 of the appendix) that the general solution of Equation (7.2.3) may be written

$$\mathbf{v}(\mathbf{x}, t) = \Sigma_{n=0}^{\infty} \Sigma_{i=1}^{m} a_{ni} \mathbf{c}_{ni}(t) F_n(\mathbf{x}) \exp(\sigma_{ni} t), \qquad (7.2.5)$$

where the a_{ni} are arbitrary constants. Here the σ_{ni} are the eigenvalues of the matrix $A_n = \alpha J^* - \lambda_n D$, which we shall refer to as the *temporal eigenvalues* of the problem when we need to distinguish them from the spatial eigenvalues λ_n. If the eigenvalue σ_{ni} of A_n is simple (non-repeated), then $\mathbf{c}_{ni}$ is a constant rather than a function of t, and is the eigenvector of A_n corresponding to the eigenvalue σ_{ni}, but in general $\mathbf{c}_{ni}$ is a polynomial in t. The eigenvalues σ_{ni} of A_n satisfy

$$\det(\sigma_n I - A_n) = \det(\sigma_n I - \alpha J^* + \lambda_n D) = 0. \qquad (7.2.6)$$

These are mth order polynomials, so that there are m eigenvalues $\sigma_{n1}, \sigma_{n2}, \cdots,$ σ_{nm} for each n. If σ_{0i} has negative real part for all i, but σ_{ni} has positive real part for some $n \neq 0$ and some i, then we say that *Turing instability* occurs. The spatially homogeneous steady state is stable to spatially homogeneous perturbations but unstable to spatially inhomogeneous mode n perturbations. A randomly chosen perturbation from the steady state, such as might happen through stochastic effects in a real system, will almost certainly not be orthogonal to F_n, or equivalently will contain a component of mode n, so that we expect to see growth of the unstable mode breaking the symmetry of the spatially homogeneous solution. The solution will begin to look like $a_{ni}\mathbf{c}_{ni}F_n(\mathbf{x})\exp(\sigma_{ni}t)$. If there is more than one unstable value of n, the one with the highest growth rate $\text{Re}\,\sigma_{ni}$ will be expected to dominate, at least while the linear approximation holds.

Example 7.1

Consider Equation (7.2.2) on $\Omega = (0, \pi)$, with homogeneous Neumann boundary conditions. Denote the eigenvalues of $A_n = \alpha J^* - \lambda_n D$ by σ_{ni}, and let σ_{ni} have positive real part for some $n = p \neq 0$ and some $i = j$, and negative real part for all other n and i. Describe the behaviour of the system after a random perturbation from the steady state.

For $\Omega = (0, \pi)$ and homogeneous Neumann boundary conditions, $\lambda_n = n^2$ for $n = 0, 1, \cdots$, and $F_n(x) = \cos nx$. There will only be one growing mode, given by $F_p(x) = \cos px$, which is spatially inhomogeneous since p is positive. Multiples $\mathbf{c}_{pj}\cos px$ of this mode (where $\mathbf{c}_{pj}$ is the eigenvector corresponding to the simple eigenvalue σ_{pj} of the matrix A_p) will grow exponentially with exponent σ_{pj} until the nonlinear terms become important. (Bifurcation theory may be used to show that, under certain conditions, the solution of the full nonlinear time-dependent problem will tend to a steady state close to a multiple of $\mathbf{c}_{pj}\cos px$.)

Often our spatial modes F are sinusoidal, given by $\cos(\mathbf{k} \cdot \mathbf{x})$, $\sin(\mathbf{k} \cdot \mathbf{x})$ or linear combinations of these. These satisfy $-\nabla^2 F = k^2 F$, where $k = |\mathbf{k}|$, and so are spatial modes with spatial eigenvalue $\lambda = k^2$ if they also satisfy the boundary conditions. An important special case is the spatial mode corresponding to $\lambda = k^2 = 0$, the constant function. Rather than dealing with combinations of $\cos$ and $\sin$ all the time, it is easier to think of these spatial modes as being $\exp(i\mathbf{k} \cdot \mathbf{x})$ for $k \geq 0$, since we can obtain any linear combination of $\cos(\mathbf{k} \cdot \mathbf{x})$

and $\sin(\mathbf{k} \cdot \mathbf{x})$ by multiplying this by a complex constant and taking the real part. For such sinusoidal functions k is known as the *wave-number*, and the distance l between consecutive peaks is known as the *wave-length*, and given by $l = 2\pi/k$.

- If all the spatial modes in a problem are sinusoidal, we test stability of a system of equations to perturbations of wave-number k by substituting a vector multiple of $\exp(i\mathbf{k}{\cdot}\mathbf{x}+\sigma t)$ into the linearised equations and determining whether any resulting eigenvalue σ has positive real part.

- If the spatial modes are given by $F(\mathbf{x})$, we test stability similarly, but substituting a vector multiple of $F(\mathbf{x})\exp(\sigma t)$ into the linearised equations.

This determination of stability is usually the most important step in solving the questions that arise in this chapter.

EXERCISES

7.1. Consider the equation

$$\frac{\partial \tilde{u}}{\partial \tilde{t}} = f(\tilde{u}) + D\frac{\partial^2 \tilde{u}}{\partial \tilde{x}^2}$$

on $(0, L)$, with homogeneous Dirichlet boundary conditions $\tilde{u}(0,t) = \tilde{u}(L,t) = 0$.

a) Rescale the space and time variables to obtain an equation of the form

$$\frac{\partial u}{\partial t} = \gamma^2 f(u) + D\frac{\partial^2 u}{\partial x^2}$$

on $(0, \pi)$, with homogeneous Dirichlet boundary conditions.

b) Derive the linearised equation

$$\frac{\partial v}{\partial t} = \gamma^2 f'(u^*)v + D\frac{\partial^2 v}{\partial x^2}$$

on $(0, \pi)$, with homogeneous Dirichlet boundary conditions.

c) Solve the problem with initial condition $v(x,0) = v_0(x) = \sin x$.

d) Fix $f'(u^*) > 0$. Is increased diffusion stabilising or de-stabilising? Is increased domain size stabilising or de-stabilising?

7.2. Consider the system (7.2.2) in plane polar co-ordinates (R, ϕ), on the annulus $a < R < a + \delta$, $0 < \phi < 2\pi$. A system of this form has been proposed as a model for the formation of tentacles in hydra, a marine coelenterate, which has an annular cross-section. Assume for simplicity that δ is so small that any variations with R may be neglected.

Show that growth of a spatial mode with n peaks (corresponding to n tentacles) is possible if the equation

$$\det\left(\sigma_n I - \alpha J^* + \frac{n^2}{a^2}D\right) = 0$$

has a solution with positive real part, where J^* is defined in the usual way.

[In plane polar co-ordinates, $\nabla^2 u = \frac{1}{R}\frac{\partial}{\partial R}\left(R\frac{\partial u}{\partial R}\right) + \frac{1}{R^2}\frac{\partial^2 u}{\partial \phi^2}$.]

7.3 Turing Bifurcations

The last section made use of the ideas of linear stability theory. It showed that the essence of determining stability was to look at the eigenvalues σ of some matrices of the form $A = \alpha J^* - \lambda D$, where the λ are the eigenvalues of a spatial operator. The eigenvalues σ are temporal eigenvalues of the problem. For Turing instability, we need these eigenvalues to be stable for $\lambda = 0$, so that the steady state is stable to spatially uniform perturbations, but not for some $\lambda = \lambda_n > 0$, a spatial eigenvalue. In this section, we shall look at systems involving parameters, and consider the possibility that a loss of stability may result from a change in parameter values. We have seen that for difference and ordinary differential equations, such a loss of stability is often accompanied by a bifurcation to new steady state or other solutions (see Sections A.2 and B.4 of the appendix). This is also true for partial differential equations, and familiar saddle-node, transcritical, pitchfork and Hopf bifurcations may occur, but the theory will not be given in this book.

   Let us look a bit more closely at Equation (7.2.6), which tells us how the spatial and temporal eigenvalues are related. We shall first analyse the equation that would result if the spatial eigenvalues could take any non-negative value,

$$\det(\sigma I - A) = \det(\sigma I - \alpha J^* + \lambda D) = 0. \tag{7.3.7}$$

For each λ this is a polynomial of degree m, and has m solutions σ_i. Let

$$\rho(\lambda) = \max_{1 \le i \le m} \mathrm{Re}\,\sigma_i(\lambda), \tag{7.3.8}$$

the real part of the eigenvalue with greatest real part. A relationship between temporal and spatial eigenvalues like this one between ρ and λ is generally called a *dispersion relation*, from its original application to waves in fluids where waves of different wavelength travel at different speeds, and hence are dispersed before arriving at their destination.

Of course ρ will depend not only on λ but also on α and the parameters in D and J^*. Let us fix all these parameters except one, which we shall call the bifurcation parameter and denote by μ, and write $\rho(\lambda; \mu)$ for $\rho(\lambda)$. Let us assume that the curve $\rho(\lambda; \mu) = 0$ in (μ, λ)-space is as shown in Figure 7.2(a). Note that we have assumed that

$$\rho(0; \mu) < 0, \qquad\qquad (7.3.9)$$

so that the steady state is stable to spatially uniform perturbations, as required. Then as μ increases the dispersion relation passes through the configurations shown in Figure 7.2(b). For $\mu < \mu_c$ all spatial modes are stable, for $\mu = \mu_c$ a spatial mode with $\lambda = \lambda_c$ would be marginally stable, while for $\mu > \mu_c$ any spatial mode with eigenvalue satisfying $\underline{\lambda}(\mu) < \lambda < \overline{\lambda}(\mu)$ would be unstable. The critical values λ_c and μ_c are found by solving the equations

$$\rho(\lambda; \mu) = \frac{d\rho}{d\lambda}(\lambda; \mu) = 0. \qquad\qquad (7.3.10)$$

Turing instability is possible for $\mu > \mu_c$, but will only in fact occur if the interval of potential instability $(\underline{\lambda}(\mu), \overline{\lambda}(\mu))$ actually contains an eigenvalue λ_n.

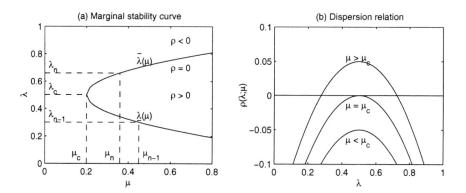

Figure 7.2   (a) The curve $\rho(\lambda; \mu) = 0$ in (μ, λ)-space, and (b) plots of the dispersion relation $\rho(\lambda; \mu)$ against λ for various values of μ, if potential Turing instability first occurs at $\mu = \mu_c$.

Now let us take into account the fact that λ must be a spatial eigenvalue. Referring to Figure 7.2, we can see that as μ passes through μ_n the nth spatial mode loses stability, and it may be shown that this is accompanied by a bifurcation. We say that a *mode-n Turing bifurcation* on the finite domain takes place at $\mu = \mu_n$, with spatial eigenvalue $\lambda = \lambda_n$. It is a *zero-eigenvalue bifurcation* of transcritical or pitchfork type if the eigenvalue with greatest real part is real (and so passes through zero when $\mu = \mu_n$), and *Hopf* if it is (they are) complex (so that a *pair* of complex conjugate temporal eigenvalues crosses the real axis when $\mu = \mu_n$.) At the bifurcation point we have

$$\rho(\lambda_n; \mu_n) = 0, \qquad (7.3.11)$$

an equation for μ_n.

If the domain is the whole of $\mathbb{R}^N$ then any non-negative λ is an eigenvalue, and Turing instability occurs for any $\mu > \mu_c$. We say that a (zero-eigenvalue or Hopf) *Turing bifurcation* occurs at $\mu = \mu_c$, with spatial eigenvalue $\lambda = \lambda_c$. Turing instability calculations are much easier in this case, although the bifurcation theory is not.

The ideas above are easily generalised to more than one bifurcation parameter. For example, if there are two bifurcation parameters μ_1 and μ_2, and there is a zero-eigenvalue Turing bifurcation of the nth mode on a finite domain, then

$$\rho(\lambda_n; \boldsymbol{\mu}) = 0 \qquad (7.3.12)$$

at $\boldsymbol{\mu} = \boldsymbol{\mu}_n$, where we have written $\boldsymbol{\mu} = (\mu_1, \mu_2)$. This defines a *curve of marginal stability* of the nth mode in (μ_1, μ_2)-space.

It is often useful to think of *zero-eigenvalue* Turing bifurcations in a rather different way. Equation (7.3.7) may be written

$$\sigma^m + a_1(\lambda; \mu)\sigma^{m-1} + \cdots + a_{m-1}(\lambda; \mu)\sigma + a_m(\lambda; \mu) = 0, \qquad (7.3.13)$$

where each a_i is an ith order polynomial in λ, and we have again distinguished a bifurcation parameter μ. Let the system enter the region of potential Turing instability as μ increases past μ_c, with critical spatial eigenvalue λ_c. Because Equation (7.3.13) has a root $\sigma = 0$ for $\mu > \mu_c$ but not for $\mu < \mu_c$, the mth order polynomial $a_m(\lambda; \mu)$ must satisfy

$$a_m(\lambda; \mu) = \frac{da_m}{d\lambda}(\lambda; \mu) = 0 \qquad (7.3.14)$$

at $(\lambda_c; \mu_c)$. Similarly, if there is a zero-eigenvalue Turing bifurcation of the nth mode on a finite domain, then

$$a_m(\lambda_n; \mu) = 0 \qquad (7.3.15)$$

at μ_n. These equations are often much easier to derive and analyse than the corresponding equations for ρ.

Bifurcation theory fleshes out the ideas of this section, and takes account of the nonlinear terms in the equations. It is much more difficult when the domain is infinite, and we shall only refer to bifurcation results in a finite domain.

Near a zero-eigenvalue Turing bifurcation with spatial eigenvalue λ_n the linear theory tells us to expect spatially inhomogeneous solutions which initially grow at a rate $\exp(\rho t)$, and which look at any point in time like a vector multiple of the spatial mode $F_n(\mathbf{x})$ corresponding to λ_n. Bifurcation theory tells us that under certain conditions the solution of the fully nonlinear problem here tends to leading order to a steady vector multiple of $F_n(\mathbf{x})$. Near a Hopf–Turing bifurcation the linear theory tells us to expect spatially and temporally inhomogeneous solutions whose time variation is like $\exp((\rho + i\omega_n)t)$, where ω_n is the imaginary part of either of the eigenvalues with greatest real part at the bifurcation point, and which look at any point in time like a vector multiple of $F_n(\mathbf{x})$. Bifurcation theory tells us that under certain conditions the solution of the fully nonlinear problem here tends to leading order to a vector multiple of $\exp(i\omega_n t)F_n(\mathbf{x})$, a periodic function of time.

7.4 Activator-inhibitor Systems

Turing instability cannot occur with a single equation, and so we shall investigate the simplest case of two reaction-diffusion equations. We shall see that Hopf–Turing bifurcations still cannot occur, so all the bifurcations in this section will be zero-eigenvalue. It will turn out that one of the reactants involved must be a short-range activator and the other a long-range inhibitor, in a sense that will be made precise later.

In the case $m = 2$, Equation (7.3.13) for σ becomes

$$Q(\sigma) = \sigma^2 + a_1\sigma + a_2 = 0. \tag{7.4.16}$$

The stability of the roots of this equation is discussed in Section B.2.2 of the appendix. From this, or from the fact that

$$\rho = \max\left\{\operatorname{Re}\sigma_\pm\right\} = \max\operatorname{Re}\left(\frac{-a_1 \pm \sqrt{a_1^2 - 4a_2}}{2}\right) \tag{7.4.17}$$

we can see that $\rho < 0$ if and only if both $a_1 > 0$ and $a_2 > 0$, and that a zero-eigenvalue bifurcation resulting in a change in stability can occur when $a_2 = 0$ and $a_1 > 0$, and a Hopf bifurcation when $a_1 = 0$ and $a_2 > 0$.

7.4.1 Conditions for Turing Instability

In this subsection we shall derive the conditions that must be satisfied for Turing instability to occur in a system of two reaction-diffusion equations.

Thus, we shall seek conditions that the spatially homogeneous steady state (u^*, v^*) is stable to spatially homogeneous but not to spatially inhomogeneous perturbations.

The system becomes

$$u_t = \alpha f(u,v) + D_1 \nabla^2 u, \quad v_t = \alpha g(u,v) + D_2 \nabla^2 v. \tag{7.4.18}$$

Defining $(\hat{u}, \hat{v}) = (u,v) - (u^*, v^*)$ and linearising,

$$\hat{u}_t = \alpha f_u^* \hat{u} + \alpha f_v^* \hat{v} + D_1 \nabla^2 \hat{u}, \quad \hat{v}_t = \alpha g_u^* \hat{u} + \alpha g_v^* \hat{v} + D_2 \nabla^2 \hat{v}, \tag{7.4.19}$$

where we have used $(\hat{u}, \hat{v})$ both for the perturbation from the steady state and for its linearised approximation, and asterisks denote evaluation at the steady state. The algebraic Equation (7.3.7) becomes

$$\begin{vmatrix} \sigma - \alpha f_u^* + \lambda D_1 & -\alpha f_v^* \\ -\alpha g_u^* & \sigma - \alpha g_v^* + \lambda D_2 \end{vmatrix} = 0,$$

or

$$Q(\sigma) = \sigma^2 + a_1(\lambda)\sigma + a_2(\lambda) = 0, \tag{7.4.20}$$

where

$$a_1(\lambda) = (D_1 + D_2)\lambda - \alpha(f_u^* + g_v^*), \tag{7.4.21}$$

$$a_2(\lambda) = D_1 D_2 \lambda^2 - \alpha(D_1 g_v^* + D_2 f_u^*)\lambda + \alpha^2(f_u^* g_v^* - f_v^* g_u^*). \tag{7.4.22}$$

Of course, a_1 and a_2 are the trace and determinant respectively of the matrix $A = \alpha J^* - \lambda D$. Any parameter(s) may be taken as the bifurcation parameter(s) here, and different applications may require different choices. However, to fix ideas we shall be explicit and choose both D_1 and D_2 as bifurcation parameters.

Both roots of equation (7.4.20) are stable if and only if both a_1 and a_2 are positive, i.e.

$$a_1(\lambda) = (D_1 + D_2)\lambda - \alpha(f_u^* + g_v^*) > 0 \tag{7.4.23}$$

and

$$a_2(\lambda) = D_1 D_2 \lambda^2 - \alpha(D_2 f_u^* + D_1 g_v^*)\lambda + \alpha^2(f_u^* g_v^* - f_v^* g_u^*) > 0. \tag{7.4.24}$$

The steady state is stable to spatially uniform perturbations if these inequalities are satisfied for $\lambda = 0$, so that $a_1(0) > 0$ and $a_2(0) > 0$, or

$$f_u^* + g_v^* = \operatorname{tr} J^* < 0, \tag{7.4.25}$$

and

$$f_u^* g_v^* - f_v^* g_u^* = \det J^* > 0. \tag{7.4.26}$$

Hence Equation (7.4.23) holds, i.e. $a_1(\lambda) > 0$ for all (positive) λ, and it follows that a Hopf bifurcation is impossible.

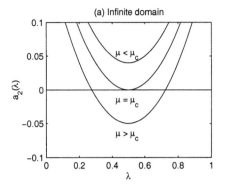
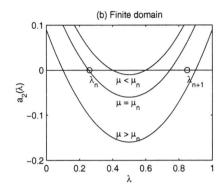

Figure 7.3   Possible configurations for $a_2(\lambda)$ when condition (7.4.26) holds. In (a), we consider potential Turing instability (or actual Turing instability in an infinite domain where all positive values of λ are eigenvalues). For the top curve, no instability can occur for any value of λ; for the middle curve, a single point of potential marginal Turing instability appears; for the bottom curve, instability occurs for any λ lying between the roots of $a_2(\lambda) = 0$. In (b), the domain is taken to be finite, and there is a countable set of eigenvalues λ_n. For the top curve, despite the fact that $a_2(\lambda) < 0$ for a range of λ, the spatially uniform steady state is not unstable because there is no eigenvalue within that range; for the middle curve, there is a point of marginal stability of the nth mode; for the bottom curve, the nth mode is unstable, (as is the $(n + 1)$th).

With the condition (7.4.26), we can sketch the possible configurations for $a_2(\lambda)$, in Figure 7.3.

From Equation (7.3.14), or from the middle curve in Figure 7.3(a), the conditions $a_2(\lambda) = a_2'(\lambda) = 0$ must be satisfied at a zero-eigenvalue bifurcation point (λ_c, μ_c) to potential Turing instability. From the second of these,

$$\lambda_c = \frac{1}{2}\alpha\left(\frac{f_u^*}{D_1} + \frac{g_v^*}{D_2}\right),\tag{7.4.27}$$

and substituting into the first,

$$(D_2 f_u^* + D_1 g_v^*)^2 = 4D_1 D_2(f_u^* g_v^* - f_v^* g_u^*).\tag{7.4.28}$$

This curve is the envelope of the curves $a_2(\lambda) = 0$ for all positive λ. Since λ_c must be positive, we must also have

$$D_2 f_u^* + D_1 g_v^* > 0.\tag{7.4.29}$$

The part of the curve (7.4.28) where the inequality (7.4.29) holds separates regions in (D_1, D_2)-space, the space of bifurcation parameters, where potential

Turing instability occurs from regions where it does not, and is called the curve of potential marginal Turing instability. If $\Omega = \mathbb{R}^N$, so that any positive λ is an eigenvalue, then any point of potential instability is in fact a point of instability, and the curve is called the curve of marginal Turing stability or Turing bifurcation curve. The unstable side of the curve is where a_2 has two real distinct positive roots, so that

$$(D_2 f_u^* + D_1 g_v^*)^2 > 4D_1 D_2 (f_u^* g_v^* - f_v^* g_u^*). \tag{7.4.30}$$

If we denote the two positive roots by $\underline{\lambda}$ and $\overline{\lambda}$, then the spatial mode with eigenvalue λ is unstable if $\underline{\lambda} < \lambda < \overline{\lambda}$. The conditions for potential Turing instability are therefore inequalities (7.4.25), (7.4.26), (7.4.29), and (7.4.30).

Now let us take account of the fact that λ must be an eigenvalue. From equation (7.3.15), or from the middle curve in Figure 7.3(b), the condition $a_2(\lambda) = 0$ must be satisfied at a zero-eigenvalue bifurcation point (λ_n, μ_n) to mode-n Turing instability. Hence

$$D_1 D_2 \lambda_n^2 - \alpha(D_2 f_u^* + D_1 g_v^*)\lambda_n + \alpha^2(f_u^* g_v^* - f_v^* g_u^*) = 0. \tag{7.4.31}$$

This curve separates regions in parameter space where the steady state is stable to mode n perturbations from regions where it is not, and is called the curve of mode-n marginal Turing instability, or the mode-n Turing bifurcation curve. The unstable side of the curve is where $a_2(\lambda_n) < 0$, so that

$$D_1 D_2 \lambda_n^2 - \alpha(D_2 f_u^* + D_1 g_v^*)\lambda_n + \alpha^2(f_u^* g_v^* - f_v^* g_u^*) < 0. \tag{7.4.32}$$

Alternatively, the nth mode is unstable whenever

$$\underline{\lambda} < \lambda_n < \overline{\lambda}, \tag{7.4.33}$$

where $\underline{\lambda}$ and $\overline{\lambda}$ are the positive roots of $a_2(\lambda) = 0$.

EXERCISES

7.3. We have shown that if certain conditions are satisfied then Turing instability occurs, but we have not checked that it is possible to choose parameters so that the conditions are satisfied. Here we show by an example that it is possible, so that diffusion can indeed de-stabilise a steady state. Consider the linear partial differential equation

$$\mathbf{v}_t = J\mathbf{v} + D\mathbf{v}_{xx}$$

where

$$J = \begin{pmatrix} 3 & 13 \\ -1 & -3 \end{pmatrix}, \quad D = \begin{pmatrix} 1 & 0 \\ 0 & 9 \end{pmatrix}.$$

a) Show that the trivial steady state solution of $\mathbf{u}_t = J\mathbf{u}$ is stable.

b) Consider the partial differential equation on the domain $0 < x < \pi$, with zero Neumann boundary conditions. Are there any unstable spatial modes for this problem? If so, which?

7.4. Consider the curve of potential marginal Turing instability defined by

$$(D_2 f_u^* + D_1 g_v^*)^2 = 4 D_1 D_2 (f_u^* g_v^* - f_v^* g_u^*) \qquad (7.4.34)$$

in (D_1, D_2)-space.

a) Show that the curve is a conic section of the form $A D_1^2 + B D_1 D_2 + C D_2^2 = 0$, with $A > 0$, $B < 0$, $C > 0$ and $B^2 > 4AC$.

b) Deduce that it determines two straight lines with positive slope through the origin, $D_2 = \underline{\beta} D_1$ and $D_2 = \overline{\beta} D_1$, say, where $0 < \underline{\beta} < \overline{\beta}$.

c) Show that the curve $a_2(\lambda_n) = 0$ given by

$$a_2(\lambda_n) = D_1 D_2 \lambda_n^2 - \alpha(D_1 g_v^* + D_2 f_u^*)\lambda_n + \alpha^2 \det J^*$$
$$= (D_1 \lambda_n - \alpha f_u^*)(D_2 \lambda_n - \alpha g_v^*) - \alpha^2 f_v^* g_u^* = 0$$

is a hyperbola which asymptotes to $D_1 = \alpha f_u^*/\lambda_n$ as $D_2 \to \infty$ and to $D_2 = \alpha g_v^*/\lambda_n$ as $D_1 \to \infty$.

d) Show that the hyperbola touches $D_2 = \underline{\beta} D_1$ (in the negative quadrant) and $D_2 = \overline{\beta} D_1$ (in the positive quadrant).

e) The (D_1, D_2)-plane is therefore as in Figure 7.4. If you were at the point marked with an asterisk in this diagram, what behaviour would you expect the reaction-diffusion system to exhibit?

7.5. The conditions $a_2 = a_2' = 0$ at the (potential) Turing bifurcation point imply that a_2 is a perfect square there (as a function of λ).

a) Deduce that

$$\lambda_c = \alpha \sqrt{\frac{\det J^*}{D_1 D_2}}. \qquad (7.4.35)$$

b) Using Equation (7.4.28), show that this is consistent with equation (7.4.27).

7.6. A simplified version of the Gierer–Meinhardt reaction-diffusion system is given by

$$u_t = \frac{u^2}{v} - bu + \nabla^2 u, \quad v_t = u^2 - v + d\nabla^2 v,$$

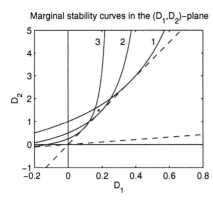

Marginal stability curves in the (D_1, D_2)-plane

Figure 7.4   Marginal Turing stability in (D_1, D_2)-space. The dotted lines are the lines $D_2 = \overline{\beta} D_1$ and $D_2 = \underline{\beta} D_1$, and the biologically realistic region of potential Turing instability is above $D_2 = \overline{\beta} D_1$ in the positive quadrant. Curves of marginal stability for modes 1, 2 and 3 are shown.

where b and d are positive constants.

a) Find the Jacobian matrix at the positive spatially uniform steady state.

b) Find the conditions for stability of the positive spatially uniform steady state solution to spatially uniform perturbations.

c) Find the curve of marginal Turing stability of the positive spatially uniform steady state for the equations with diffusion on the whole of $\mathbb{R}^N$.

d) Hence show that the parameter domain for Turing instability of the positive spatially uniform steady state for the equations with diffusion on the whole of $\mathbb{R}^N$ is given by $0 < b < 1$, $bd > 3 + 2\sqrt{2}$, and sketch the region of (b, d) parameter space in which such instability may occur.

e) Show further that on the curve of marginal stability the critical eigenvalue λ_c is given by $\lambda_c = (1 + \sqrt{2})/d$.

7.4.2 Short-range Activation, Long-range Inhibition

In this subsection we investigate the consequences of the conditions (7.4.25), (7.4.26), (7.4.29) and (7.4.30) for a system of two reaction-diffusion systems to exhibit Turing instability. It will become clear that the sign pattern of the Jacobian matrix must be of one of two kinds, and the sign pattern can be interpreted in terms of which chemical is an activator and which an inhibitor. Both sign patterns may be interpreted as activator-inhibitor systems, one using cross-activation and one pure activation.

The reactant u is a *self-activator/self-inhibitor* if an increase in u results in an increase/decrease in u_t according to the kinetic equations. This occurs if f_u is positive/negative. It *activates/inhibits* v if an increase in u results in an increase/decrease in v_t according to the kinetic equations. This occurs if g_u is positive/negative. u is an *activator/inhibitor* if it activates/inhibits both itself and v. Self-activation, otherwise known as *auto-catalysis*, results in positive feedback.

From Equations (7.4.25) and (7.4.29), $f_u^* + g_v^* < 0 < D_2 f_u^* + D_1 g_v^*$. These imply

– that the diffusion coefficients D_1 and D_2 cannot be equal, and

– that f_u^* and g_v^* must be of opposite sign.

Let us take $f_u^* > 0$, $g_v^* < 0$ without loss of generality. Then inequality (7.4.26) implies that $f_v^* g_u^* < f_u^* g_v^* < 0$, so that f_v^* and g_u^* are also of opposite sign. The possible sign patterns for the Jacobian matrix at the steady state are

$$\begin{pmatrix} + & - \\ + & - \end{pmatrix} \quad \text{or} \quad \begin{pmatrix} + & + \\ - & - \end{pmatrix}.$$

The first represents an *activator-inhibitor* system, or *pure* activator-inhibitor system, with u the activator and v the inhibitor.

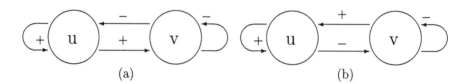

(a)   (b)

Figure 7.5   Diagrammatic representation of (a) pure activation-inhibition and (b) cross-activation-inhibition.

The second is sometimes called a *cross-activator-inhibitor* system. If we define $w = -v$ in this case, the linearised equations become

$$u_t = \alpha f_u^* u - \alpha f_v^* w + D_1 \nabla^2 u, \quad w_t = -\alpha g_u^* u + \alpha g_v^* w + D_2 \nabla^2 w,$$

with Jacobian sign pattern

$$\begin{pmatrix} + & - \\ + & - \end{pmatrix},$$

a pure activator-inhibitor system with u the activator and w the inhibitor. Hence any system with Turing instability may be thought of as a pure activator-inhibitor system.

The constants f_u^* and g_v^* are the self-activation or auto-catalysis rates of u and v at the steady state. (In fact g_v^* is negative so this is self-inhibition, at rate $|g_v^*|$.) These represent the exponential rates of growth of u if $v = 0$ and of v if $u = 0$ in the absence of diffusion. Since the self-inhibition rate $|g_v^*|$ of v is greater than the self-activation rate f_u^* of u, ($|g_v^*| = -g_v^* > f_u^*$), how does the activator u manage to produce the instability?

If activation acts over a short range, and inhibition over a long range, it is possible that even if inhibition is globally stronger than activation, activation may be locally stronger (more intense) than inhibition. This may allow activation to overcome inhibition locally, and instigate a pattern-forming process.

Let us define $r_1 = \sqrt{2D_1/(\alpha f_u^*)}$ to be the *range* of u. It has dimensions of length, and represents a typical (root-mean-square) distance travelled by a molecule of u in the time it would take for the self-activation to multiply the concentration of u by a factor e. Similarly, we define $r_2 = \sqrt{2D_2/(\alpha|g_v^*|)}$ to be the range of v, a typical distance travelled by a molecule of v in the time it would take for the self-inhibition to multiply the concentration of v by a factor $1/e$. Using inequality (7.4.29) we obtain $r_1 < r_2$, so that the range of activation is less than the range of inhibition. We have *short-range activation, long-range inhibition*, and the activation is more focused than the inhibition. The process of pattern formation is as follows.

– Activator u increases locally through stochasticity.

– Consequently, activator u and inhibitor v both increase locally.

– Inhibitor diffuses further than activator.

– Inhibitor therefore (a) loses control of peak and (b) initiates trough at a distance.

What do we expect the wave-length of the pattern to be? From Equation (7.4.27),

$$\lambda_c = \frac{1}{r_1^2} - \frac{1}{r_2^2},$$

(emphasising that we must have $r_2 > r_1$, the range of the inhibitor greater than the range of the activator, for Turing instability). In $\mathbb{R}^N$ we expect to see this mode after the bifurcation has taken place; in a finite domain we would expect a mode with λ_n close to λ_c. The critical wave-number is $k_c = \sqrt{\lambda_c}$, and critical wave-length $l_c = 2\pi/k_c$, given by

$$l_c = 2\pi r_1 \left(1 - \frac{r_1^2}{r_2^2}\right)^{-\frac{1}{2}}.$$

If $r_2 \gg r_1$, then this is given approximately by $l_c \approx 2\pi r_1$, dominated by the (short) range r_1 of the activator.

EXERCISES

7.7. Some prey-predator systems, notably in plankton, exhibit patchy spatial distributions of the species. The Mimura–Murray system is a model for such ecological patchiness. It is given by

$$u_t = u(f(u) - v) + D_1 \nabla^2 u, \quad v_t = v(u - g(v)) + D_2 \nabla^2 v,$$

where g is a positive increasing function, while f increases from $f(0) > 0$ and then decreases. (In the terminology of Chapter 2, the prey u exhibits non-critical depensation.)

a) Sketch a possible phase plane for the system in the absence of diffusion.

b) Show that the system is a cross-activator-inhibitor system if the nullcline $u = g(v)$ cuts the nullcline $v = f(u)$ to the left of the hump.

7.8. The Gierer–Meinhardt reaction-diffusion system is given by

$$u_t = a - bu + \frac{u^c}{v} + D_1 \nabla^2 u, \quad v_t = u^c - v + D_2 \nabla^2 v,$$

where $a \geq 0$ and b, c, D_1 and D_2 are positive constants. Find conditions under which this is a pure activator-inhibitor system near the non-trivial spatially uniform steady state.

7.9. The Schnakenberg reaction-diffusion system is given by

$$u_t = a - u + u^2 v + D_1 \nabla^2 u, \quad v_t = b - u^2 v + D_2 \nabla^2 v.$$

a) Find the non-trivial spatially uniform steady state solution (u^*, v^*) of this system.

b) Show that it is a cross-activator-inhibitor system near (u^*, v^*) under certain conditions on the parameters.

c) Find the ranges r_1 and r_2 of u and v.

d) If $r_1 \ll r_2$, give an approximation to the critical wave-length of the pattern formed.

7.4.3 Do Activator-inhibitor Systems Explain Biological Pattern Formation?

We have looked in some detail at a mechanism that could potentially produce spatial patterns in chemical concentrations, but too many processes that are not merely chemical have been neglected for this to be a good explanation of biological pattern formation, let alone morphogenesis, in the vast majority of cases. We shall discuss the effects of growth, population (including cell) movement, and mechanical forces in later sections. Even from the purely chemical point of view there are shortcomings, and it was nearly forty years after the publication of Turing's paper before the existence of Turing patterns was finally demonstrated experimentally. The main stumbling block is the requirement that both $f_u^* + g_v^* < 0$ (7.4.25) and $D_2 f_u^* + D_1 g_v^* > 0$ (7.4.29) must be satisfied. As we have shown, the diffusion coefficients of the activator and inhibitor must be different, with $D_2 > D_1$ (short-range activation and long-range inhibition), but it is worse than this; unless D_1 and D_2 are very different the mechanism cannot be robust. Both $f_u^* + g_v^*$ and $D_2 f_u^* + D_1 g_v^*$ are close to zero, which implies that both the stability to spatially uniform perturbations and the instability to spatially non-uniform perturbations are easily disrupted. The behaviour is unlikely to be observed experimentally, and even less likely to be the basis for biological pattern formation, where robustness is a necessity. Diffusion coefficients of molecules dissolved in water tend not to be very different; Einstein showed that they are proportional to the square root of molecular size, so even with small molecules compared to large ones, it is difficult to obtain the required difference. However, real living systems do not typically consist of molecules dissolved in water. The activator could bind reversibly to a stationary or nearly stationary third element, such as a gel-like substance, and its diffusion would then be substantially slowed down. In the first experiments to exhibit Turing patterns, the chlorite-iodide-malonic-acid (CIMA) reaction, the activator iodide binds to an immobile starch indicator, and this has been suggested as the basis for the pattern formation. The analysis of such activator-inhibitor-immobiliser or $(2 + 1)$ systems has shown that they are much more likely to exhibit Turing instability than the corresponding system without the immobiliser. They also admit the possibility of Hopf–Turing bifurcations. The immobiliser often stabilises the spatially uniform steady state rather than shortening the range of the activator, but whatever the mechanism such additional reactants are likely to be important in real systems.

7.5 Bifurcations with Domain Size

One of the features of most biological systems where patterns are formed is that they grow. In this section we shall look at the effects of domains of different sizes. This seems to be crucial for robust generation of pattern in many cases.

By sketching marginal stability curves in (D_1, D_2)-space we have essentially been thinking of D_1 and D_2 as bifurcation parameters. We now wish to think of α, representing domain size, as the bifurcation parameter. The first thing to notice is that the curve (7.4.28) of potential marginal Turing stability is independent of α, so that changing α cannot move us across this curve. Let us fix D_1 and D_2 somewhere in the potentially unstable region, so that inequalities (7.4.29) and (7.4.30) hold, and allow α to vary.

Does size matter? The only thing it can affect is which modes are unstable. We know that mode n is unstable if the inequality (7.4.33) holds. It is easy to see from Equation (7.4.22) that the real positive roots $\underline{\lambda}$ and $\overline{\lambda}$ of $a_2(\lambda) = 0$ may be written as $\underline{\lambda} = \alpha\underline{\kappa}$ and $\overline{\lambda} = \alpha\overline{\kappa}$, where $\underline{\kappa}$ and $\overline{\kappa}$ are constants (independent of α). Hence if

- $\alpha\underline{\kappa} < \lambda < \alpha\overline{\kappa}$, and if

- λ is an eigenvalue, $\lambda = \lambda_n$ for some n,

then we have instability of the nth mode. The first of these conditions says that the point (α, λ) must lie between two straight lines in the (α, λ)-plane, shown in Figure 7.6(a), and the second that it must also lie on one of the lines $\lambda = \lambda_n$. The α-interval for which each mode is unstable may then be seen, and mode n is unstable for

$$\frac{\lambda_n}{\overline{\kappa}} < \alpha < \frac{\lambda_n}{\underline{\kappa}}.$$

There is no Turing instability for sufficiently small domains, $\alpha < \lambda_1/\overline{\kappa}$. As the size of the domain increases, α increases, and each mode in turn becomes unstable and then re-stabilises. Eventually, the unstable α-intervals overlap, and for α sufficiently large the trivial steady state never re-stabilises. The bifurcation diagram is sketched in Figure 7.6(b). Any point in the region of potential Turing instability exhibits Turing instability for some α, or some domain size. Spatial modes F_n with higher and higher n become unstable, and then re-stabilise. These tend to be more and more spatially complex as n increases, so that we get more and more peaks in the solution as α increases.

As an example of the effect of domain size and shape, we shall ask whether a single reaction-diffusion mechanism could be responsible for generating practically all the common mammalian coat patterns observed. We shall make predictions using this hypothesis, first drawing some general conclusions about the kind of patterns to be expected from such systems, and then giving the

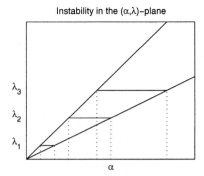

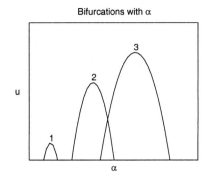

Figure 7.6   (a) A diagram of instability regions for the first three modes in the (α, λ)-plane; α represents domain size. Modes 1, 2 and 3, with eigenvalues λ_1, λ_2 and λ_3, are unstable for the values of α between the appropriate dotted lines. (b) Bifurcation diagram with bifurcation parameter α, corresponding to the instabilities in (a). The numbers denote the unstable modes, and the mode 2 and mode 3 unstable intervals overlap.

results of some numerical computations on a particular system. In particular, we shall make predictions about tail patterns, by first looking at solutions of reaction-diffusion equations on the surface of a cylinder. Finally we shall test them against real coat patterns.

Development of colour pattern on the coat of mammals occurs towards the end of embryogenesis, but may reflect a pre-pattern laid down much earlier. The steps are as follows; melanocytes are pigment cells, and melanoblasts their precursors.

– Genetic determination of melanoblasts.

– Migration of melanoblasts.

– Melanoblasts become melanocytes.

– Melanocytes respond to their chemical environment, which decides whether they are to produce melanin or not.

We shall model the chemical environment of the melanocytes.

We have no data on what this mechanism might be, but suggest that a Turing instability is the basis of its pattern-forming ability. We shall take as a model an activator-inhibitor system of two reaction-diffusion equations, given by Equations (7.4.18). We shall assume that the chemicals diffuse in a thin surface layer of the embryo, so that the equations are to be solved on the surface of a region in space.

Writing $\mathbf{v}$ as the linearised approximation to $\hat{\mathbf{u}} = (\hat{u}, \hat{v})^T$, the linearised system may be written

$$\mathbf{v}_t = \alpha J^* \mathbf{v} + D\nabla^2 \mathbf{v}. \tag{7.5.36}$$

The correct domain to use to consider tail patterns can be approximated by the surface of a cone, but to get a general idea of the processes involved we shall look at the surface of a cylinder, radius a. We shall therefore use cylindrical polars R, ϕ, z, with $R = a$, $0 < \phi < 2\pi$, $0 < z < h$. The boundary conditions are

$$\mathbf{v}_z(\phi, 0, t) = \mathbf{v}_z(\phi, h, t) = 0, \tag{7.5.37}$$

zero flux conditions, and

$$\mathbf{v}(0, z, t) = \mathbf{v}(2\pi, z, t), \quad \mathbf{v}_\phi(0, z, t) = \mathbf{v}_\phi(2\pi, z, t), \tag{7.5.38}$$

periodic boundary conditions, ensuring that $\mathbf{v}$ and the flux of $\mathbf{v}$ are continuous on the line $\phi = 0$ or 2π.

The temporal eigenvalues are given by Equation (7.3.7), $\det(\sigma - \alpha J^* + \lambda D) = 0$, where λ is an eigenvalue of the spatial problem. This spatial eigenvalue problem is $-\nabla^2 F = \lambda F$, with $F_z(\phi, 0) = F_z(\phi, h) = 0$, $F(0, z) = F(2\pi, z)$, and $F_\phi(0, z) = F_\phi(2\pi, z)$, and is considered in Section C.6 of the appendix. It has the solution

$$\lambda_{mn} = -\frac{1}{a^2}\frac{P_n''(\phi)}{P_n(\phi)} - \frac{Q_m''(z)}{Q_m(z)} = \frac{n^2}{a^2} + \frac{m^2\pi^2}{h^2},$$

m and n non-negative integers, with eigenfunctions $F_{mn}(\phi, z) = P_n(\phi)Q_m(z) = \cos(m\pi z/h)\exp(in\phi)$ for each (m, n).

The temporal eigenvalues for the (m, n)th mode are given by

$$\det(\sigma_{mn}I - \alpha J^* + \lambda_{mn}D) = 0.$$

Familiar arguments, as in Section 7.4.1, imply that this can only be unstable if

$$a_2(\lambda) = D_1 D_2 \lambda^2 - \alpha(D_2 f_u^* + D_1 g_v^*)\lambda + \alpha^2 \det J^* < 0$$

for some λ_{mn}, i.e. if $\alpha\underline{\kappa} < \lambda_{mn} < \alpha\overline{\kappa}$, where $\alpha\underline{\kappa}$ and $\alpha\overline{\kappa}$ are the roots of $a_2(\lambda) = 0$, as in Section 7.5. The condition for instability of the (m, n)th mode is thus

$$\alpha\underline{\kappa} < \frac{n^2}{a^2} + \frac{m^2\pi^2}{h^2} < \alpha\overline{\kappa}.$$

Now recall from Section 7.2 that $\alpha = \gamma^2$, where γ represents a scaling of the linear dimensions of the domain. It follows that if the cylinder is thin, $a\gamma$ is small. If $a^2\gamma^2 < \frac{1}{\overline{\kappa}}$, the first circumferential mode ($n = 1$) and all others with $n > 1$ are outside the unstable range. The only possible unstable modes are $(m, 0)$th modes, where $F(\phi, z) = \cos(m\pi z/h)$, depending on z but not on ϕ; these are therefore stripes. Predictions are as follows.

- Tails whose patterns are laid down early in embryogenesis, or which are thin, cannot exhibit spots but can exhibit stripes.

- Tails whose patterns are laid down later, and which are thick, can exhibit spots but are likely to have stripes near the end, where they are thin.

- If we think of the body of an animal as approximating a larger cylinder and the tail a smaller one at the time patterns are laid down, we conclude that spotted animals can have striped tails, but not vice versa.

The pictures below bear out the predictions well. We should be wary of concluding that the mechanism involved is therefore indeed a combination of reaction and diffusion, as other models could very well give similar predictions. However, the model has passed this test, and we can be more confident that we are on the right track.

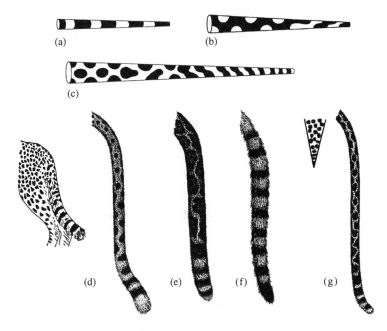

Figure 7.7 (a)–(c) Numerical solutions of an activator-inhibitor reaction-diffusion system on a tapering cylinder. Note the transition from spots to stripes in (c). Tail patterns in the (d) adult cheetah (*Acinonyx jubatis*), (e) adult jaguar (*Panthera onca*), (f) pre-natal genet (*Genetta genetta*) and (g) adult leopard (*Panthera pardus*). From Murray (*Mathematical Biology*, Springer, 2003).

EXERCISES

7.10. Consider the general system

$$u_t = \alpha f(u,v) + D_1 \nabla^2 u, \quad v_t = \alpha g(u,v) + D_2 \nabla^2 v,$$

in a domain $\Omega \subset \mathbb{R}^N$ with zero-flux boundary conditions. Let the Jacobian matrix J^* of the linearisation of the system about the spatially uniform steady state (u^*, v^*) be given by

$$J^* = \begin{pmatrix} 2 & -4 \\ 4 & -6 \end{pmatrix},$$

let $D_1 = 1$, $D_2 = 10$, and let $\Omega = (0, \pi)$.

a) Derive the condition for instability of the nth mode.

b) Sketch the bifurcation diagram with bifurcation parameter α, showing bifurcation points up to $\alpha = 10$.

7.11. We have seen that domain geometry as well as domain size is important in determining the bifurcations that occur. Consider the general system

$$u_t = f(u,v) + D_1 \nabla^2 u, \quad v_t = g(u,v) + D_2 \nabla^2 v,$$

in the rectangle $\{(x,y)|0 < x < a, 0 < y < b\}$, with zero-flux boundary conditions.

a) The spatial modes for this problem are given by $F_{mn}(x,y) = \cos(m\pi x/a)\cos(n\pi y/b)$. Show that the corresponding spatial eigenvalues are given by

$$\lambda_{mn} = \frac{m^2 \pi^2}{a^2} + \frac{n^2 \pi^2}{b^2}.$$

b) Fix D_1 and D_2 so that we are at a point of potential marginal Turing instability in (D_1, D_2)-space, with critical spatial eigenvalue $\lambda_c = k\pi^2/a^2$, where k is not a perfect square. Fix a and let $b = \gamma a$, and consider γ as a bifurcation parameter. Show that marginal stability of the (m,n)th mode occurs when

$$\gamma^2 = \gamma_{mn}^2 = \frac{n^2}{k - m^2}.$$

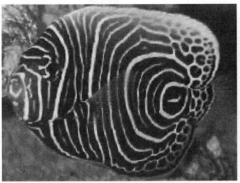 

Figure 7.8   Some examples of biological patterns where movement seems to be necessary. (a) Angel fish and (b) patterns in bacteria, from E. Ben-Jacob, O. Schochet, I. Cohen, A. Tenenbaum, A. Czirok and T. Vicsek, *Fractals* **3**, 849–868, Figure 1(d), 1995.

7.6 Incorporating Biological Movement

In a Turing system the pattern is produced passively, but in most biological pattern-forming systems there is an active response to the pattern which contributes to its formation and to subsequent morphogenesis. Cell movement seems to be important in angel-fish, where new stripes are formed between existing ones as the fish grows, and probably in mammalian coat pattern as well. There is a lot of work on patterns in colonies of micro-organisms that suggests that individual cell movement is crucial. The trigger for such movement is most often chemotaxis, introduced in Chapter 5, usually motion up a gradient of an attracting chemical, although other taxes may also play a role. In this section we shall just look at the spatial pattern-forming abilities of systems including chemotaxis. Further analysis of most of the systems here would require an investigation of the close relationship between spatial pattern formation and the propagation of signals through travelling waves.

Rather than look at standard examples from microbiology, some of which will be investigated in Exercises 7.14 and 7.15, we shall take an example from entomology.

Let us consider the nest-building behaviour of termites. Termites build nests by regurgitating material from their guts and applying it either to the surface where the nest will be or on previously applied nest material. As they do so they emit a pheromone, a volatile chemical that attracts other termites.

Let the density of termites be $n(\mathbf{x}, t)$, and the concentration of pheromone $p(\mathbf{x}, t)$. Pheromone is assumed to be emitted at a rate proportional to the density n, decays at a constant specific rate and diffuses with diffusion coefficient D_n. Using the chemotaxis model of Chapter 5, we have

$$\frac{\partial n}{\partial t} = rn\left(1 - \frac{N}{K}\right) - \nabla \cdot (\chi n \nabla p) + D_n \nabla^2 n,$$

$$(7.6.39)$$

$$\frac{\partial p}{\partial t} = \alpha n - \beta p + D_p \nabla^2 p,$$

where the term $rn(1-N/K)$ is not a growth term but accounts for the fact that the termites tend to attain a typical density K in the absence of pheromones. Let us neglect the effects of boundaries for simplicity, and look at the stability of the steady state $(n^*, p^*) = (K, \alpha K/\beta)$. Taking $\hat{n} = n - n^*$ and $\hat{p} = p - p^*$ and linearising about (n^*, p^*), we obtain

$$\frac{\partial \hat{n}}{\partial t} = -r\hat{n} - \chi K \nabla^2 \hat{p} + D_n \nabla^2 \hat{n},$$

$$(7.6.40)$$

$$\frac{\partial \hat{p}}{\partial t} = \alpha \hat{n} - \beta \hat{p} + D_p \nabla^2 \hat{p}.$$

The only spatial differential operator is $-\nabla^2$ on $\mathbb{R}^N$, with eigenvalues $\lambda = k^2 \geq 0$, and sinusoidal eigenfunctions. Stability can therefore be determined by testing stability to perturbations of wave-number k, substituting a vector multiple of $\exp(i\mathbf{k}\cdot\mathbf{x} + \sigma t)$ into the linearised Equations (7.6.40). The temporal eigenvalues σ satisfy

$$\begin{vmatrix} \sigma + r + \lambda D_n & -\chi K \lambda \\ -\alpha & \sigma + \beta + \lambda D_p \end{vmatrix} = 0,$$

or

$$Q(\sigma) = \sigma^2 + a_1(\lambda)\sigma + a_2(\lambda) = 0, \qquad (7.6.41)$$

where

$$a_1(\lambda) = (D_n + D_p)\lambda + r + \beta, \qquad (7.6.42)$$

$$a_2(\lambda) = D_n D_p \lambda^2 + (\beta D_n + r D_p - \alpha \chi K)\lambda + r\beta. \qquad (7.6.43)$$

It is clear that $a_1(\lambda) > 0$ for all λ; instability can only occur if $a_2(\lambda)$ becomes negative for some values of λ, so that Equation (7.6.41) for σ has one positive and one negative root. Let us think of the chemotaxis parameter χ as the bifurcation parameter. A sketch of $a_2(\lambda) = 0$ in the (λ, χ)-plane, Figure 7.9, immediately gives us all the qualitative information we need to know. There is a Turing bifurcation point at (λ_c, χ_c), so that $a_2(\lambda) > 0$ for all λ if $\chi < \chi_c$, but $a_2(\lambda) < 0$ for a range of λ, $\underline{\lambda} < \lambda < \overline{\lambda}$, if $\chi > \chi_c$. The value of λ_c may easily be found since a_2 is a perfect square at the bifurcation point, so $\lambda_c = \sqrt{r\beta/(D_n D_p)}$, by a similar argument to that leading to Equation (7.4.35).

As long as the chemotaxis parameter is sufficiently high, the system leads to a pattern of aggregation of termites. The nest starts off as a number of pillars with a characteristic spacing between them. Close to the bifurcation point, the characteristic spacing is close to $l_c = 2\pi/k_c$, where $k_c^2 = \lambda_c$. Eventually arcs are formed between the pillars, a cover is put over the whole structure and the complex architecture of the nest is built up.

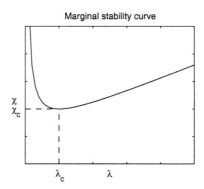

Figure 7.9   A sketch of $a_2(\lambda) = 0$ in the (λ, χ)-plane.

The importance of the chemotaxis term is that it leads to a $\nabla^2 \hat{p}$ term in the $\hat{n}$ equation, so-called *cross-diffusion*. This removes the need for D_n and D_p to be sufficiently different in order to form patterns. The effect of the pheromone is both to increase the local density of termites and to decrease the density further away, since this is where the termites are attracted from, so it plays a dual role as a short-range activator and a long-range inhibitor. This dual role is only possible through the movement of the termites.

EXERCISES

7.12. Colonies of bacteria grown on a Petri dish often produce very beautiful and intricate patterns (Figure 7.8). One model for this situation is the following. Let n be the bacterial concentration and c the concentration of a chemical that attracts the bacteria. Then a model for colony growth is given by

$$\frac{\partial n}{\partial t} = \gamma n - n^3 - \chi \nabla \cdot (n \nabla c) + D_n \nabla^2 n,$$

$$\frac{\partial c}{\partial t} = \theta(c) n - \beta c + D_c \nabla^2 c,$$

(7.6.44)

where θ is a threshold function, $\theta(c) = \theta_0$ for $c > c_0$, zero otherwise, and γ, β, χ, D_n and D_c are positive parameters.

a) Find the non-trivial spatially uniform steady state of these equations.

b) Derive the equation for the temporal eigenvalues σ.

c) Show that there is a band of eigenvalues λ with real positive σ as long as χ is sufficiently large and a further condition, which you should state, is satisfied.

d) Find the critical wave-length of the patterns.

7.13. The slime mould *Dictyostelium discoideum* is an organism first studied comprehensively from the 1940s onward by the microbiologist John Tyler Bonner because of his interests in development (then studied almost exclusively by zoologists), and in what were then called the lower plants: fungi and algae. It has a fascinating life history, and is now studied widely, both experimentally and theoretically. Starting as free-living amoebae, the cells of the slime mould move around in the soil (or on a Petri dish) and eat bacteria until a shortage of food causes a new stage of development, where they both initiate and respond to a chemotactic signal and aggregate into a multi-cellular 'slug'. This slug crawls around until it finds an appropriate place to produce a fruiting body of spore cells on top of a stalk made of non-spore cells. (It is an interesting evolutionary question to ask why some of the cells "agree" to be stalk cells, giving up any chance of reproduction.) The spore cells are then dispersed by the wind, and the life cycle starts again. A simple model of the aggregation phase is given by

$$\frac{\partial n}{\partial t} = -\nabla \cdot (\chi n \nabla p) + D_n \nabla^2 n,$$

$$\frac{\partial p}{\partial t} = \alpha n - \beta p + D_p \nabla^2 p, \tag{7.6.45}$$

almost identical to the model of this section for termite nest-building. Here n is the density of the amoebae, p the density of the chemo-attractant that they produce, and the parameters α, β, D_n, D_p and χ are all positive. There is a family of non-trivial spatially uniform steady states, and the one that occurs depends on the initial conditions.

a) Derive the conditions for potential instability of the spatially uniform steady state.

b) It has been observed experimentally that the chemotaxis parameter χ increases with time. Discuss the bifurcation(s) that might occur in an infinite domain with bifurcation parameter χ.

c) Consider the problem in a one-dimensional domain $(0, L)$ with zero-flux boundary conditions. Describe the bifurcations that occur in the system with L as the bifurcation parameter, and determine the critical wave-length when the system bifurcates to spatially inhomogeneous solutions.

7.7 Mechanochemical Models

The Turing approach to pattern formation is purely chemical. The idea is that a pre-pattern is set up by chemicals reacting and diffusing, while growth, and therefore morphogenesis, simply follows this pre-pattern. In contrast, in the mechanochemical approach, pattern formation and morphogenesis take place simultaneously, each influencing the other. The increased feedback possible with such a system is a strong argument in its favour, and certainly mechanical forces are ubiquitous in developing organisms. However, this approach requires a knowledge of continuum mechanics which is beyond the scope of this book. For those with such knowledge, an idea of the approach is given on the website, where we look at a model of wound healing. The resulting equations are rather complicated, but despite this there are sinusoidal solutions of the linearised equations, and stability is determined by substituting a vector multiple of $\exp(i\mathbf{k} \cdot \mathbf{x} + \sigma t)$ into the equations and testing whether σ might have positive real part. Unstable modes can occur, which may be related to scar formation.

7.8 Conclusions

- In biology symmetry-breaking bifurcations are an essential part of many developmental processes.

- Much information may be obtained from a linear stability analysis, but more rigorous results require bifurcation theory.

- If the relevant spatial differential operators are such that we expect sinusoidal solutions of a system, we test for stability of a system of equations by substituting in a vector multiple of $\exp(i\mathbf{k} \cdot \mathbf{x} + \sigma t)$ into the linearised equations and determining whether any resulting eigenvalue σ has positive real part. Otherwise, substitute in a vector multiple of $F(\mathbf{x})\exp(\sigma t)$, where F is a spatial mode.

- Diffusion cannot de-stabilise the spatially uniform steady state of a single reaction-diffusion equation, but it may de-stabilise that of a system of reaction-diffusion equations. This is called diffusion-driven or Turing instability.

- In a zero-eigenvalue Turing bifurcation from a spatially uniform steady state, a spatially inhomogeneous steady state solution is produced, while in a Hopf–Turing bifurcation, a temporally periodic spatially inhomogeneous solution is produced.

- In the simplest case of two equations, the conditions for Turing instability are $f_u^* + g_v^* < 0$, $D_2 f_u^* + D_1 g_v^* > 0$, $(D_2 f_u^* + D_1 g_v^*)^2 > 4 D_1 D_2 (f_u^* g_v^* - f_v^* g_u^*) > 0$.

- For these conditions to hold, one of the chemicals must activate and the other inhibit its own production. The system must be a pure activator-inhibitor or a cross-activator-inhibitor system. The process must involve short-range activation and long-range inhibition.

- Activator-inhibitor systems are likely to be supplemented by other chemicals in real biological examples of Turing pattern formation.

- As the domain grows, if Turing instability is possible at all, then successive modes become unstable and then re-stabilise. In real life, however, mode-doubling rather than successive modes are more often seen.

- If a Turing mechanism is responsible for mammalian coat patterns, then spotted animals can have striped tails, and spotted tails can have striped tips, but not vice versa. The patterns depend on the size (and shape) of the domain at the time that the pre-pattern is laid down.

- Chemotaxis and other mechanisms of biological movement are likely to be responsible, at least in part, for many biological pattern formation phenomena.

- Mechanical forces are often crucial in morphogenesis. Mechanochemical models include these forces as well as chemical effects. They tend to be rather complex and to require a knowledge of continuum mechanics for a deep understanding.

8
Tumour Modelling

- Biological processes such as cell proliferation are normally extremely tightly controlled through feedback processes that are mainly chemically mediated. This chapter focuses on cell populations that escape from such controls through mutations that allow them to manipulate their local environment.

- The mutations that cancer cells undergo may be sufficient to allow the immune system to recognise them as foreign, and hence to mount a defence against them.

8.1 Introduction

In the prosperous countries of the world, about one person in five will die of cancer. But why is it so common for cells to act in such a way as to kill their host, and therefore themselves, in apparent contradiction to the principles of natural selection? Natural selection works in two ways here. On the one hand, it acts on the population of cells within the animal. Although these cells are initially genetically identical, natural selection is strong because of high mutation rates leading to genetic variation for it to work on, high rates of reproduction, and high selective advantages. On the other hand, natural selection acts on the population of animals, and provides them with a battery of controls to guard against cancer. This is not as strong as might be expected because the heritable component of most cancers is small, as they are usually the result of a cascade of random somatic mutations, and because cancer is often a disease of old age, so that the selective advantages of not having it are small. Since there are about 10^{10} mutations per gene in a typical human lifetime, the fact that more cells do not escape from the controls that prevent them becoming cancerous shows how strong these controls are.

The numbers and distribution of normal cells throughout the body are regulated by ecological processes such as birth and death, and foreign cells in the body are controlled by the immune system (Chapter 6). Cancer cells must undergo many changes to escape the body's control and defence mechanisms. Some of their distinctive features are as follows.

– Escape from control of birth and death processes.

– Escape from control of maturation and differentiation processes.

– Non-self characteristics, and an ability to overcome the resulting immune response.

– Ability to stimulate production of their own nutrient supply.

– Poor control of genetic processes.

– Escape from control of migration processes.

These changes are the result of successive random genetic mutations or other rare events. Chemical carcinogens may cause simple local changes in DNA sequence, ionising radiation may cause chromosome breaks and translocations, or viruses may introduce foreign DNA into the cell, beginning to provide genetic variation for natural selection to work on. As the changes proceed, the feature of worsening control of genetic processes provides more variation. The changes do not occur all at once, and tumours usually start as a mild disorder of cell behaviour that slowly develops through well-characterised stages into full cancer, a process known as *tumour progression*. Even after such a cataclysmic carcinogenic event as the Hiroshima bomb, leukaemia rates in the city did not begin to rise significantly for five years. Only a small number of genes, known as oncogenes, are involved in the development of cancer. These oncogenes are normally involved in the control of proliferation and differentiation, and carcinogenic mutations often occur in the genes that control the activity of these oncogenes.

Cells that normally form part of multiplying cell populations are much more likely to become cancerous than others. Cancers of the epithelial cells, which form the linings of the body, the skin, and the surface of the gut, are known as carcinomas and constitute 90% of all cancers. Those of the haematopoietic (blood cell production) system are known as leukaemias, and those of the immune system as lymphomas. Such populations normally regulate how many stem cells are produced, and non-stem cells undergo a strict maturation process, passing through a fixed number of divisions and then dying. Mutations can weaken the control on stem cell production, or the maturation and differentiation process, or the death process, and such populations may then grow out of control, forming a tumour.

Often tumour cells do not multiply much faster than normal cells, but carry on multiplying in circumstances where normal cells would cease to do so. The normal cell requires a balanced salt solution and an energy source to survive. However, it will not grow and multiply unless special proteins, known as growth factors, are also present. These factors stimulate growth by attaching themselves to receptors on the cell surface. Tumour cells may produce their own growth factors, or their receptors may have altered in such a way that growth is possible even when no growth factor is present.

So-called *avascular* or *prevascular* tumours cannot grow indefinitely, and are normally too small to be seen *in vivo*. Oncologist Judah Folkman of Harvard University offers a holiday for two in the best hotel in Miami to any one of his research students who can grow a tumour in the laboratory to more than two millimetres in diameter. But growth is diffusion-limited, and sufficient nutrient cannot be delivered to the tumour by diffusion in order for it to grow to such a size. Real tumours enter the *vascular* stage by stimulating blood-vessel formation, or *angiogenesis*, in order to grow, which they do by secreting a substance known as tumour angiogenesis factor. After blood supply is ensured, tumours can grow to 16000 times their original volume in a few weeks. One promising avenue in cancer research focuses on ways to control the pathway to angiogenesis and thus prevent vascularisation. There is a final step that a tumour must take to be recognised as cancer in the clinical sense. Its cells must break free from the usual controls on cell migration, allowing them to spread from the original site to numerous distant sites, a process known as *metastasis*.

In this chapter we shall survey some mathematical models for the various stages of tumour progression, starting with phenomenological and mechanistic models for prevascular growth, looking at the effects of growth promoters and inhibitors, and then briefly at vascularisation and metastasis. We finish with a model for the effect of the immune system response.

8.2 Phenomenological Models

A tumour, at least in its early stages, has a sigmoid growth curve, first accelerating and then decelerating to an apparent limit. For this reason the logistic equation

$$\frac{dN}{dt} = rN\left(1 - \frac{N}{K}\right),$$

(8.2.1)

discussed in Chapter 1, has been used as a model for tumour growth. Here N is the size of the tumour, usually measured as a number of cells or as a volume. Of course $N(t) \to K$ in this model. Generalising this model, von Bertalanffy used the equation

$$\frac{dN}{dt} = f(N) = \alpha N^\lambda - \beta N^\mu$$

(8.2.2)

to represent tumour growth, where α, β, λ and μ are positive parameters with $\mu > \lambda$.

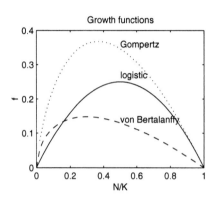

Figure 8.1   Comparison of the logistic, von Bertalanffy (with $\lambda = 2/3$ and $\mu = 1$) and Gompertz functions for tumour growth.

The function f is sketched in Figure 8.1, and for $N(0) > 0$, $N(t) \to K$ as $t \to \infty$, where $K = (\alpha/\beta)^{1/(\mu-\lambda)}$. As stated in the introduction, tumours do tend to a steady state size in the diffusion-limited phase of their growth, before they stimulate blood-vessel production (angiogenesis) that allows them to improve their nutrient supply. A particular case of the von Bertalanffy equation is the surface rule model, which states that growth is proportional to surface area (since nutrients have to enter through the surface) while decay is proportional to size. Then for a tumour of constant shape we recover (8.2.2) with $\lambda = 2/3$ and $\mu = 1$.

Now let $\mu = 1$ (assuming decay is proportional to size), and define $a = \alpha - \beta$, $b = \beta(\mu - \lambda) = \beta(1 - \lambda)$. The von Bertalanffy Equation (8.2.2) becomes

$$\frac{dN}{dt} = aN^\lambda - bN^\lambda\left(\frac{N^{1-\lambda} - 1}{1 - \lambda}\right).$$

(8.2.3)

Now take the limit as $\lambda \to 1-$, and this becomes

$$\frac{dN}{dt} = aN - bN\log N = -bN\log(\frac{N}{K}),$$

(8.2.4)

where $K = \exp(a/b)$. This is known as the *Gompertz equation*, which we have already met in Chapter 1 in quite a different context, as a model for the increase in mortality rate with age in a human population. It provides an excellent fit to empirical growth curves for avascular tumours and vascular tumours in their early stages, often much better than the more intuitive von Bertalanffy equation, but why should this be so? Interpretations in terms of $-\log(N/K)$ as the proliferative fraction of cells in the tumour cell population and derivations in terms of the entropy of the system have been proposed, but a satisfactory answer to the question has yet to be found.

The phenomenological models that we have discussed in this section are of limited use on their own, being merely descriptive rather than explanatory, but they are often used as the basis on which to build models of phenomena that occur later in the natural history of the tumour.

EXERCISES

8.1. Using the substitution $u = \log(N/K)$, solve the Gompertz Equation (8.2.4) with initial condition $N(0) = N_0$ to obtain the solution $N(t) = K\exp(-Ae^{-bt})$, where $A = -\log(N_0/K)$.

8.2. If either $\lambda = 1$ or $\mu = 1$, the von Bertalanffy equation is a *Bernoulli equation* and may be solved by standard substitutions.

a) Show that the von Bertalanffy Equation (8.2.2) with $\mu = 1$ may be written
$$\frac{dN}{dt} = \beta N\left\{\left(\frac{N}{K}\right)^{\nu} - 1\right\},$$
where $\nu = \mu - \lambda = 1 - \lambda$.

b) Using the Bernoulli transformation $u = (N/K)^{-\nu}$, show that the solution of this equation with initial condition $N(0) = N_0$ is given by
$$N(t) = K\left(\frac{N_0^{\nu}}{N_0^{\nu}(1 - e^{-\beta\nu t}) + K^{\nu}e^{-\beta\nu t}}\right)^{1/\nu}.$$

c) Confirm your result by using the transformation $u = \log(N/K)$.

8.3 Nutrients: the Diffusion-limited Stage

In order to grow a tumour requires oxygen and other nutrients. Normal tissues have blood vessels passing through them, and nutrients in the blood pass into the tissues through the vessel walls. In the early stages of development tumours have no such blood supply, and rely on nutrients diffusing from the adjacent normal tissue. As the tumour grows, diffusion can no longer provide sufficient nutrient, nutrient concentrations near its centre fall and cells die, resulting in a *necrotic* core. The tumour can grow no further and reaches a *diffusion-limited steady state*. A similar situation can occur after vascularisation, i.e. after the tumour has triggered production of its own blood supply, if the pressure in the tumour gets high enough to collapse the blood vessels in the tumour. In this section we shall investigate the conditions under which a necrotic core is produced by modelling the nutrient concentration in this steady state. We shall assume that the problem is spherically symmetric, leading to simplifications in the equations which are summarised in Section C.2.3 of the appendix. Let $c(r)$ be the concentration of the limiting nutrient, which we assume for definiteness to be oxygen, at radius r. Let the radius of the necrotic core (when it exists) be r_1 and that of the tumour be r_2. We shall take r_2 to be given and seek information on r_1, so we are addressing the question of how large the necrotic core would be if the tumour were of a given size, rather than how large the tumour will become. Let c satisfy the steady-state diffusion equation

$$0 = -k + D\nabla^2 c = -k + D\frac{1}{r^2}\frac{d}{dr}\left(r^2\frac{dc}{dr}\right) \tag{8.3.5}$$

for $r_1 < r < r_2$, where k is a constant representing the rate of uptake of oxygen and D is the constant diffusion coefficient. The oxygen is only taken up by living cells, so that

$$0 = D\nabla^2 c = D\frac{1}{r^2}\frac{d}{dr}\left(r^2\frac{dc}{dr}\right) \tag{8.3.6}$$

for $r < r_1$. Let c_2 be the concentration in the normal tissue, provided by the perfusing blood vessels, and c_1 the concentration at or below which cells die. The fact that r_1 is unknown, so that the domains of Equations (8.3.5) and (8.3.6) are not given *a priori*, is a difficulty to be overcome. Such problems are known as *free boundary problems*.

Let us first consider small tumours, so small that there is no necrotic core. Then $r_1 = 0$ and the boundary conditions are

$$\frac{dc}{dr}(0) = 0, \quad c(r_2) = c_2, \tag{8.3.7}$$

using symmetry. Multiplying Equation (8.3.5) through by r^2, integrating once, dividing through by r^2 and integrating again, we obtain

$$c(r) = \frac{1}{6}\frac{k}{D}r^2 + \frac{A}{r} + B, \qquad (8.3.8)$$

where A and B are the two constants of integration, or using the boundary conditions (8.3.7),

$$c(r) = -\frac{1}{6}\frac{k}{D}(r_2^2 - r^2) + c_2. \qquad (8.3.9)$$

This is valid as long as $c(0) \geq c_1$,

$$r_2^2 \leq r_c^2 = 6(c_2 - c_1)\frac{D}{k}. \qquad (8.3.10)$$

Now assume $r_2 > r_c$, and is known, so that $r_1 > 0$, and is to be found. We may integrate Equation (8.3.5) in the necrotic core to deduce that c must be constant there, $c = \hat{c}$, say. By definition $\hat{c} \leq c_1$, but since no consumption of nutrient occurs for $c \leq c_1$ then we must have $\hat{c} = c_1$. (This may be made mathematically rigorous by using a maximum principle.) The boundary conditions for the region of living cells are now

$$c(r_1) = c_1, \quad J(r_1) = 0, \quad c(r_2) = c_2. \qquad (8.3.11)$$

where $J = -D\frac{dc}{dr}$ is the (radial) flux of nutrient. The condition at $r = r_2$ is as before; the conditions at $r = r_1$ ensure continuity of concentration and flux at the boundary with the necrotic core. (There is no flux at all in the necrotic core, since $c = c_1$, constant, and so by continuity there can be none at r_1+.) Note that there are three boundary conditions, although Equation (8.3.5) is only a second order differential equation. The extra condition is crucial in allowing us to determine r_1. *Continuity conditions are often the key to a determination of an unknown boundary in a free boundary problem.*

We again integrate Equation (8.3.5) to obtain

$$c(r) = \frac{1}{6}\frac{k}{D}r^2 + \frac{A}{r} + B. \qquad (8.3.12)$$

Applying the boundary conditions (8.3.11), we obtain

$$c_1 = \frac{1}{6}\frac{k}{D}r_1^2 + \frac{A}{r_1} + B, \quad c_2 = \frac{1}{6}\frac{k}{D}r_2^2 + \frac{A}{r_2} + B, \quad 0 = \frac{1}{3}\frac{k}{D}r_1 - \frac{A}{r_1^2}. \qquad (8.3.13)$$

Subtracting the first of these from the second, and substituting in the value of A obtained from the third, we obtain

$$c_2 - c_1 = \frac{1}{6}\frac{k}{D}\left(r_2^2 - r_1^2 - 2r_1^3(\frac{1}{r_1} - \frac{1}{r_2})\right)$$

$$= \frac{1}{6}\frac{k}{D}r_2^2(1 + 2\frac{r_1}{r_2})(1 - \frac{r_1}{r_2})^2 = \frac{1}{6}\frac{k}{D}(1 + 2\frac{r_1}{r_2})(r_2 - r_1)^2, \qquad (8.3.14)$$

from which r_1 may be found. If $r_2 \to \infty$, then from the second of these equalities $r_1/r_2 \to 1$, so from the third $r_2 - r_1 \to h$, a constant, where

$$h^2 = 2\frac{D}{k}(c_2 - c_1). \qquad (8.3.15)$$

In a large tumour there is a shell of proliferating cells, whose thickness depends on the excess nutrient concentration above a threshold, how fast the nutrient is consumed and how fast it diffuses, but not on the size of the tumour itself.

The condition $r_2 \to \infty$ above is a condition on a parameter of the problem, and should not be confused with the possible behaviour of the radius of the tumour as a function of time.

EXERCISES

8.3. Some tumours are better approximated by circular cylinders than by spheres.

  a) For such a tumour, show that the analogue of Equation (8.3.12) is given by

$$c(R) = \frac{1}{4}\frac{k}{D}R^2 + A\log R + B.$$

  b) Find the critical value of the outer radius R_2 above which a necrotic core begins to form.

  c) Find the relationships between A, B and the necrotic radius R_1.

  d) Show that if the external radius of the cylinder is large, $R_2 - R_1 \to h$, where h is a constant to be found.

8.4. There is evidence in some tumours that intermediate levels of nutrient are sufficient for the cells to survive but not for them to proliferate, so that there is a layer of quiescent cells between the necrotic core and the outer proliferative layer, which consume nutrient at a lower rate than the proliferative cells. Set up a mathematical model for this situation. (You are not asked to solve it.) What conditions would you apply at the boundaries between the layers?

8.4 Moving Boundary Problems

In Section 8.3, we could not investigate the tumour growth because we neglected a fundamental principle, that of conservation of mass, which we shall now include. We shall also include kinetics, so we now have $r_1 = r_1(t)$, $r_2 = r_2(t)$. The necrotic core occupies $0 \le r < r_1(t)$ and the living cells $r_1(t) < r < r_2(t)$. Apart from this the problem for c is essentially unchanged if we make the reasonable (quasi-steady-state, see Chapter 6) assumption that the oxygen diffusion time-scale is much shorter than the tumour growth time-scale, so that Equation (8.3.5) still holds. A necrotic core forms for $r_2 > r_c$, and c is given by Equation (8.3.9) if $r_2 < r_c$, Equations (8.3.12) and (8.3.13) if $r_2 > r_c$. From now on we shall only consider the more difficult case $r_2 > r_c$, and leave the case $r_2 < r_c$ as an exercise. We must include the effects of proliferation of live cells and degradation of dead ones. We shall assume that all cells outside the necrotic core are proliferating, although there is some evidence that cells go through an intermediate non-proliferative stage before dying through lack of nutrients. The assumption on proliferation is that cell volume is produced by living cells at a rate P. In general this will depend on the nutrient concentration c. The assumption on degradation is that cell volume is lost at a rate L as necrotic cells and are broken down and their waste products removed. Let $\rho(r, t)$ be the density of the tumour, and $\mathbf{v}(r, t)$ the velocity field in the tumour, at radius r and time t. Then conservation of mass gives

$$\frac{\partial \rho}{\partial t} = -\rho L - \nabla \cdot \mathbf{J} = -\rho L - \nabla \cdot (\rho \mathbf{v}) \qquad (8.4.16)$$

in $0 < r < r_1(t)$ and

$$\frac{\partial \rho}{\partial t} = \rho P - \nabla \cdot \mathbf{J} = \rho P - \nabla \cdot (\rho \mathbf{v}). \qquad (8.4.17)$$

in $r_1(t) < r < r_2(t)$. Here $\mathbf{J}$ is the mass flux in the tumour, which is due simply to advection, so that $\mathbf{J} = \rho \mathbf{v}$ (see Chapter 5). This is a kinetic free boundary problem. Both r_1 and r_2 are unknown, although we know the relationship between them from the problem for the nutrient. We shall require an extra condition to determine them.

Let us assume that the density in the tumour is constant. Quoting Section C.2.3 of the appendix, the conservation of mass equations become

$$\nabla \cdot \mathbf{v} = \frac{1}{r^2} \frac{\partial}{\partial r}(r^2 v) = -L, \quad \nabla \cdot \mathbf{v} = \frac{1}{r^2} \frac{\partial}{\partial r}(r^2 v) = P, \qquad (8.4.18)$$

in $0 < r < r_1(t)$ and $r_1(t) < r < r_2(t)$ respectively, where v is the radial velocity. Integrating and applying continuity of the velocity field at $r = r_1(t)$,

$$v = -\frac{1}{3}Lr, \quad v = \frac{1}{3}Pr - \frac{1}{3}(P + L)\frac{r_1^3}{r^2} \qquad (8.4.19)$$

in $0 < r < r_1(t)$ and $r_1(t) < r < r_2(t)$ respectively.

So far in this section we have introduced a new unknown v and found a solution for it that would apply whatever r_1 and r_2 were. For the last piece in the jigsaw we need to connect the v problem with r_1 or r_2. We use the fact that the outermost cells in the tumour are moving at the velocity of expansion of the tumour, $\frac{dr_2}{dt}(t) = v(r_2(t))$. Hence

$$\frac{dr_2}{dt} = \frac{1}{3} P r_2 \left(1 - \frac{P + L}{P} \frac{r_1^3}{r_2^3} \right). \tag{8.4.20}$$

Hence r_2 increases until $\frac{r_2^3}{r_1^3}$ becomes equal to $\frac{P+L}{P}$. If $L \ll P$ then $(P+L)/P \approx 1$, so $r_2/r_1 \approx 1$, and we can use the approximation that $r_2 - r_1 = h$ given in Equation (8.3.15) to obtain $r_2 \approx 3hP/L$.

An alternative approach to this problem is to integrate Equations (8.4.16) and (8.4.17) over the volume $V(t)$ occupied by the tumour, again taking ρ to be constant. We obtain

$$\int_{V(t)} \nabla \cdot \mathbf{J} dV = \int_{V(t)} \nabla \cdot (\rho \mathbf{v}) dV = - \int_{V_1(t)} \rho L dV + \int_{V_2(t)} \rho P dV, \tag{8.4.21}$$

where $V_1(t)$ is the necrotic core and $V_2(t)$ the living cells. Using the divergence theorem,

$$\int_{V(t)} \nabla \cdot (\rho \mathbf{v}) dV = \int_{S(t)} \rho \mathbf{v} \cdot \mathbf{n} dS,$$

where $S(t)$ is the surface of $V(t)$, and since $\mathbf{v} \cdot \mathbf{n}$ is the normal component of velocity on the surface, this integral is equal to $\rho \frac{dV}{dt}$. (Alternatively, for those familiar with fluid dynamics, this is clear from the interpretation of $\nabla \cdot \mathbf{v}$ as the rate of dilatation.) The equation reduces to

$$\frac{dV}{dt} = - \int_{V_1} L dV + \int_{V_2} P dV \tag{8.4.22}$$

The interpretation of the terms in this equation is clear. For P and L constant it gives

$$\frac{dV}{dt} = P(V - V_1) - L V_1, \tag{8.4.23}$$

where V_1 is the volume of the necrotic core. In terms of the radii r_1 and r_2,

$$r_2^2 \frac{dr_2}{dt} = \frac{1}{3} P(r_2^3 - r_1^3) - \frac{1}{3} L r_1^3,$$

and we have recovered Equation (8.4.20). In principle, we can now solve this equation by substituting in the formula obtained from Equation (8.3.14) for r_1 in terms of r_2, and then integrating, but in practice this has to be done numerically.

However, we can obtain some useful information from the equation. Usually the loss rate L is much smaller than the proliferation rate P, and we can see that if $L \ll P$ the outside radius r_2 of the tumour satisfies $r_2(t) \to 3hP/L$ as $t \to \infty$. In other words, the tumour cannot grow beyond a certain size while its nutrient supply is diffusion-limited.

EXERCISES

8.5. Show that the model of this section with P constant predicts exponential growth of the tumour while $r_2 < r_c$.

8.6. Extend the model by including a layer of quiescent cells between the necrotic core and the outer proliferative layer.

8.7. In this exercise we analyse the steady state size r_2^* of the tumour model discussed in the last two sections.

  a) Find an expression for the steady state size r_2^* of the tumour as a function of the parameters c_1, c_2, k, D, P and L, all taken to be constant.

  b) Confirm that $r_2^* \to 3hP/L$ as $L/P \to 0$, where h is given by Equation (8.3.15).

8.5 Growth Promoters and Inhibitors

A crucial feature of tissues is that they produce chemical substances that control (activate or inhibit) the growth of the surrounding tissue. These substances are known as local control or *paracrine* factors. One of the characteristic properties of tumour cells is their ability to escape from these local controls. Let us consider a homogeneous spherical tumour of radius R that secretes an inhibitory paracrine factor (growth inhibitor) c. We shall perform a spherically symmetric steady state analysis, as in Section 8.3. The equation satisfied by the growth inhibitor is

$$0 = \lambda - \mu c + D\nabla^2 c = \lambda - \mu c + D \frac{1}{r^2} \frac{d}{dr}\left(r^2 \frac{dc}{dr}\right) \qquad (8.5.24)$$

in $0 < r < R$, where λ is the rate at which the chemical is secreted, μ the specific rate at which it is depleted, and D a constant diffusion coefficient. The boundary conditions are given by

$$\frac{dc}{dr}(0) = 0, \quad -D\frac{dc}{dr}(R) = Pc(R). \qquad (8.5.25)$$

The first of these arises from symmetry, the second states that the flux of chemical out of the tumour is proportional to the concentration difference between the inside and the outside, assuming that the concentration outside is negligible. The constant P is known as the *permeability* of the interface between the tumour and the normal tissue. The equation (8.5.24) is linear, and hence may be solved by the method of complementary function plus particular integral, with complementary function $A\exp(\alpha r)/r + B\exp(-\alpha r)/r$, where $\alpha = \sqrt{\mu/D}$, and particular integral λ/μ. The general solution satisfying the boundary condition at 0 is given by

$$c = \frac{\lambda}{\mu} + A\frac{\sinh(\alpha r)}{r} \tag{8.5.26}$$

in $0 < r < R$. The constant of integration A is then determined by the boundary condition at R, and we obtain

$$c(r) = \frac{\lambda}{\mu}\left(1 - \frac{R\sinh(\alpha r)}{r\sinh(\alpha R)f(\alpha R)}\right) \tag{8.5.27}$$

in $0 < r < R$, where

$$f(x) = 1 + \beta\left(\coth x - \frac{1}{x}\right) \tag{8.5.28}$$

and $\beta = \sqrt{\mu D}/P = \alpha D/P$. The function $c(r)$ is a decreasing function of r, shown in Figure 8.2.

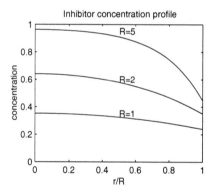

Figure 8.2   The inhibitor profile. If growth were completely inhibited for $c > c_1 = 0.4$, say, then growth would occur throughout the tumour for $R = 1$, only in a thin outer shell for $R = 2$, and nowhere in the tumour for $R = 5$ (so that the tumour would in fact never reach this size).

As the radius R of the tumour increases, then it is easy (although algebraically messy) to show that $c(0)$ increases from 0 to λ/μ, and $c(R)$ increases from 0 to $\frac{\lambda\beta}{\mu(1+\beta)}$. The concentration of the inhibitor inside the tumour increases as the tumour grows.

Now if we suppose that growth is very precisely controlled by the inhibitor, in the manner of a strongly cooperative modifier as discussed in Chapter 6, we can assume that growth only occurs where $c < c_1$. Hence growth ceases everywhere in the tumour as long as $c(R) > c_1$,

$$\frac{\lambda}{\mu}\left(1 - \frac{1}{f(\alpha R)}\right) > c_1. \tag{8.5.29}$$

If αR is small, this can never hold, whereas if αR is large, it approximates to $c_1 < \frac{\lambda\beta}{\mu(1+\beta)}$. If $c_1 < \frac{\lambda\beta}{\mu(1+\beta)}$, the tumour will cease growing at some finite value of R, but if $c_1 > \frac{\lambda\beta}{\mu(1+\beta)}$, tumour growth cannot be controlled by the inhibitor. (It may of course still be controlled by some other mechanism, such as lack of nutrient.) We suppose that normal tissue is sensitive to inhibitor, so c_1 is relatively low and tissue growth is under control, but that a tumour is relatively insensitive, so c_1 is larger and it escapes inhibitory control.

EXERCISES

8.8. On the assumption that there is no other growth control mechanism, sketch the bifurcation diagram for the steady state tumour radius R^* with bifurcation parameter c_1.

8.9. Let the concentration c of an inhibitor satisfy $0 = \lambda - \mu c + D\nabla^2 c$ within a tumour, $0 = -\mu c + D\nabla^2 c$ outside it.

a) What is the rationale behind this model?

b) What boundary conditions would you impose at the boundary between normal and tumour tissue?

c) Solve the problem in one spatial dimension (for algebraic simplicity), assuming the tumour occupies the region $-L < x < L$.

d) Sketch the bifurcation diagram for the steady state tumour size $2L^*$ with bifurcation parameter c_1.

8.6 Vascularisation

In order to grow beyond the diffusion-limited stage, tumours have to have a blood supply. They seem to achieve this by secreting a *tumour angiogenesis factor* (TAF) which diffuses across the tissue between the tumour and a blood vessel, activates new blood vessel formation (angiogenesis), and attracts these vessels towards the tumour. Let c be the concentration of TAF in the region between the tumour and the target blood vessel. Then it can be modelled by

$$\frac{\partial c}{\partial t} = -f(c)g(n) - h(c) + D_c\nabla^2 c. \tag{8.6.30}$$

Here $h(c)$ represents the rate of decay of the TAF and $f(c)g(n)$ the rate of take-up by the cells n which make up the new blood vessels. The equation for these cells is

$$\frac{\partial n}{\partial t} = F(n)G(c) - H(n) - \nabla \cdot \mathbf{J}, \tag{8.6.31}$$

where $H(n)$ represents their rate loss in the absence of TAF, and $F(n)G(c)$ their rate of production when stimulated by TAF at concentration c. The cell flux $\mathbf{J}$ of is made up of two parts,

$$\mathbf{J} = \mathbf{J}_{\text{diff}} + \mathbf{J}_{\text{chemo}} = -D_n \nabla n + \mathbf{J}_{\text{chemo}}, \tag{8.6.32}$$

since the cells both move about at random and are attracted towards the tumour by TAF. The chemotactic flux is typically modelled by

$$\mathbf{J}_{\text{chemo}} = n\chi(c)\nabla c, \tag{8.6.33}$$

as in Chapter 5. This states that cells flow up a TAF gradient at a speed proportional to the magnitude of the gradient; the constant of proportionality is $\chi(c)$, generally dependent on c but usually taken to be constant, known as the *chemotaxis coefficient*.

The forms taken for the various functions are as follows:

$$h(c) = dc, \quad f(c) = \frac{V_m c}{K_m + c}, \quad g(n) = \frac{n}{n_0}, \tag{8.6.34}$$

$$F(n) = rn\left(1 - \frac{n}{n_0}\right), \quad H(n) = k_p n, \tag{8.6.35}$$

and

$$G(c) = \begin{cases} 0, & c \le c^*, \\ \frac{c - c^*}{c_b}, & c > c^*, \end{cases} \tag{8.6.36}$$

where $c^* \le c_b$.

The model is too complicated for much progress to be made analytically. Numerical simulations with biologically realistic parameter values show cells in a distant blood vessel proliferating and migrating towards the tumour, as occurs in empirical studies.

8.7 Metastasis

A tumour that grows as a sphere, however large it grows, is not malignant. One of the defining characteristics of malignant tumours is that they metastasise. In metastasis, cells break off and are transported to other parts of the body, where secondary tumours may be initiated.

It seems that all tissues secrete factors to inhibit their own growth, and we have already considered the effect of these on tumour growth in Section 8.5. But there is also evidence that tumours can produce their own growth-promoting substances. In this section we shall consider the effect of the interaction between these two chemicals, and interpret the results from the point of view of metastasis.

Let u be the concentration of the activator and v the concentration of the inhibitor in the tumour, and let $\mathbf{u} = (u, v)^T$ be the column vector of concentrations. Then if the activator and the inhibitor react and diffuse inside the tumour, the equations to be satisfied are those of Chapter 7,

$$\frac{\partial \mathbf{u}}{\partial t} = \mathbf{f}(\mathbf{u}) + D\nabla^2 \mathbf{u}, \tag{8.7.37}$$

where $\mathbf{f} = (f, g)^T$ is the column vector of the reaction kinetics and D is the diagonal matrix of diffusion coefficients of u and v.

Just as in Chapter 7, for certain reaction kinetics and $D_2 > D_1$ there exists a range of unstable spatial eigenvalues $\underline{\lambda} < \lambda < \overline{\lambda}$ for which the real part $\operatorname{Re}\sigma$ of the temporal eigenvalue is positive. Within this unstable range $\operatorname{Re}\sigma$ will have a maximum which indicates a fastest-growing mode which will eventually dominate over time, at least until nonlinear effects become important. This will give rise to a spatially heterogeneous pattern, which may be a precursor of local invasion and eventual metastasis.

8.8 Immune System Response

Now let us consider the immune response to the cancers. Let us assume that the tumour cells have undergone some mutation on the road to malignancy that allows the immune system to recognise them as foreign. Then the immune system mounts a two-pronged attack, as discussed in Section 6.5, a humoral response (antibodies in the bodily fluids) and a cytotoxic response (killer cells primed to recognise the cancer cells). For definiteness let us consider the cytotoxic response. Let $X(\mathbf{r}, t)$ be the concentration of tumour cells (cells per unit volume) at position $\mathbf{r}$ and time t, and let $E(\mathbf{r}, t)$ be the concentration of effector cells constituting the cytotoxic response. We assume that in the absence of an immune response the tumour grows according to the equation

$$\frac{\partial X}{\partial \tau} = rX\left(1 - \frac{X}{K}\right) + D\nabla^2 X, \tag{8.8.38}$$

a combination of logistic growth and diffusion of the cells, by an argument similar to that of Chapter 5. The effector cells work by combining with the tumour cells and destroying them by *lysis* (splitting),

$$E + X \overset{k_1}{\to} C \overset{k_2}{\to} E + P, \tag{8.8.39}$$

(cf Chapter 6), where C is the complex of effector cell and tumour cell and P is the product of lysis, waste that will be removed by the body. The equations for the effector cells and the complex may then be derived using the law of mass action, as in Chapter 6, and are given by

$$\frac{dE}{d\tau} = -k_1 EX + k_2 C, \quad \frac{dC}{d\tau} = k_1 EX - k_2 C, \tag{8.8.40}$$

neglecting spatial variation for the cells of the immune system. Since lysis is expected to occur much faster than the other processes in the volume element considered, we make the quasi-steady-state hypothesis

$$k_1 EX - k_2 C = 0 \tag{8.8.41}$$

Adding Equations (8.8.40), we obtain

$$\frac{d}{d\tau}(E + C) = 0, \tag{8.8.42}$$

so that $E + C = E_0$. Then the quasi-steady-state hypothesis gives

$$E = \frac{k_2 E_0}{k_2 + k_1 X}. \tag{8.8.43}$$

The rate at which X is destroyed by the immune cells is given by

$$-k_1 EX = -\frac{k_2 k_1 E_0 X}{k_2 + k_1 X}, \tag{8.8.44}$$

a version of the Michaelis–Menten equation familiar from Chapter 6. Thus the equation for X including the immune response is given by

$$\frac{\partial X}{\partial \tau} = rX\left(1 - \frac{X}{K}\right) - \frac{k_2 k_1 E_0 X}{k_2 + k_1 X} + D\nabla^2 X = f(X) + D\nabla^2 X. \tag{8.8.45}$$

The function f satisfies $f(0) = 0$. If $r > k_1 E_0$ it has a single positive zero $X = X^*$, whereas if $r < k_1 E_0$, $f(X) < 0$ for all $X > 0$. We know from Chapter 5 that a travelling wave solution exists for $r > k_1 E_0$, but that $X \to 0$ for $r < k_1 E_0$. A strong immune response can drive the tumour to extinction, but a weak one allows it to grow indefinitely.

EXERCISES

8.10. In this exercise we shall analyse the spatially uniform steady states of Equation (8.8.45).

a) Let $x = k_1 X/k_2$, and show that spatially uniform steady states x^* satisfy
$$(\theta - x^*)(1 + x^*) = \beta,$$
where $\beta = k_1^2 E_0 K/(k_2 r)$, $\theta = k_1 K/k_2$.

b) Sketch bifurcation diagrams of x^* with bifurcation parameter β in the cases $\theta < 1$ and $\theta > 1$.

c) If β falls slowly, how do you expect the behaviour of the system to differ depending on whether θ is less than or greater than 1?

8.9 Conclusions

– Early in its development a tumour has a sigmoid growth curve, described phenomenologically by a logistic equation, a von Bertalanffy equation or a Gompertz equation. These phenomenological models are often used as the basis on which to build models of phenomena that occur later in the natural history of the tumour.

– An avascular tumour cannot grow above a certain size because the rate at which nutrients are provided is diffusion-limited. Cells in the core die, and proliferative cells occupy a shell close to the surface whose thickness tends to a constant.

– Modelling of tumour growth is complicated by the fact that we do not know *a priori* the region occupied by the tumour, and in fact this is the most important piece of information that we are trying to find. Mathematical problems of this kind are called free boundary problems. The principle that is used in formulating boundary conditions for such problems is that there is a velocity field within the tumour, and that the cells on the surface of the tumour are moving with the velocity with which the tumour is expanding.

– Normal cells produce chemicals that inhibit their own proliferation, providing negative feedback. Such chemicals will prevent bunches of cells from growing above a certain limiting size. Tumour cells, by becoming insensitive to such an inhibitor, can break free completely from such controls.

– Tumours overcome the diffusion-limited stage by stimulating *angiogenesis*, the growth of blood vessels to perfuse the tumour. This is accomplished by a chemical known as tumour angiogenesis factor, and the chemotactic effect of this chemical on blood vessel cells is crucial to the growth of the network of blood vessels.

– In metastasis cells break free from a tumour and are transported to other parts of the body. The causes are complicated, including changes in cell adhesion properties, but are preceded by the tumour losing its spherical shape. This instability of the spherical shape may be triggered by a diffusive instability between activators and inhibitors produced by the tumour.

– In a cytotoxic immune response effector cells bind with cancer cells and cause them to break apart (*lyse*). It may be modelled in a similar way to binding of an enzyme with a substrate, giving a Michaelis–Menten-type loss term. A strong response may prevent the growth of a tumour completely; a weak response may slow its growth but still allow it to grow indefinitely.

Further Reading

Many books on Mathematical Biology include a section on single-species population dynamics, sometimes with and sometimes without age structure. More mathematical texts are L. Edelstein-Keshet (*Mathematical models in biology*, Random House, 1988), F. C. Hoppensteadt (*Mathematical methods of population biology*, Cambridge, 1982), R. M. May (*Theoretical ecology*, Princeton, 1981), J. D. Murray (*Mathematical biology*, Springer, 1989, new edition, 2002), and M. Kot (*Elements of mathematical ecology*, Cambridge, 2001), and more biological ones are J. Maynard Smith (*Models in ecology*, Cambridge, 1974), N. J. Gotelli (*A primer of ecology*, Sinauer, 1995) and M. Begon, M. Mortimer and D. J. Thompson (*Population ecology*, Blackwell, 1996). H. Caswell (*Matrix population models*, Sinauer, 2001) discusses matrix models in detail. Evolutionary questions in age-structured populations are dealt with by B. Charlesworth (*Evolution in age-structured populations*, Cambridge, 1980) and S. C. Stearns (*The evolution of life histories*, Oxford, 1992). N. Keyfitz (*Applied mathematical demography*, second edition, Springer 1985) contains applications to human demography. C. W. Clark (*Mathematical Bioeconomics*, Wiley, 1990) is the standard text on natural resource management, and includes economic questions that we have not dealt with here. In this context costs as well as benefits (such as yield) may be included in the utility to be maximised. Metapopulation ecology is covered extensively in I. Hanski (*Metapopulation ecology*, Oxford, 1999). The assumption that the sites are identical and identically isolated may be relaxed, by using the incidence function method explained there.

Readable biologically based introductions to the material in Chapter 2 are by M. Begon, J. L. Harper and C. R. Townsend (*Ecology*, Blackwell, 1996) and, for the host-parasitoid material in particular, M. P. Hassell (*The spatial and temporal dynamics of host-parasitoid interactions*, Oxford, 2000). Maynard Smith (1974) and J. M. Emlen (*Population biology*, Macmillan, 1984) are good more mathematical treatments. Edelstein-Keshet (1988), Murray (1989, 2002) and Kot (2001) are more mathematical still, but do not lose sight of the biology. A recent book featuring case studies, which we do not have space for here, is W. S. C. Gurney and R. M. Nisbet (*Ecological dynamics*, Oxford, 1998). The ecosystems section is based on this book. The original works of A. J. Lotka (*Elements of physical biology*, Williams and Wilkins, 1925, reprinted as *Elements of mathematical biology*, Dover, 1956) and V. Volterra (various dates, depending on whether you read French, Italian, or would like an English translation; try looking in F. Oliveira-Pinto and B. W. Conolly, *Applicable mathematics of non-physical phenomena*, Ellis Horwood, 1982) are worth a read. R. M. May (*Stability and complexity in model ecosystems*, Princeton, 1976) looks at the relationship between stability and complexity. Hanski (1999) is worth reading for more on metapopulation models.

An excellent book which includes both the mathematics and the biological background on a variety of diseases is R. M. Anderson and R. M. May (*Infectious diseases of humans*, Oxford, 1991). More mathematical books are D. J. Daley and J. Gani (*Epidemic modelling*, Cambridge, 1999), which includes the very important stochastic approach, F. Brauer and C. Castillo-Chavez (*Mathematical models in population biology and epidemiology*, Springer, 2001), and O. Diekmann and J. A. P. Heesterbeek (*Mathematical epidemiology of infectious diseases*, Wiley, 2000), which includes extensions of the basic theory in various directions. The early papers of W. O. Kermack and A. G. McKendrick (see Oliveira-Pinto and Conolly, 1982) are worth reading.

Charles Darwin (*The Origin of Species*, John Murray, 1859, or many later editions) is required reading for anyone with an interest in evolution. D. J. Futuyma (*Evolutionary biology*, Sinauer, 1986) is a good modern text. Good references for mathematical population genetics are J. Roughgarden (*Theory of population genetics and evolutionary ecology*, Prentice Hall, 1996) and A. W. F. Edwards (*Foundations of mathematical genetics*, Cambridge, 2000), the former being more biologically and the latter more mathematically oriented. J. Maynard Smith (*Evolutionary genetics*, Oxford, 1998) is also good. For background genetics, D. L. Hartl and E. W. Jones (*Essential genetics*, Jones and Bartlett, 1999) is very readable. Non-technical references for game theory are K. Sigmund (*Games of life*, Oxford, 1993) and, specifically for the evolution of cooperation, R. Axelrod (*The evolution of cooperation*, Basic Books, 1984). A more mathematical treatment is given by J. Maynard Smith (*Evolution and the theory of games*, Cambridge, 1982). For full rigour, try J. Hofbauer and K. Sigmund (*Evolutionary games and population dynamics*, Oxford, 1998), who include proofs of results only stated here.

Much of the material in this chapter is basic to the study of almost any biological phenomenon where spatial variation is important, and so is treated in most books on mathematical biology that go beyond purely kinetic phenomena. Two excellent examples are Edelstein-Keshet (1988) and Murray (1989, 2002). Texts concentrating on molecular and cellular applications are L. A. Segel (*Mathematical models in molecular and cellular biology*, Cambridge, 1980) and S. I. Rubinow (*Introduction to mathematical biology*, Wiley, 1975), while biological invasions are covered by N. Shigesada and K. Kawasaki (*Biological invasions*, Oxford, 1997). A. Okubo (*Diffusion and ecological problems*, Springer, 1980, new edition with S. A. Levin 2001) gives a very readable account of applications in ecology emphasising the mathematical modelling aspects. An interesting book giving the background on random walks, emphasising the physics of the process but written from a biologist's point of view, is H. C. Berg (*Random walks in biology*, Princeton, 1993).

Many texts treat biochemical kinetics from a mathematical perspective, including Segel (1980), Edelstein-Keshet (1988), Murray (1989, 2002), and J. P. Keener and J. Sneyd (*Mathematical physiology*, Springer, 1998). The last three of these also include neural modelling, and all four include examples of excitable and oscillatory behaviour, which A. Goldbeter (*Biochemical oscillations and cellular rhythms*, Cambridge, 1996) treats exclusively. The place to go for extensions of these ideas to systems physiology as well as further applications to cellular physiology is Keener and Sneyd (1998). The treatment of immunology as a problem in population dynamics is elucidated in M. A. Nowak and R. M. May (*Virus dynamics*, Oxford, 2000). Other mathematical approaches to immunology are dealt with in A. S. Perelson's chapter in Segel (1980). We have been precluded by space considerations from including many fascinating areas of molecular and cellular biology, including all aspects of genetic information utilisation, control and replication, models of protein folding, analysis of pattern and sequence, continuum mechanics, combinatorics, neural networks, control theory and evolution, systems physiology, etc. Some aspects of dynamics have also been ignored, such as bursting, chaos, and dynamic diseases.

A. M. Turing's original paper (The chemical basis of morphogenesis, *Philosophical transactions of the Royal Society of London B*, **237**, 37–72, 1952) is beautifully written and well worth reading. A classic from the early part of the last century is D'Arcy Thompson's *On Growth and Form* (Cambridge, 1917 and several later editions). H. Meinhardt (*Models of biological pattern formation*, Academic, 1982) did early work on reaction-diffusion models, and has also published a book on sea-shell patterns (H. Meinhardt, *The algorithmic beauty of sea shells*, Springer, 1998). The books by Edelstein-Keshet (1988) and Murray (1989, 2002) have extended sections on pattern formation.

Biological background on cancer may be found in L. Wolpert (*The triumph of the embryo*, Oxford, 1991), who gives a short non-technical introduction, or

B. Alberts *et al* (*Molecular biology of the cell*, Garland, fourth edition 2002), who give more detail. Mathematical models of tumours are given in J. A. Adam and N. Bellomo (*A survey of models for tumor-immune system dynamics*, Birkhäuser, 1997), who include interactions with the immune system, and T. E. Wheldon (*Mathematical models in cancer research*, Hilger, 1988), who includes models of treatment régimes.

A
Some Techniques for Difference Equations

A.1 First-order Equations

We shall consider the first-order difference Equation (1.2.3),

$$N_{t+1} = f(N_t), \tag{A.1.1}$$

also called a *recurrence equation* or *map*, to be solved with the initial condition N_0 given. This defines a sequence $N_0, N_1, N_2, \cdots$, called a solution of the equation with the initial condition. It is *stable* if another solution $N_0', N_1', N_2', \cdots$ remains close to the first solution whenever it starts close, $|N_t - N_t'|$ is small for all t whenever $|N_0 - N_0'|$ is small, and *asymptotically stable* if also $|N_t - N_t'| \to 0$ as $t \to \infty$. It is *neutrally stable* if it is stable but not asymptotically stable. It is a *steady state* (or *fixed point* or *equilibrium*) solution N^* if $N_t = N^*$ for all t; it is clear from Equation (A.1.1) that the condition for N^* to be a steady state is that $N^* = f(N^*)$. It is *periodic of period p* if $N_{t+p} = N_t$ for all t, but $N_{t+q} \neq N_t$ for any t and any $q < p$, and *aperiodic* if it is not periodic.

A.1.1 Graphical Analysis

We wish to answer qualitative questions about the solution of Equation (A.1.1) with the initial condition N_0 given. For example, does the solution tend to a steady state, does it tend to a periodic solution or is it more complex than that? If the equation represents population growth then clearly $f(0) = 0$ and there is a steady state at zero; we can investigate the existence of others graphically by sketching $N_{t+1} = N_t$ and $N_{t+1} = f(N_t)$ together in the (N_t, N_{t+1})-plane; any intersection of these graphs is a steady state.

We can then investigate the behaviour by the method of *cobwebbing*. The idea is as follows.

- Choose a starting value N_0, and begin at the point (N_0, N_0) in the (N_t, N_{t+1})-plane.

- Draw a vertical line to the curve $N_{t+1} = f(N_t)$; this reaches the curve at the point $(N_0, f(N_0)) = (N_0, N_1)$.

- Draw a horizontal line to the diagonal $N_{t+1} = N_t$; this reaches the diagonal at the point (N_1, N_1).

- Repeat the process to arrive at (N_2, N_2), and then indefinitely until the behaviour of the equation with this starting value becomes clear.

- If necessary, do the same with other starting values.

Some examples of the process are shown in Figure A.1.

A.1.2 Linearisation

It is plausible but not quite obvious that the dividing line between the oscillatorily stable and oscillatorily unstable behaviour shown in Figure A.1 is $f'(N^*) = -1$. Let us check this by defining $n = N - N^*$; then subtracting $N^* = f(N^*)$ from Equation (A.1.1) gives

$$n_{t+1} = f(N^* + n_t) - f(N^*) = f'(N^*)n_t + h.o.t.,$$

where *h.o.t.* stands for higher order terms. Let us assume that for n_t sufficiently small the higher order terms are negligible. Then we may infer that the solution of Equation (A.1.1) behaves similarly to that of the approximating equation

$$n_{t+1} = f'(N^*)n_t. \tag{A.1.2}$$

This is known as the *linearised equation*. The solution is $n_t = n_0 f'(N^*)^t$, and so the trivial steady state is oscillatorily unstable, oscillatorily asymptotically stable, monotonically asymptotically stable or monotonically unstable according to whether $\lambda = f'(N^*)$ satisfies $\lambda < -1$, $-1 < \lambda < 0$, $0 < \lambda < 1$ or $1 < \lambda$ respectively. The condition for asymptotic stability is

$$|\lambda| = |f'(N^*)| < 1,$$

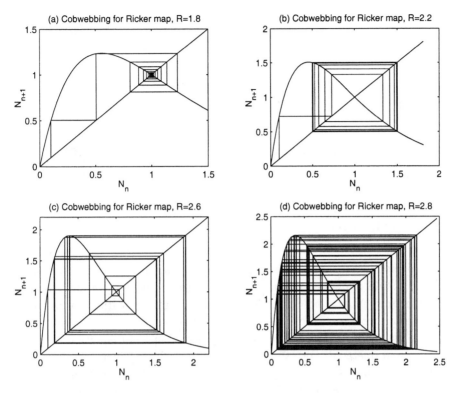

Figure A.1   Some examples of cobweb maps, as a parameter of the equation increases. The equation is essentially irrelevant, as long as it has a hump. In (a), the solution tends oscillatorily to a steady state. In (b) and (c), it tends to a period-2 and a period-4 solution respectively. In (d), there is chaotic behaviour. This illustrates a typical period-doubling cascade to chaos.

and if $|\lambda| = 1$ the steady state is stable but not asymptotically stable. The use of the notation λ reflects the fact that the place of $f'(N^*)$ will be taken by the eigenvalues of a matrix for systems of equations, as we shall see, and we shall often refer to $f'(N^*)$ itself as an eigenvalue. The neglect of the higher order terms can be shown to be justified sufficiently close to the steady state as long as we are not on the borderline between two kinds of behaviour, i.e. as long as $f'(N^*) \neq 0$ (when the nonlinear terms determine monotonicity) or ± 1 (when they determine stability). We summarise the behaviour of the linearised equation in the diagram below.

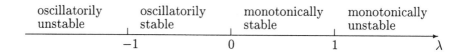

Figure A.2   Behaviour of solutions of Equation (A.1.2).

A.2 Bifurcations and Chaos for First-order Equations

Now let us assume that there is a parameter μ in the equations, and consider equations of the form

$$x_{t+1} = f(x_t, \mu).$$

We shall not necessarily interpret this as an equation of population dynamics, and so will allow x_t to take positive or negative values; we have changed notation from N to x to emphasise this. Often, in applications, x will be a perturbation from a steady state.

It is clear that the behaviour of solutions of such an equation can vary with the parameter μ. For example, if the eigenvalue f_x^* increases through 1 as μ increases past some value μ_c, we expect the steady state to change character from monotonically stable to monotonically unstable, according to the linearised analysis of the last section. A diagram of the solution behaviour (showing the steady states and periodic orbits, their stability, etc) against the parameter μ is known as a *bifurcation diagram*, and the points where the solution behaviour changes as *bifurcation points*. All the bifurcations that can occur for first-order equations are described below.

A.2.1 Saddle-node Bifurcations

A typical example of the saddle-node bifurcation is

$$x_{t+1} = f(x_t, \mu) = x_t + \mu - x_t^2.$$

The equation has no steady states for $\mu < 0$, and two ($x^* = \pm\sqrt{\mu}$) for $\mu > 0$. The positive square root is stable and the negative square root unstable. The bifurcation point is at $(x, \mu) = (x_c, \mu_c) = (0, 0)$, where $f(0, 0) = 0$, and the eigenvalue at the bifurcation point is $f_x(0, 0) = 1$. More generally, a saddle-node bifurcation is said to occur when, near the bifurcation point, the equation $x_{t+1} = f(x_t, \mu)$ possesses a unique curve of fixed points in the (x, μ)-plane, which passes through the bifurcation point (x_c, μ_c), and lies on one side of the line $\mu = \mu_c$. Conditions for such a bifurcation to occur at (x_c, μ_c) are $f(x_c, \mu_c) = 0$, $f_x(x_c, \mu_c) = 1$, $f_\mu(x_c, \mu_c) \neq 0$, $f_{xx}(x_c, \mu_c) \neq 0$. Such bifurcations occur, for example, in models for insect pests where outbreaks may occur. A typical bifurcation diagram for such a model is sketched below.

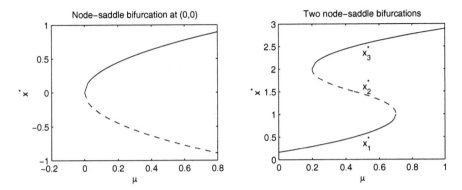

Figure A.3   (a) The prototype saddle-node bifurcation. (b) Typical saddle-node bifurcations in models of insect pests. Here x_1^* is a stable endemic steady state, x_2^* an unstable intermediate steady state, and x_3^* a stable outbreak steady state x_3^*. The intermediate and outbreak states appear through a saddle-node bifurcation at μ_1. There is a hysteresis effect; the insect population will not reach the outbreak state x_3^* until the endemic and intermediate steady states disappear through a second saddle-node bifurcation point at μ_2, but will then remain at outbreak levels unless μ subsequently decreases past μ_1.

A.2.2 Transcritical Bifurcations

The prototype equation for the transcritical bifurcation is

$$x_{t+1} = x_t + \mu x_t - x_t^2.$$

The equation has two steady states, the trivial one $x^* = 0$ and the non-trivial one $x^* = \mu$, $x^* = 0$ being stable for $\mu < 0$ and $x^* = \mu$ for $\mu > 0$. The bifurcation point is at $(x, \mu) = (x_c, \mu_c) = (0, 0)$, and the eigenvalue at the bifurcation point is $f_x(0, 0) = 1$. More generally, a transcritical bifurcation is said to occur when, near the bifurcation point, the equation $x_{t+1} = f(x_t, \mu)$ possesses two curves of fixed points in the (x, μ)-plane, each of which passes through the bifurcation point (x_c, μ_c) and exists on both sides of the line $\mu = \mu_c$. An exchange of stability between the two curves of steady states takes place at the bifurcation point. Without loss of generality (by redefining x if necessary), we take one of these curves to be the line $x = 0$. (If we do not do this the following conditions are more unwieldy.) Conditions for such a bifurcation to occur at (x_c, μ_c) are then $f(x_c, \mu_c) = 0$, $f_x(x_c, \mu_c) = 1$, $f_\mu(x_c, \mu_c) = 0$, $f_{x\mu}(x_c, \mu_c) \neq 0$, $f_{xx}(x_c, \mu_c) \neq 0$. Such bifurcations occur, for example, from the trivial steady state in models for population dynamics with intra-specific competition when the basic reproductive ratio R_0 increases past 1.

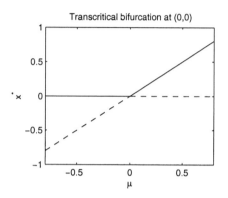

Figure A.4   A transcritical bifurcation.

A.2.3 Pitchfork Bifurcations

The prototype equation for the pitchfork bifurcation is

$$x_{t+1} = f(x_t, \mu) = x_t + \mu x_t - x_t^3.$$

The equation has one steady state for $\mu < 0$, the trivial one $x^* = 0$, and three steady states for $\mu > 0$, the trivial one $x^* = 0$ and the non-trivial ones $x^* = \pm\sqrt{\mu}$, the trivial one being stable for $\mu < 0$ and both non-trivial ones being stable for $\mu > 0$. The bifurcation point is at $(x, \mu) = (x_c, \mu_c) = (0, 0)$, and the eigenvalue at the bifurcation point is $f_x(0, 0) = 1$. More generally, a pitchfork bifurcation is said to occur when, near the bifurcation point, the equation $x_{t+1} = f(x_t, \mu)$ possesses two curves of fixed points in the (x, μ)-plane, each of which passes through the bifurcation point (x_c, μ_c), one of which exists on one side and the other on both sides of the line $\mu = \mu_c$. Without loss of generality (by redefining x if necessary), we take the curve that exists on both sides of $\mu = \mu_c$ to be the line $x = 0$, which again simplifies the conditions for the bifurcation. The trivial steady state is stable on one side of $\mu = \mu_c$, and both non-trivial solutions are stable if $x = 0$ is unstable and unstable if $x = 0$ is stable. Conditions for such a bifurcation to occur at (x_c, μ_c) are then $f(x_c, \mu_c) = 0$, $f_x(x_c, \mu_c) = 1$, $f_\mu(x_c, \mu_c) = 0$, $f_{xx}(x_c, \mu_c) = 0$, $f_{x\mu}(x_c, \mu_c) \neq 0$, $f_{xxx}(x_c, \mu_c) \neq 0$. These bifurcations are most important in mathematical biology in their role in period-doubling bifurcations, which we shall describe next.

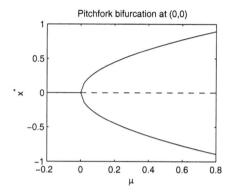

Figure A.5   A pitchfork bifurcation.

A.2.4 Period-doubling or Flip Bifurcations

The prototype equation for the period-doubling or flip bifurcation is

$$x_{t+1} = f(x_t, \mu) = -x_t - \mu x_t + x_t^3.$$

One steady state is the trivial one $x^* = 0$, which exists for all μ. There is also a curve of steady states given by $x^2 = \mu + 2$, which arises through a pitchfork bifurcation from the trivial steady state at $(x, \mu) = (0, -2)$; we are not interested in this bifurcation. The bifurcation point of interest is at $(x, \mu) = (x_c, \mu_c) = (0, 0)$, with eigenvalue $f_x(0, 0) = -1$. Here the trivial steady state loses its stability, moving from oscillatorily stable to oscillatorily unstable, but there is *no* stable steady state for $\mu > \mu_c$. What happens to solutions of the equation here? The oscillatory nature of the instability gives us a clue, and we can answer this question by considering the second iterate $f^2 = f \circ f$ of f, which is equivalent to considering every other term in the sequence x_t. It is easy to see that as μ increases past 0, and the derivative f_x on the trivial branch decreases past -1, the derivative $(f^2)_x$ on that branch increases past 1, and (with a bit more work) that f^2 undergoes a pitchfork bifurcation there. Two new steady states x_1^* and x_2^* of f^2 appear which are not steady states of f. The only possibility is that these correspond to period-2 solutions of f, oscillating between x_1^* and x_2^*. A stable period-2 orbit bifurcates from the trivial steady state at $(0,0)$. More generally, a period-doubling bifurcation is said to occur when, near the bifurcation point, the equation $x_{t+1} = f(x_t, \mu)$ possesses a single curve of fixed points in the (x, μ)-plane, while the second iterate f^2 undergoes a pitchfork bifurcation at (x_c, μ_c). Without loss of generality (by redefining x if necessary), we take the curve of fixed points to be the line $x = 0$, which again leads to simpler bifurcation conditions. This curve is stable on one side of $\mu = \mu_c$, and the period-2 solution is unstable if it occurs where $x^* = 0$ is stable, stable if it occurs where $x^* = 0$ is unstable. Conditions for such a bifurcation to occur at (x_c, μ_c) are then $f(x_c, \mu_c) = 0$, $f_x(x_c, \mu_c) = -1$, $f_\mu^2(x_c, \mu_c) = 0$, $f_{xx}^2(x_c, \mu_c) = 0$, $f_{x\mu}^2(x_c, \mu_c) \neq 0$, $f_{xxx}^2(x_c, \mu_c) \neq 0$.

But more can be shown. Under quite general conditions, the stable steady states x_1^* and x_2^* of f^2 suffer exactly the same fate as the trivial steady state of f, producing stable steady states of f^4, then f^8, and so on. The bifurcation of a stable steady state to an unstable steady state and a stable orbit of period two is then followed by a cascade of such period-doubling bifurcations, leading to orbits of period 4, 8 and so on. This cascade accumulates at some value μ_∞ of the bifurcation parameter. For values of μ greater than μ_∞ much more complicated behaviour is possible, including *chaos*. There are many definitions of chaos, and a full discussion of them would lead us too far afield. For our purposes, chaotic behaviour in Equation (A.1.1) is characterised by the following.

– There are aperiodic solutions.

– The *butterfly effect* occurs. By this we mean that there is sensitive dependence on the initial condition, so that a small error in specifying the initial condition can lead to large differences in the predictions of the model.

A chaotic solution is almost indistinguishable from random behaviour, despite being derived from a deterministic equation.

The period-doubling route to chaos typically occurs in models for population growth with humped functions f, with bifurcation parameter the basic reproductive ratio, so that there is reason to believe that chaotic behaviour may occur in ecological systems. However with real data, where stochastic effects inevitably play a part, it is even more difficult to tell whether unpredictable behaviour arises from chaos or merely from stochasticity, and there is still controversy over whether chaos is observed in ecological and other biological systems.

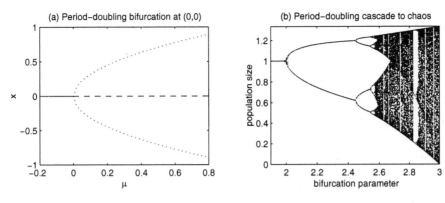

Figure A.6   (a) A single period-doubling bifurcation. The trivial steady state loses its stability for $\mu > 0$, and the dotted lines represent stable period-2 solutions taking alternately the upper and lower values on the fork. (b) A cascade of period-doubling bifurcations to chaos.

Example A.1

A function $f : [0, \infty) \to [0, \infty)$ modelling population dynamics with intra-specific competition exhibits exact compensation (Section 1.2.2) if it is mono-tonic increasing, $f(0) = 0$, and $f(N) \to N_{max}$, a constant, as $N \to \infty$.

If f exhibits exact compensation, solutions of Equation (A.1.1) tend to a steady state as $t \to \infty$.

The result is obvious by looking at cobweb maps, although a rigorous proof is slightly more demanding.

Example A.2

A function $f : [0, \infty) \to [0, \infty)$ modelling population dynamics with intra-specific competition exhibits over-compensation (Section 1.2.2) if $f(0) = 0$, $f(N) \to 0$ as $N \to \infty$. Such a function is *uni-modal* if there exists a $N^{\dagger}$ such that f is monotonic increasing for $N < N^{\dagger}$ and decreasing for $N > N^{\dagger}$.

Typically, solutions of Equation (A.1.1) with f a uni-modal function ex-hibit a cascade of period-doubling bifurcations to chaos if an increase in the bifurcation parameter results in steepening of the function.

Remark A.3

We can consider such a function to be a function $f : [0, \hat{N}] \to [0, \hat{N}]$ with $f(0) = 0$, $f(\hat{N}) = 0$, the usual definition of a uni-modal function, by choosing $\hat{N} > f(N^{\dagger})$ and redefining f on $[f(N^{\dagger}), \hat{N}]$; this makes no difference to the dynamics from time 1 onwards.

A.3 Systems of Linear Equations: Jury Conditions

In Section 1.8, on Fibonacci's model for rabbit population growth, we derived the second-order linear difference Equation (1.8.22), $y_{n+2} = y_{n+1} + y_n$. Linear difference equations of the mth order may be analysed by writing them as m equations of the first order. In this case, defining $x_n = y_{n+1}$, equation (1.8.22) may be written $x_{n+1} = x_n + y_n$, $y_{n+1} = x_n$, and the initial conditions (1.8.23) become $x_0 = 1$, $y_0 = 1$. Generalising, we may have to consider

$$\mathbf{z}_{n+1} = M\mathbf{z}_n, \tag{A.3.3}$$

where $\mathbf{z}_n$ is an m-vector and M an $m \times m$-matrix, with $\mathbf{z}_0$ given. Let us look for a solution in the form $\mathbf{z}_n = \lambda^n \mathbf{c}$, where $\mathbf{c}$ is an m-vector. Substituting into Equation (A.3.3) and cancelling λ^n,

$$\lambda \mathbf{c} = M\mathbf{c} \tag{A.3.4}$$

or

$$(M - \lambda I)\mathbf{c} = \mathbf{0}.$$

For any λ this has a solution $\mathbf{c} = \mathbf{0}$, but if λ is an eigenvalue of the matrix with eigenvector $\mathbf{c}$ then this is a non-trivial solution. For this to happen $M - \lambda I$ must be singular,

$$\det(M - \lambda I) = 0 \tag{A.3.5}$$

This is a polynomial of the mth degree and has m roots. If they are all distinct (the usual case in Mathematical Biology) then the general solution of Equation (A.3.3) is

$$\mathbf{z}_n = \Sigma_{i=1}^m A_i \lambda_i^n \mathbf{c}_i,$$

where $\mathbf{c}_i$ is the eigenvector corresponding to the eigenvalue λ_i. The A_i are arbitrary constants that are determined by the initial conditions. Note that

- if each $|\lambda_i| < 1$ then $|\mathbf{z}_n| \to 0$ as $n \to \infty$, and

- if there exists i such that $|\lambda_i| > 1$, and if $A_i \neq 0$, then $|\mathbf{z}_n| \to \infty$ as $n \to \infty$.

In the case $m = 2$ the eigenvalue Equation (A.3.5) is given by

$$\lambda^2 + a_1 \lambda + a_2 = 0,$$

where $a_1 = -\text{tr}\, M$, $a_2 = \det M$. The necessary and sufficient conditions for asymptotic stability, $|\lambda_i| < 1$ for $i = 1, 2$, are the *Jury conditions*

$$|a_1| < a_2 + 1, \quad a_2 < 1. \tag{A.3.6}$$

If $1 + a_1 + a_2 = 0$ ($\text{tr}\, M = 1 + \det M$) there is an eigenvalue $\lambda = 1$, if $a_1 = 1 + a_2$ ($1 + \text{tr}\, M + \det M = 0$) there is an eigenvalue $\lambda = -1$, and if $|a_1| < 1 + a_2$ and $a_2 = 1$ ($|\text{tr}\, M| < 1 + \det M$ and $\det M = 1$) there is a pair of complex conjugate eigenvalues on the unit circle.

Jury conditions may be derived for $m > 2$, but they get rapidly more complicated. For $m = 3$, with eigenvalue equation $\lambda^3 + a_1 \lambda^2 + a_2 \lambda + a_3 = 0$, they are

$$|a_1 + a_3| < a_2 + 1, \quad |a_3| < 1, \quad |a_2 - a_3 a_1| < |1 - a_3^2|. \tag{A.3.7}$$

A.4 Systems of Nonlinear Difference Equations

Throughout this section we shall consider second-order systems of the form

$$N_{t+1} = f(N_t, P_t), \quad P_{t+1} = g(N_t, P_t), \tag{A.4.8}$$

although the results may be extended to systems of higher order. Graphical analysis is much more difficult than for the single equation, although it is often helpful to plot solutions in (N, P)-space, but linearisation and bifurcation analyses are still available, and we look at these methods in this section.

Some new kinds of behaviour occur here, and we need some definitions. An *invariant curve* is a curve Γ in (N, P)-space such that if $(N_0, P_0) \in \Gamma$, then $(N_t, P_t) \in \Gamma$ for all $t > 0$. Such a curve is *stable* (or *orbitally stable*) if a solution remains close to it whenever it starts close to it, and *asymptotically (orbitally) stable* if the distance between such a solution and the curve tends to zero as $t \to \infty$. A solution N_t which starts and therefore remains on a closed invariant curve Γ may either return to its starting point after a finite number of steps, or not. We say it has rational or irrational *rotation number*, respectively.

A.4.1 Linearisation of Systems

Let us assume that there exists a steady state (N^*, P^*) of this system; it satisfies

$$N^* = f(N^*, P^*), \quad P^* = g(N^*, P^*).$$

Perturbations from this steady state may be defined by $(n, p) = (N, P) - (N^*, P^*)$. Linearising about the steady state, in the same way as was done for the first-order equation, we obtain the approximate equations (the linearised equations)

$$n_{t+1} = \frac{\partial f}{\partial N}(N^*, P^*)n_t + \frac{\partial f}{\partial P}(N^*, P^*)p_t, \tag{A.4.9}$$

$$p_{t+1} = \frac{\partial g}{\partial N}(N^*, P^*)n_t + \frac{\partial g}{\partial P}(N^*, P^*)p_t, \tag{A.4.10}$$

or

$$\mathbf{n}_{t+1} = J^*\mathbf{n}_t, \tag{A.4.11}$$

where $\mathbf{n}$ is the column vector $(n, p)^T$, J is the *Jacobian* of the transformation, viz

$$J(N, P) = \begin{pmatrix} f_N(N, P) & f_P(N, P) \\ g_N(N, P) & g_P(N, P) \end{pmatrix},$$

and a star denotes evaluation at the steady state. Comparing Equation (A.4.11) with Equation (A.3.3) and using the Jury conditions (A.3.6) given in the last section, we infer asymptotic stability of the steady state if

$$|\operatorname{tr} J^*| < \det J^* + 1, \quad \det J^* < 1. \tag{A.4.12}$$

A.4.2 Bifurcation for Systems

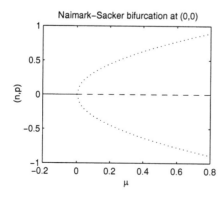

Figure A.7   A Naimark–Sacker bifurcation. The trivial steady state loses its stability at the origin. The dotted line represents the invariant closed curve in (n, p)-space.

The bifurcations of first-order difference equations described in the last section also occur for systems as an eigenvalue of J^* passes through ± 1, but there is also a new possibility. This is the Naimark–Sacker bifurcation, often referred to as a Hopf bifurcation (see Appendix B) for difference equations. At such a bifurcation point, the Jacobian matrix has two complex conjugate eigenvalues of modulus 1. This may occur for systems of any order greater than 1, and the results are similar, but we shall consider second-order systems for simplicity. Then, in terms of the Jacobian matrix, the conditions at the bifurcation point are $|\text{tr } J^*| < \det J^* + 1$, $\det J^* = 1$. Consider the case where, as a bifurcation parameter μ increases past a bifurcation value μ_c, the two complex conjugate eigenvalues cross out of the unit disc, so that $\det J^*$ increases past 1 and the steady state therefore loses its stability for $\mu > \mu_c$. Assume also that, at the bifurcation point, $\lambda^k \neq 1$ for $k = 1, 2, 3, 4$. Then there are two possibilities.

– In the subcritical case, there exists an unstable closed invariant curve in **n**-space for $\mu < \mu_c$.

– In the supercritical case, there exists a stable closed invariant curve in **n**-space for $\mu > \mu_c$.

Solutions on these invariant curves may have a rational or irrational rotation number. Therefore, if we have a system such as (A.4.8) and we know its solutions are bounded, and if we also know that the steady state becomes unstable through two complex conjugate eigenvalues crossing out of the unit disc, we expect to see solutions of the system on a closed invariant curve, with rational or irrational rotation number, oscillating about the steady state in the phase plane. This is typical of realistic host-parasitoid models such as those of Chapter 2.

B

Some Techniques for Ordinary Differential Equations

B.1 First-order Ordinary Differential Equations

Consider the first-order ordinary differential equation,

$$\dot{N} = f(N), \tag{B.1.1}$$

to be solved with initial condition

$$N(0) = N_0. \tag{B.1.2}$$

Definitions of steady states and their stability are similar to the difference equation case, Appendix A, and are omitted. The condition for N^* to be a steady state is now $f(N^*) = 0$.

B.1.1 Geometric Analysis

Now note that if N currently takes a value where $f(N)$ is positive, it will subsequently increase (since $\dot{N} = f(N) > 0$), and if $f(N)$ is negative it will decrease. A sketch of the graph of f tells us all we need to know about the steady states (points N^* where $f(N^*) = 0$) and their stability (stable if $f'(N^*) < 0$, unstable if $f'(N^*) > 0$), and the asymptotic (long-term) behaviour of the solution. It is clear that as $t \to \infty$, $N \to \pm\infty$ or to a steady state.

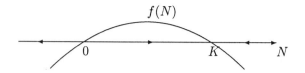

Figure B.1   A typical function $f(N)$ with an unstable steady state at 0 and a stable steady state at K.

B.1.2 Integration

Equation (B.1.1) with initial condition (B.1.2) may be integrated by separating the variables, to obtain the implicit solution

$$t = \int_0^t dt = \int_{N_0}^N \frac{dN}{f(N)}.$$

This confirms that as $t \to \infty$, $N \to \pm\infty$ or to a steady state (a zero of f), but also tells us how fast it does so.

B.1.3 Linearisation

Let us assume that there is a steady state solution N^*. Defining $n = N - N^*$ and using the fact that $f(N^*) = 0$, we obtain

$$\dot{n} = f(N^* + n) = f'(N^*)n + h.o.t., \qquad (B.1.3)$$

where $h.o.t.$ is an abbreviation for *higher order terms*. Let us assume that for n sufficiently small the higher order terms are negligible. Then we may infer that the solution of Equation (B.1.3) behaves similarly to that of the approximating *linearised* equation

$$\dot{n} = f'(N^*)n, \qquad (B.1.4)$$

which has solution $n(t) = n_0 \exp(f'(N^*)t)$. The neglect of the higher order terms can be shown to be justified sufficiently close to the steady state as long as $f'(N^*) \neq 0$. It follows that the steady state is exponentially stable if $f'(N^*) < 0$ and exponentially unstable if $f'(N^*) > 0$, while if $f'(N^*) = 0$ the nonlinear terms determine stability.

B.2 Second-order Ordinary Differential Equations

A second-order system of ordinary differential equations is given by

$$\dot{U} = f(U, V), \quad \dot{V} = g(U, V), \tag{B.2.5}$$

to be solved with two initial conditions

$$U(0) = U_0, \quad V(0) = V_0. \tag{B.2.6}$$

Definitions of periodic solutions and their stability and orbital stability are similar to the difference equation case, and are omitted. A periodic solution which is the limit as $t \to \pm\infty$ of other solutions is known as a *limit cycle*.

B.2.1 Geometric Analysis (Phase Plane)

A lot of information about the solutions of such systems for general initial conditions may be obtained by sketching the (U, V)-plane, known as the *phase plane*, together with the *solution trajectories*. These solution trajectories represent solutions of the ordinary differential equations as curves in the (U, V)-plane, with time as a parameter. The procedure is as follows.

– Find out where the *nullcline* $f = 0$ is, where $f < 0$ and where $f > 0$. Do the same for g.

– There are *steady states* where $f = g = 0$, i.e. where the nullclines cross. Mark these.

– In the region(s) where $f > 0$ and $g > 0$, both U and V are increasing. Mark them with an arrow pointing rightwards and upwards. Mark other regions with an appropriate arrow.

– On the nullcline $f = 0$, U is neither increasing or decreasing. Mark it with an upward-pointing arrow where $g > 0$, downward where $g < 0$. Mark $g = 0$ similarly.

– Sketch in the solution trajectories following the arrows.

In some cases it will be necessary or useful to do more than this. For example, it might be necessary to analyse the behaviour near the steady states (see Section B.2.2 below). There might also be some trajectories that move along axes or do something else special. These tend to be important, and should be marked.

Example B.1

Sketch the phase plane for the system

$$\dot{U} = U(1 - U), \quad \dot{V} = -V.$$

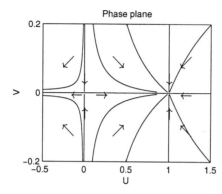

Figure B.2 Phase plane for $\dot{U} = U(1 - U)$, $\dot{V} = -V$. Note the trajectories along the axes and the line $U = 1$.

B.2.2 Linearisation

Let (U^*, V^*) be a steady state of Equations (B.2.5), so that $f(U^*, V^*) = g(U^*, V^*) = 0$. Defining $u = U - U^*$, $v = V - V^*$, we may proceed as we did for the first-order Equation (B.1.1). Assuming that we may neglect higher order terms if u and v are sufficiently small, we obtain the approximate (linearised) equations

$$\begin{aligned}
\dot{u} &= f_U(U^*, V^*)u + f_V(U^*, V^*)v, \\
\dot{v} &= g_U(U^*, V^*)u + g_V(U^*, V^*)v,
\end{aligned} \tag{B.2.7}$$

or, defining the Jacobian matrix $J(U, V)$ in the usual way,

$$\dot{\mathbf{w}} = J^*\mathbf{w}, \tag{B.2.8}$$

where $\mathbf{w}$ is the column vector $(u, v)^T$, and a star denotes evaluation at the steady state. The behaviour of the system near (U^*, V^*) depends on the eigenvalues of the matrix $J^* = J(U^*, V^*)$.

It can be shown that the neglect of higher order terms is valid, and the nonlinear system behaves like the linear system near the steady state, as long as neither of the eigenvalues of J^* has zero real part.

Making the definitions $\beta = \mathrm{tr}\, J^*$, $\gamma = \det J^*$, $\delta = \mathrm{disc}\, J^*$, the eigenvalue equation is $\lambda^2 - \beta\lambda + \gamma = 0$, and we may determine the character of the steady state from the signs of these.

Theorem B.2 (Steady States and Eigenvalues)

– If $\gamma < 0$ the (trivial) steady state of the second-order system (B.2.8) is a saddle point. Both eigenvalues are real, one positive and one negative.

– If $\gamma > 0$, $\delta > 0$, $\beta < 0$, it is a stable node. Both eigenvalues are real and negative.

– If $\gamma > 0$, $\delta > 0$, $\beta > 0$, it is an unstable node. Both eigenvalues are real and positive.

– If $\gamma > 0$, $\delta < 0$, $\beta < 0$, it is a stable focus. The eigenvalues are complex conjugates, with negative real part.

– If $\gamma > 0$, $\delta < 0$, $\beta > 0$, it is an unstable focus. The eigenvalues are complex conjugates, with positive real part.

– If $\gamma > 0$, $\delta < 0$, $\beta = 0$, it is a centre. The eigenvalues are complex conjugates, and purely imaginary.

The proof follows from the formula for solutions of the quadratic $\lambda^2 - \beta\lambda + \gamma = 0$.

In all cases above except the last, the same is true of the steady state of the nonlinear system (B.2.5). In the last case, it may be a centre or a stable or unstable focus depending on the nonlinear terms in the equation.

Theorem B.3 (Routh–Hurwitz Criteria for Second-order Systems)

Necessary and sufficient conditions for both roots of the quadratic

$$\lambda^2 + a_1\lambda + a_2 = 0 \tag{B.2.9}$$

to have negative real parts are

$$a_1 > 0, \quad a_2 > 0. \tag{B.2.10}$$

If $a_2 = 0$ there is an eigenvalue $\lambda = 0$ while if $a_1 = 0$ and $a_2 > 0$ there is a pair of complex conjugate eigenvalues on the real axis. It follows that necessary and sufficient conditions for asymptotic stability of the trivial steady state of the second-order linearised system (B.2.8) are given by

$$\beta < 0, \quad \gamma > 0, \tag{B.2.11}$$

where $\beta = \operatorname{tr} J^*$, $\gamma = \det J^*$. (This also follows from Theorem B.2.) If either of these inequalities is strictly violated, then it is unstable.

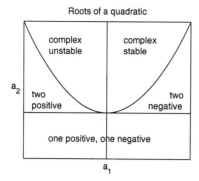

Roots of a quadratic

Figure B.3   Properties of the roots of $\lambda^2 + a_1\lambda + a_2 = 0$. The roots are stable in the positive quadrant. From there, zero-eigenvalue bifurcations may take place if the (positive) a_1-axis is crossed, and Hopf bifurcations if the (positive) a_2-axis is crossed. See Section B.4.

Figure B.3 summarises the information we have about the roots of the quadratic Equation (B.2.9).

Example B.4

Sketch the phase plane for the system

$$\dot{U} = V, \quad \dot{V} = -U(1-U) + cV,$$

where c is a positive constant. The nullclines are $V = 0$ and $V = U(1-U)/c$. If $V > 0$, then $\dot{U} > 0$; if $V < 0$, then $\dot{U} < 0$. If $V > U(1-U)/c$ then $\dot{V} > 0$; and vice versa. The steady states are at $(0,0)$ and $(1,0)$. The Jacobian matrix is

$$J(U,V) = \begin{pmatrix} 0 & 1 \\ -1 + 2U & c \end{pmatrix}.$$

At $(0,0)$ we have $\gamma = \det J^* = 1$, $\beta = \operatorname{tr} J^* = c > 0$ and $\delta = \operatorname{disc} J^* = c^2 - 4$. The character of the critical point depends on c; for $c \geq 2$ it is an unstable node whereas for $0 < c < 2$ it is an unstable focus. At $(1,0)$ we have $\gamma = \det J^* = -1$, so that the critical point is a saddle point. The phase plane is sketched below.

B.2.3 Poincaré–Bendixson Theory

Theorem B.5 (Poincaré–Bendixson)

Let (U,V) satisfy Equations (B.2.5) with initial conditions (B.2.6). Let f and g be Lipschitz continuous. Let (U,V) be bounded as $t \to \infty$.

Then either (U,V) is or tends to a critical point as $t \to \infty$, or it is or tends to a periodic solution. The same result holds as $t \to -\infty$.

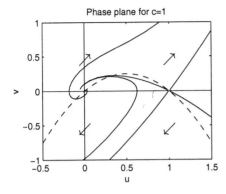

Figure B.4   Phase plane for the system $\dot{U} = V$, $\dot{V} = -U(1-U)+cV$, with $c = 1$. The point $(0,0)$ is an unstable focus and the point $(1,0)$ a saddle point. The dashed line is the nullcline $V = U(1-U)/c$.

The result follows from the fact that solution trajectories cannot cross. Then the topology of the plane gives the result.

Theorem B.6 (Dulac Criterion)

Let Ω be a simply connected region of the plane. Let the functions f and g be in $C^1(\Omega)$. Let $B \in C^1(\Omega)$ be such that the expression $\frac{\partial(Bf)}{\partial U} + \frac{\partial(Bg)}{\partial V}$ is not identically zero and does not change sign in Ω.

Then there are no periodic orbits of Equation (B.2.5) in Ω.

Theorem B.7 (Bendixson Criterion)

This is a particular case of the Dulac criterion with $B(U,V) = 1$ for all $(U,V) \in \Omega$.

These results follow from Green's theorem for integrals in the plane. Note that they are *negative criteria*. We can never deduce the existence of a periodic solution from them.

B.3 Some Results and Techniques for mth Order Systems

Consider the system

$$\dot{\mathbf{x}} = \mathbf{f}(\mathbf{x}),\qquad\qquad(B.3.12)$$

to be solved with initial conditions

$$\mathbf{x}(0) = \mathbf{x_0}.\qquad\qquad(B.3.13)$$

Here $\mathbf{f}$ is a column vector of functions f_i.

B.3.1 Linearisation

Linearisation may be applied to systems of any order. A steady state is asymptotically stable if all eigenvalues of the linearisation have negative real part. This may sometimes be analysed using the Routh–Hurwitz criteria, which may be derived for general m, but these get rather complicated as m increases. The conditions for $m = 3$ are as follows. Let the eigenvalue equation be given by

$$\lambda^3 + a_1\lambda^2 + a_2\lambda + a_3 = 0.$$

Then all the roots of this cubic have negative real part if and only if all the following inequalities hold:

$$a_1 > 0, \quad a_3 > 0, \quad a_1a_2 - a_3 > 0. \qquad (\text{B.3.14})$$

Another test that is sometimes useful in determining stability and related questions for higher-order systems is Descartes' rule of signs. Let the polynomial p be given by

$$p(\lambda) = a_0\lambda^n + a_1\lambda^{n-1} + \cdots + a_n = 0. \qquad (\text{B.3.15})$$

Let k be the number of sign changes in the sequence of coefficients $a_0, a_1, \cdots, a_n$, ignoring any zeroes, and let m be the number of real positive roots of the polynomial (B.3.15). Then $m \le k$, and k and m have the same parity (even or odd). Setting $\mu = -\lambda$ and applying the rule again, we may obtain information on the number of real negative roots.

B.3.2 Lyapunov Functions

Definitions: a function $\Phi : \mathbb{R}^m \to \mathbb{R}$ is *positive definite* in $\Omega \subset \mathbb{R}^m$ about a point $\mathbf{x} = \mathbf{x}^*$ if (a) $\Phi(\mathbf{x}^*) = 0$, and (b) $\Phi(\mathbf{x}) > 0$ for all $\mathbf{x} \in \Omega \backslash \{\mathbf{x}^*\}$. A function Ψ is *negative definite* if $-\Psi$ is positive definite. For a system $\dot{\mathbf{x}} = \mathbf{f}(\mathbf{x})$ and a function $\Phi \in C^1(\mathbb{R}^m, \mathbb{R})$ we may define a derivative, the *derivative of the function along trajectories of the system*, by

$$\dot{\Phi}(\mathbf{x}) \equiv \sum_{i=1}^{m} \frac{\partial\Phi}{\partial x_i}(\mathbf{x})f_i(\mathbf{x}).$$

A *Lyapunov function* $\Phi : \mathbb{R}^m \to \mathbb{R}$ for the system $\dot{\mathbf{x}} = \mathbf{f}(\mathbf{x})$ is a continuously differentiable positive definite function Φ in Ω whose derivative along trajectories of the system satisfies $\dot{\Phi}(\mathbf{x}) \le 0$ in Ω. If a Lyapunov function exists for a system then $\mathbf{x}^*$ is a stable steady state of the system. If also $\dot{\Phi}$ is negative definite in Ω then $\mathbf{x}^*$ is globally asymptotically stable in Ω, that is all solutions $\mathbf{x}$ of the system with initial conditions in Ω satisfy $\mathbf{x}(t) \to \mathbf{x}^*$ as $t \to \infty$.

B.3.3 Some Miscellaneous Facts

- Let each component function f_i in Equation (B.3.12) be Lipschitz continuous. Then system (B.3.12) with initial conditions (B.3.13) has a solution in a neighbourhood of $t = 0$. Moreover the solution is unique.

- Under the same conditions, the only way the solution can cease to exist is by blow-up in finite time.

- Hence *a priori* bounds can give existence for all time. These are usually obtained by finding *positively invariant sets*. (A set $D \subset \mathbb{R}^m$ is said to be *positively invariant* for Equation (B.3.12) if whenever $\mathbf{x}(0) \in D$ then $\mathbf{x}(t) \in D$ for all $t > 0$; it is *negatively invariant* if the same is true for all $t < 0$.)

B.4 Bifurcation Theory for Ordinary Differential Equations

The stability of a steady state solution of a system of ordinary differential equations

$$\dot{\mathbf{x}} = \mathbf{f}(\mathbf{x}, \mu)$$

depends on the eigenvalues of the Jacobian matrix there; it is asymptotically stable if all the eigenvalues have negative real part, and unstable if at least one of them has positive real part. As for difference equations, we use bifurcation theory to study the qualitative changes in solution behaviour that may occur as the parameter μ varies.

B.4.1 Bifurcations with Eigenvalue Zero

The description and analysis of bifurcations with eigenvalue 0 is almost identical to that of bifurcations with eigenvalue 1 for difference equations. It is easy to see why this is so if we compare the difference equation $x_{t+1} = x_t + f(x_t, \mu)$ with the differential equation

$$\dot{x} = f(x, \mu). \tag{B.4.16}$$

The steady states of each are the same, so the bifurcation diagrams for steady state solutions are the same. Moreover the eigenvalue of the difference equation is greater by 1 than the eigenvalue of the differential equation. We can therefore translate all the results of Section A.2 on saddle-node, transcritical and pitchfork bifurcations directly to the differential equation case. The conditions for each of these bifurcations of Equation (B.4.16) to occur at a bifurcation point (x_c, μ_c) are as follows. Again we have simplified the conditions by taking the transcritical and pitchfork bifurcations to be from the trivial solution.

- Saddle-node bifurcation: $f(x_c, \mu_c) = 0$, $f_x(x_c, \mu_c) = 0$, $f_\mu(x_c, \mu_c) \neq 0$, $f_{xx}(x_c, \mu_c) \neq 0$.

- Transcritical bifurcation, with one branch of solutions $x = 0$: $f(x_c, \mu_c) = 0$, $f_x(x_c, \mu_c) = 0$, $f_\mu(x_c, \mu_c) = 0$, $f_{x\mu}(x_c, \mu_c) \neq 0$, $f_{xx}(x_c, \mu_c) \neq 0$.

- Pitchfork bifurcation, with one branch of solutions $x = 0$: $f(x_c, \mu_c) = 0$, $f_x(x_c, \mu_c) = 0$, $f_\mu(x_c, \mu_c) = 0$, $f_{xx}(x_c, \mu_c) = 0$, $f_{x\mu}(x_c, \mu_c) \neq 0$, $f_{xxx}(x_c, \mu_c) \neq 0$.

These three bifurcations of steady states are the only ones that are possible for first order ordinary differential equations; there is no counterpart of the period-doubling bifurcation in difference equations (which has eigenvalue –1).

B.4.2 Hopf Bifurcations

In dimensions higher than 1, there is another way for a steady state to lose stability, by a pair of complex conjugate eigenvalues crossing the imaginary axis into the right half plane. The bifurcation associated with this loss of stability is usually called after Hopf (who analysed it for mth order systems in 1942), although others point to the fact that the bifurcation appeared 50 years earlier in Poincaré's work, and was analysed (for second-order systems) by Andronov in 1929, and call it the Poincaré–Andronov–Hopf bifurcation. The prototype for the Hopf bifurcation is

$$\dot{x} = \mu x - \omega y - x(x^2 + y^2), \quad \dot{y} = \omega x + \mu y - y(x^2 + y^2), \tag{B.4.17}$$

where ω is a constant. The bifurcation point is $(x, y, \mu) = (0, 0, 0)$, and the Jacobian matrix at $(0, 0, \mu)$ has eigenvalues $\mu \pm i\omega$. Transforming to polar coordinates (R, ϕ) by taking $R^2 = x^2 + y^2$, $\phi = \arctan(y/x)$, the equations become

$$\dot{R} = \mu R - R^3, \quad \dot{\phi} = \omega. \tag{B.4.18}$$

This has the trivial solution $R = 0$ for all values of μ, and a periodic solution $R = \sqrt{\mu}$, $\phi = \omega t$ for $\mu > 0$. The trivial solution loses stability as μ increases past 0, and the periodic solution is stable where it exists. More generally, let $\mathbf{x}^* = 0$ be a solution of a system of ordinary differential equations for all μ, and, for μ near μ_c, let the Jacobian matrix J^* of the system have two complex conjugate eigenvalues $\lambda(\mu)$ and $\overline{\lambda}(\mu)$ which are on the imaginary axis at $\mu = \mu_c$, all other eigenvalues having negative real part. For second-order systems this occurs if $\operatorname{tr} J^* = 0$ at $\mu = \mu_c$ while $\det J^* > 0$. Assume also that the complex conjugate eigenvalues cross the imaginary axis into the right half plane as μ increases past μ_c, $\operatorname{Re}\lambda'(\mu_c) > 0$. Then there exists a periodic solution, unique up to phase shifts, for every μ in a one-sided neighbourhood of μ_c. There are two possibilities.

– In the subcritical case, an unstable periodic solution exists for $\mu < \mu_c$, where the trivial solution is stable.

– In the supercritical case, a stable periodic solution exists for $\mu > \mu_c$, where the trivial solution is unstable.

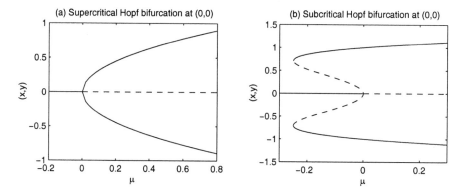

Figure B.5   (a) Super- and (b) subcritical Hopf bifurcations. The supercritical bifurcation is stable; the subcritical bifurcation is unstable near the bifurcation point, but frequently such bifurcations become stable further along the branch through a saddle-node bifurcation, as shown here. In this case, we expect to see the appearance of a large amplitude oscillation as μ increases through zero.

The condition for sub- or supercriticality is algebraically hairy. Make a transformation so that the bifurcation point is at the origin $(x, y, \mu) = (0, 0, 0)$, and so that the system with $\mu = 0$ is given by

$$\begin{pmatrix} \dot{x} \\ \dot{y} \end{pmatrix} = \begin{pmatrix} 0 & -\omega \\ \omega & 0 \end{pmatrix} \begin{pmatrix} x \\ y \end{pmatrix} + \begin{pmatrix} f(x, y, 0) \\ g(x, y, 0) \end{pmatrix}.$$

Define a by

$$a = \frac{1}{16}(f_{xxx} + f_{xyy} + g_{xxy} + g_{yyy})$$
$$+ \frac{1}{16\omega}(f_{xy}(f_{xx} + f_{yy}) - g_{xy}(g_{xx} + g_{yy}) - f_{xx}g_{xx} + f_{yy}g_{yy}), \quad \text{(B.4.19)}$$

evaluated at the origin. The condition for supercriticality is $a < 0$.

Typical bifurcation diagrams are given in Figure B.5. Since periodic behaviour is common in biology, the Hopf bifurcation is a useful tool. It arises, for example, in models for the propagation of a train of nerve impulses, in oscillatory metabolic processes, and in predator-prey systems.

C

Some Techniques for Partial Differential Equations

C.1 First-order Partial Differential Equations and Characteristics

Consider McKendrick's partial differential equation

$$\frac{\partial P}{\partial a} + \frac{\partial P}{\partial t} = -\mu P. \tag{C.1.1}$$

This is to be solved in the positive quadrant of the (a, t)-plane, and conditions are given at $a = 0$ and $t = 0$. Now think of travelling in this plane along one of the lines of the Lexis diagram (Figure 1.12), a straight line of slope 1, $t = t(a) = a + c$, for c constant. What happens to P as we follow one of these lines? On the line, P may be given in terms of a only, $P(a, t) = P(a, t(a)) = P(a, a + c)$. Taking the total derivative with respect to a,

$$\frac{dP}{da} = \frac{\partial P}{\partial a} + \frac{\partial P}{\partial t}\frac{dt}{da} = -\mu P,$$

using Equation (C.1.1). The partial differential equation for P reduces to an ordinary differential equation, which may be solved by separation of variables. A condition on $t = 0$ or $a = 0$ is required to complete the solution, depending on which is hit first by the line we are following in the Lexis diagram, i.e. depending whether $c > 0$ or $c < 0$.

Any curve along which a partial differential equation reduces to an ordinary differential equation is known as a *characteristic curve* (or just a *characteristic*) for the equation. More generally, consider a partial differential equation of the form

$$f \frac{\partial u}{\partial x} + g \frac{\partial u}{\partial y} = h.$$

The curve $(x(s), y(s))$, given parametrically, is a characteristic curve for this partial differential equation if the equation reduces to an ordinary differential equation on it. But since

$$\frac{du}{ds} = \frac{\partial u}{\partial x} \frac{dx}{ds} + \frac{\partial u}{\partial y} \frac{dy}{ds},$$

we need

$$\frac{dx}{ds} = f, \quad \frac{dy}{ds} = g, \quad \frac{du}{ds} = h.$$

These are the equations of the characteristics.

C.2 Some Results and Techniques for the Diffusion Equation

C.2.1 The Fundamental Solution

Consider the one-dimensional version of the diffusion Equation (5.2.4) with $f = 0$,

$$\frac{\partial u}{\partial t} = D \frac{\partial^2 u}{\partial x^2} \tag{C.2.2}$$

on the whole of $\mathbb{R}$, with initial condition

$$u(x, 0) = \delta(x). \tag{C.2.3}$$

Here δ is the Dirac delta function which is zero everywhere except at $x = 0$ and whose integral is unity. We claim that the solution of this initial-value problem is given by

$$u(x, t) = \frac{1}{\sqrt{4\pi D t}} \exp\left(-\frac{x^2}{4Dt}\right), \tag{C.2.4}$$

called the fundamental solution of the diffusion equation on $\mathbb{R}$. To verify this assertion, we need to check two things.

- The function satisfies Equation (C.2.2). This is an exercise in partial differentiation. Some algebra can be avoided by using the fact that $\log u = -\frac{1}{2}\log(4\pi Dt) - x^2/4Dt$, so that

$$\frac{\partial u}{\partial t} = \left(-\frac{1}{2t} + \frac{x^2}{4Dt^2}\right)u.$$

- The function satisfies the initial condition (C.2.3). Since $\delta(x)$ is not a true function this needs to be interpreted as meaning that $u(x,t) \to \delta(x)$ as $t \to 0+$, i.e. that

 - $u(x,0+) = 0$ for $x \neq 0$,

 - $\int_{-\infty}^{\infty} u(x,0+)dx = 1$.

The first of these is clear because of the exponential in u; for the second, we need the standard result that, for σ^2 independent of x,

$$\int_{-\infty}^{\infty} \exp\left(-\frac{x^2}{2\sigma^2}\right)dx = \sqrt{2\pi\sigma^2}.$$

Then, for any t,

$$\int_{-\infty}^{\infty} u(x,t)dx = \frac{1}{\sqrt{4\pi Dt}}\int_{-\infty}^{\infty}\exp\left(-\frac{x^2}{4Dt}\right)dx = \frac{1}{\sqrt{4\pi Dt}}\sqrt{2\pi(2Dt)} = 1.$$

(C.2.5)

Taking the limit as $t \to 0+$, and assuming that we can interchange limits and integrals, the result follows, and our proof is complete.

It is then easy to show that the solution of

$$\frac{\partial u}{\partial t} = D\left(\frac{\partial^2 u}{\partial x^2} + \frac{\partial^2 u}{\partial y^2}\right) \tag{C.2.6}$$

on the whole of $\mathbb{R}^2$, with initial conditions

$$u(x,y,0) = \delta(x)\delta(y) \tag{C.2.7}$$

is given by

$$u(x,y,t) = \frac{1}{4\pi Dt}\exp\left(-\frac{x^2}{4Dt}\right)\exp\left(-\frac{y^2}{4Dt}\right), \tag{C.2.8}$$

called the fundamental solution of the diffusion equation in two dimensions. This may be generalised in the obvious way to n dimensions,

$$u(\mathbf{x},t) = \frac{1}{(4\pi Dt)^{n/2}}\exp\left(-\frac{|\mathbf{x}|^2}{4Dt}\right). \tag{C.2.9}$$

The fundamental solution in two dimensions may be written

$$u(R, t) = \frac{1}{4\pi Dt} \exp\left(-\frac{R^2}{4Dt}\right),$$

(C.2.10)

and in three dimensions,

$$u(r, t) = \frac{1}{(4\pi Dt)^{3/2}} \exp\left(-\frac{r^2}{4Dt}\right).$$

(C.2.11)

The fundamental solutions may be used to solve more general initial-value problems.

Example C.1

Let $\hat{u}$ be the fundamental solution of the diffusion equation in one dimension. Show that the solution of the initial-value problem

$$\frac{\partial u}{\partial t} = D\frac{\partial^2 u}{\partial x^2}$$

(C.2.12)

on $\mathbb{R}$,

$$u(x, 0) = f(x),$$

(C.2.13)

is given by

$$u(x, t) = \int_{-\infty}^{\infty} \hat{u}(x - y, t)f(y)dy,$$

(C.2.14)

or equivalently, by a change of variables,

$$u(x, t) = \int_{-\infty}^{\infty} \hat{u}(y, t)f(x - y)dy.$$

(C.2.15)

Idea of proof: we have

$$(u_t - Du_{xx})(x, t) = \int_{-\infty}^{\infty} (\hat{u}_t - D\hat{u}_{xx})(y, t)f(x - y)dy = 0,$$

(C.2.16)

so that the given function u satisfies the diffusion equation, and

$$\lim_{t\to 0} u(x, t) = \lim_{t\to 0} \int_{-\infty}^{\infty} \hat{u}(x - y, t)f(y)dy = \int_{-\infty}^{\infty} \delta(x - y)f(y)dy = f(x),$$

(C.2.17)

so that it also satisfies the initial condition.

Now consider equations where there is a linear source of u as well as diffusion,

$$\frac{\partial u}{\partial t} = \alpha u + D\frac{\partial^2 u}{\partial x^2} \tag{C.2.18}$$

on $\mathbb{R}$,

$$u(x, 0) = f(x). \tag{C.2.19}$$

Define a new function v by $v(x, t) = u(x, t)e^{-\alpha t}$. Then

$$\frac{\partial v}{\partial t} = \frac{\partial u}{\partial t}e^{-\alpha t} - \alpha u e^{-\alpha t}, \tag{C.2.20}$$

$$\frac{\partial^2 v}{\partial x^2} = \frac{\partial^2 u}{\partial x^2}e^{-\alpha t}, \tag{C.2.21}$$

and

$$\frac{\partial v}{\partial t} - D\frac{\partial^2 v}{\partial x^2} = \left(\frac{\partial u}{\partial t} - \alpha u - D\frac{\partial^2 u}{\partial x^2}\right)e^{-\alpha t} = 0. \tag{C.2.22}$$

Moreover,

$$v(x, 0) = u(x, 0) = f(x), \tag{C.2.23}$$

so we can solve for v.

If $f(x) = \delta(x)$, v is the fundamental solution of the diffusion equation on $\mathbb{R}$, and

$$u(x, t) = e^{\alpha t}v(x, t) = \frac{1}{\sqrt{4\pi Dt}}\exp\left(\alpha t - \frac{x^2}{4Dt}\right). \tag{C.2.24}$$

It satisfies the integral condition

$$\int_{-\infty}^{\infty} u(x, t)dx = e^{\alpha t}. \tag{C.2.25}$$

C.2.2 Connection with Probabilities

On the website, we show that the right-hand side of Equation (C.2.4) is the probability density function at time t of the position of a particle performing a diffusion random walk in one dimension starting at the origin.

The normal (Gaussian) distribution in one dimension with mean zero and variance σ^2 is given by

$$N(0, \sigma^2) \sim \frac{1}{\sqrt{2\pi\sigma^2}}\exp\left(-\frac{x^2}{2\sigma^2}\right). \tag{C.2.26}$$

Hence the particle probability density function is normally distributed with mean zero, variance $2Dt$. Its root mean square distance from the origin is the standard deviation $\sqrt{2Dt}$. We may interpret this as meaning that the distance of a diffusing particle from its starting point after a time t is given on average by $\sqrt{2Dt}$.

The right-hand side of Equation (C.2.9) is the probability density function at time t of the position of a particle performing a diffusion random walk in n dimensions starting at the origin. Each component of the position of the particle is normally distributed with mean zero and variance $2Dt$. Its root mean square distance from the origin is $\sqrt{2nDt}$, by Pythagoras' theorem.

C.2.3 Other Coordinate Systems

In cylindrical polar coordinates R, ϕ, z,

$$\nabla u = \left(\frac{\partial u}{\partial R}, \frac{1}{R} \frac{\partial u}{\partial \phi}, \frac{\partial u}{\partial z} \right), \tag{C.2.27}$$

$$\nabla \cdot \mathbf{J} = \frac{1}{R} \frac{\partial}{\partial R}(R J_R) + \frac{1}{R} \frac{\partial J_\phi}{\partial \phi} + \frac{\partial J_z}{\partial z}, \tag{C.2.28}$$

and

$$\nabla^2 u = \frac{1}{R} \frac{\partial}{\partial R} \left(R \frac{\partial u}{\partial R} \right) + \frac{1}{R^2} \frac{\partial^2 u}{\partial \phi^2} + \frac{\partial^2 u}{\partial z^2}. \tag{C.2.29}$$

For cylindrically symmetric flow, all ϕ- and z-derivatives are zero. The formulae for plane polar coordinates are the same without the z-components and derivatives.

In spherical polar coordinates r, θ, ϕ,

$$\nabla u = \left(\frac{\partial u}{\partial r}, \frac{1}{r} \frac{\partial u}{\partial \theta}, \frac{1}{r \sin \theta} \frac{\partial u}{\partial \phi} \right), \tag{C.2.30}$$

$$\nabla \cdot \mathbf{J} = \frac{1}{r^2} \frac{\partial}{\partial r}(r^2 J_r) + \frac{1}{r \sin \theta} \frac{\partial}{\partial \theta}(\sin \theta J_\theta) + \frac{1}{r \sin \theta} \frac{\partial J_\phi}{\partial \phi}, \tag{C.2.31}$$

and

$$\nabla^2 u = \frac{1}{r^2} \frac{\partial}{\partial r} \left(r^2 \frac{\partial u}{\partial r} \right) + \frac{1}{r^2 \sin \theta} \frac{\partial}{\partial \theta} \left(\sin \theta \frac{\partial u}{\partial \theta} \right) + \frac{1}{r^2 \sin^2 \theta} \frac{\partial^2 u}{\partial \phi^2}. \tag{C.2.32}$$

For spherically symmetric flow, all θ- and ϕ-derivatives are zero.

C.3 Some Spectral Theory for Laplace's Equation

The *spectrum* of a differential operator is the set of its eigenvalues. We give some facts about the spectra for the Laplacian operator with various boundary conditions. Let $\Omega \in \mathbb{R}^n$ be a well-behaved finite domain with boundary $\partial\Omega$. Any sufficiently smooth domain is well-behaved, as are such domains as rectangles in $\mathbb{R}^2$. The equation

$$-\nabla^2 F = \lambda F$$

in Ω, with homogeneous Dirichlet boundary conditions $F = 0$ on $\partial\Omega$, has solution $F = 0$ for *any* value of λ, but for certain values of λ it has non-trivial solutions. These values of λ are the *eigenvalues* of the operator $-\nabla^2$ on Ω with Dirichlet boundary conditions, and the corresponding solutions are the *eigenfunctions*. The operator has an infinite sequence F_n of eigenfunctions forming an orthogonal basis for the Hilbert space consisting of square-integrable functions on Ω, and the corresponding eigenvalues λ_n are real and satisfy

$$0 < \lambda_0 < \lambda_1 \leq \lambda_2 \leq \cdots, \tag{C.3.33}$$

and $\lambda_n \to \infty$ as $n \to \infty$. If Dirichlet conditions are replaced by Neumann (zero-flux) conditions $\mathbf{n} \cdot \nabla F = 0$, also written $\frac{\partial F}{\partial n} = 0$, where $\mathbf{n}$ is the outward-pointing normal on $\partial\Omega$, the same conclusions hold, except now

$$0 = \lambda_0 < \lambda_1 \leq \lambda_2 \leq \cdots. \tag{C.3.34}$$

These results are useful because they allow us to write any function u of $\mathbf{x}$ satisfying the boundary conditions (and residing in the correct function space) as a linear combination of the appropriate eigenfunctions, $u(\mathbf{x}) = \sum_{n=0}^{\infty} a_n F_n(\mathbf{x})$. This is a (generalised) Fourier series. Moreover, since any function v of $\mathbf{x}$ and t is just a different function of $\mathbf{x}$ at each time t, we can write any such function satisfying the boundary conditions (and residing in the correct function space) as a linear combination of the appropriate eigenfunctions, the combination changing with t,

$$v(\mathbf{x}, t) = \sum_{n=0}^{\infty} G_n(t) F_n(\mathbf{x}). \tag{C.3.35}$$

We shall use this many times when analysing linearised reaction-diffusion equations. We give some examples below that appear repeatedly.

Example C.2

Find the spectrum and the eigenfunctions of $-\nabla^2$ on $\Omega = (0, L)$, with homogeneous Dirichlet boundary conditions.

In one dimension, $\nabla^2 = \frac{d^2}{dx^2}$. We need to solve the problem

$$-\frac{d^2 F}{dx^2} = \lambda F \text{ in } (0, L), \quad F(0) = F(L) = 0.$$

If $\lambda = -k^2 < 0$, then $F(x) = Ae^{kx} + Be^{-kx}$, which cannot satisfy the boundary conditions unless $A = B = 0$. If $\lambda = 0$, then $F(x) = A + Bx$, which also cannot satisfy the boundary conditions unless $A = B = 0$. If $\lambda = k^2 > 0$, then $F(x) = A \sin kx + B \cos kx$. The boundary condition $F(0) = 0$ implies $B = 0$, and the boundary condition $F(L) = 0$ then implies that $A \sin kL = 0$, so $A = 0$ (which we reject as trivial) or $kL = n\pi$ for some positive integer n. Indexing these so that we start with $n = 0$, we have $k_n = \frac{(n+1)\pi}{L}$, $\lambda_n = \frac{(n+1)^2 \pi^2}{L^2}$, $F_n(x) = \sin \frac{(n+1)\pi x}{L}$, taking $A = 1$ since we recognise that any constant multiple of an eigenfunction is also an eigenfunction. Here and generally where we have sinusoidal solutions, k_n is called the nth *wave-number*. The *wave-length* of F_n is $\frac{2\pi}{k} = \frac{2L}{n}$.

Example C.3

Find the spectrum and the eigenfunctions of $-\nabla^2$ on $\Omega = (0, L)$, with homogeneous Neumann (zero-flux) boundary conditions.

We need to solve the problem

$$-\frac{d^2 F}{dx^2} = \lambda F \text{ in } (0, L), \quad F'(0) = F'(L) = 0.$$

Using a similar argument to that of Example C.2, but now noting that $\lambda = 0$ does give a non-trivial solution, we have $k_n = \frac{n\pi}{L}$, $\lambda_n = \frac{n^2 \pi^2}{L^2}$, and $F_n = \cos \frac{n\pi x}{L}$.

Example C.4

Find the spectrum and the eigenfunctions of $-\nabla^2$ on $\Omega = (0, 2\pi)$, with periodic boundary conditions.

Periodic boundary conditions occur naturally in examples. Usually the dependent variable is an angular variable such as ϕ in cylindrical polars, and the line $\phi = 0$ is identical to the line $\phi = 2\pi$. In order that the function be continuously differentiable, it and its derivative must be the same on the two lines. Hence we need to solve the problem

$$-\frac{d^2 F}{d\phi^2} = \lambda F \text{ in } (0, 2\pi), \quad F(0) = F(2\pi), \quad F'(0) = F'(2\pi).$$

Using a similar argument to that of Example C.2, we have $k_n = n$, $\lambda_n = n^2$, and $F_0(\phi) = 1$, $F_n(\phi) = \cos n\phi$ or $F_n(\phi) = \sin n\phi$ if $n \geq 1$. We write $F_n(\phi) = e^{in\phi}$, where it is understood that both the real and the imaginary parts of F_n are eigenfunctions.

Example C.5

Find the spectrum and eigenfunctions of $-\nabla^2$ when $\Omega = \mathbb{R}^n$, and boundary conditions are replaced by the requirement that the function be bounded as $|\mathbf{x}| \to \infty$.

We need to find bounded solutions of the equation

$$-\nabla^2 F = \lambda R \text{ in } \mathbb{R}^n.$$

Then F given by $F(\mathbf{x}) = \exp(i\mathbf{k} \cdot \mathbf{x})$ is an eigenfunction for any constant vector $\mathbf{k} \in \mathbb{R}^n$, called a *wave-number vector*, and $\lambda = \mathbf{k} \cdot \mathbf{k}$ is the corresponding eigenvalue. Again, it is understood that the real and imaginary parts of F are both eigenfunctions. Hence any non-negative real number λ is an eigenvalue in this case, rather than just a countable set $\{\lambda_n\}$. In this case the sum (C.3.35) has to be replaced by an integral,

$$v(\mathbf{x}, t) = \int_{\mathbb{R}^n} G(\mathbf{k}, t) \exp(i\mathbf{k} \cdot \mathbf{x}) d\mathbf{k}, \tag{C.3.36}$$

a Fourier transform.

C.4 Separation of Variables in Partial Differential Equations

The method of separation of variables is a method of constructing solutions of various partial differential equations. It is of limited scope, being in particular only applicable to linear problems, but is nevertheless important in applications, including mathematical biology. It is used, for example, in determining whether spatially homogeneous solutions of reaction-diffusion equations are stable.

The diffusion operator $\frac{\partial}{\partial t} - D\nabla^2$ is always separable, in the sense that the time variable is always separable from the space variables, if the spatial domain is fixed in time and conditions are given at $t = 0$ and on the boundary of the spatial domain.

In this section we demonstrate the technique of separation of variables in n space dimensions. The result allows us to determine the asymptotic behaviour of solutions of the diffusion equation as $t \to \infty$.

Let u satisfy

$$\frac{\partial u}{\partial t} = D\nabla^2 u \qquad (C.4.37)$$

on a well-behaved bounded domain $\Omega \subset \mathbb{R}^n$, with Neumann boundary condition

$$\frac{\partial u}{\partial n} = \mathbf{n} \cdot \nabla u = 0 \qquad (C.4.38)$$

on the boundary $\partial\Omega$ of Ω, and initial condition

$$u(\mathbf{x}, 0) = u_0(\mathbf{x}) \qquad (C.4.39)$$

(Here as before $\mathbf{n}$ is the outward pointing normal to Ω at points of $\partial\Omega$.)

The strategy is simple. First, we look for functions of the form $u(\mathbf{x}, t) = F(\mathbf{x})G(t)$ which satisfy both the differential Equation (C.4.37) and the boundary condition (C.4.38). We shall find a whole set of them. Since the problem is linear, any linear combination of them will also satisfy Equations (C.4.37) and (C.4.38), by the principle of superposition. We then determine which linear combination also satisfies the initial condition (C.4.39).

Substituting $u(\mathbf{x}, t) = F(\mathbf{x})G(t)$ into Equation (C.4.37), we obtain

$$F(\mathbf{x})G'(t) = D\nabla^2 F(\mathbf{x})G(t),$$

so that

$$\frac{\nabla^2 F(\mathbf{x})}{F(\mathbf{x})} = \frac{G'(t)}{DG(t)} = -\lambda, \qquad (C.4.40)$$

say. The next part of the argument is crucial. Note that since $\lambda = \nabla^2 F(\mathbf{x})/F(\mathbf{x})$, it is independent of t, but since $\lambda = G'(t)/(DG(t))$, it is independent of $\mathbf{x}$. Hence λ is *constant*. The differential operator is separable. Hence

$$-\nabla^2 F = \lambda F, \quad G' = -\lambda DG. \qquad (C.4.41)$$

The boundary condition implies

$$\frac{\partial F}{\partial n} = \mathbf{n} \cdot \nabla F = 0. \qquad (C.4.42)$$

The first of Equations (C.4.41) with boundary conditions (C.4.42) is the eigenvalue problem discussed in Section C.3, and so we know about the spatial eigenvalues λ_n and spatial eigenfunctions F_n. In particular, the eigenvalues satisfy Equation (C.3.34). In this context of separation of time and space variation, the spatial eigenfunctions are known as the *spatial modes*. The corresponding G_n may now be found from the second of Equations (C.4.41), and are given by $G_n(t) = \exp(-\lambda_n Dt)$. This gives functions $F_n(\mathbf{x})G_n(t)$ which satisfy the diffusion Equation (C.4.37) and the boundary condition (C.4.38). The general solution of these is given by

$$u(\mathbf{x}, t) = \sum_{n=0}^{\infty} a_n F_n(\mathbf{x}) \exp(-\lambda_n Dt). \qquad (C.4.43)$$

This satisfies the initial condition (C.4.39) as well if

$$u_0(\mathbf{x}) = \sum_{n=0}^{\infty} a_n F_n(\mathbf{x}).$$

Since the F_n may be taken to be *orthonormal*, $\int_\Omega F_n(\mathbf{x})F_m(\mathbf{x})d\mathbf{x} = 0$ for $m \neq n$, $\int_\Omega F_n^2(\mathbf{x})d\mathbf{x} = 1$, and the coefficients a_n are given by

$$a_n = \int_\Omega u_0(\mathbf{x})F_n(\mathbf{x})d\mathbf{x}. \qquad (C.4.44)$$

From Equation (C.4.43) using Equation (C.3.34), $u(\mathbf{x}, t) \to a_0 F_0(\mathbf{x}) = \text{constant}$ as $t \to \infty$. For Dirichlet boundary conditions $u = 0$ on $\partial\Omega$, Equation (C.3.34) is replaced by Equation (C.3.33), and $u(\mathbf{x}, t) \to 0$ as $t \to \infty$.

Example C.6

Give the general solution of the diffusion equation in the one-dimensional domain $\Omega = (0, L)$ with homogeneous Neumann boundary conditions.

The spatial eigenvalue problem is that of example C.3, for which we can write down explicitly the eigenvalues $\lambda_0 = 0$, $\lambda_n = \frac{n^2\pi^2}{L^2}$, and the eigenfunctions $F_0(x) = 1$, $F_n(x) = \cos\frac{n\pi x}{L}$. The general solution is

$$u(x, t) = \frac{1}{2}a_0 + \sum_{n=1}^{\infty} a_n \cos\left(\frac{n\pi x}{L}\right) \exp\left(-\frac{n^2\pi^2}{L^2}Dt\right). \qquad (C.4.45)$$

(The factor $\frac{1}{2}$ in the first term is conventional; without it, the formula (C.4.44) for a_n requires a minor change for $n = 0$.)

Example C.7

Give the general solution of the diffusion equation in the one-dimensional domain $\Omega = (0, L)$ with homogeneous Dirichlet boundary conditions.

By a similar argument, we obtain

$$u(x, t) = \sum_{n=1}^{\infty} a_n \sin\left(\frac{n\pi x}{L}\right) \exp\left(-\frac{n^2\pi^2}{L^2}Dt\right). \qquad (C.4.46)$$

Example C.8

Write down the general solution of

$$\frac{\partial u}{\partial t} = \alpha u + D\nabla^2 u \qquad (C.4.47)$$

with homogeneous Neumann or Dirichlet boundary conditions, where α is a constant. Discuss the asymptotic behaviour as $t \to \infty$ if $\alpha \leq 0$.

Using the same transformation as in Section C.2.1, the general solution is given by

$$u(\mathbf{x}, t) = \sum_{n=0}^{\infty} a_n F_n(\mathbf{x}) \exp(\alpha t - \lambda_n Dt). \qquad (C.4.48)$$

For Neumann boundary conditions, when Equation (C.3.34) holds, all solutions of Equation (C.4.47) with $\alpha < 0$ decay exponentially with time. If $\alpha = 0$, the solution tends to a constant. For Dirichlet boundary conditions, when Equation (C.3.33) holds, all solutions with $\alpha \leq 0$ decay exponentially with time.

There are various conditions that must be satisfied for separation of variables to work, as follows.

- There must be a step like (C.4.40) where we can separate the t variation and the x variation in the partial differential equation. We say that the linear differential operator must be *separable*.

- All initial and boundary conditions must be on lines of constant t and x. We could not, for example, solve a problem on a growing domain directly by separation of variables.

- There are restrictions on the boundary operators that can be used. Neumann, Dirichlet and periodic boundary conditions are suitable, as are Robin boundary conditions, $au + b\frac{\partial u}{\partial n} = 0$, with $a \geq 0$, $b \geq 0$.

All of these conditions are easily violated.

C.5 Systems of Diffusion Equations with Linear Kinetics

The method of separation of variables may be extended to systems of diffusion equations. Although the separation of variables process itself is virtually unchanged, there are some differences in writing down the solutions, so we present the method here.

Consider the system of m equations on a domain Ω given by

$$\mathbf{u}_t = J\mathbf{u} + D\nabla^2\mathbf{u}, \qquad (C.5.49)$$

where J is a matrix with constant coefficients, and D is a constant matrix, usually a diagonal matrix of diffusion coefficients, with Neumann boundary conditions

$$\mathbf{n} \cdot \nabla\mathbf{u} = \frac{\partial\mathbf{u}}{\partial n} = 0 \qquad (C.5.50)$$

(by which we mean that $\mathbf{n} \cdot \nabla u_i = 0$ for each i) on $\partial\Omega$ and initial condition

$$\mathbf{u}(\mathbf{x}, 0) = \mathbf{u}_0(\mathbf{x}) \qquad (C.5.51)$$

in Ω.

Now look for a solution of (C.5.49) with (C.5.50) in separated form, $\mathbf{u}(\mathbf{x}, t) = \mathbf{c}F(\mathbf{x})G(t)$, where $\mathbf{c}$ is a constant vector. Then

$-$ F satisfies the Neumann boundary conditions, and

$-$ $\mathbf{u} = \mathbf{c}FG$ satisfies (C.5.49),

i.e.

$$\mathbf{c}F(\mathbf{x})G'(t) = J\mathbf{c}F(\mathbf{x})G(t) + D\mathbf{c}\nabla^2 F(\mathbf{x})G(t),$$

or, dividing through by the scalar functions F and G,

$$\mathbf{c}\frac{G'(t)}{G(t)} = J\mathbf{c} + D\mathbf{c}\frac{\nabla^2 F(\mathbf{x})}{F(\mathbf{x})}.$$

The left hand side of this equation is independent of $\mathbf{x}$, so the right hand side must also be independent of $\mathbf{x}$, so $\frac{\nabla^2 F(\mathbf{x})}{F(\mathbf{x})}$ must be constant, $-\lambda$, say. The operator is still separable. This leads to exactly the same eigenvalue problem for F that we had in the single equation case, so we have a set of eigenvalues λ_n of $-\nabla^2$ on Ω with Neumann boundary conditions and the corresponding eigenfunctions or spatial modes F_n.

The equation for the temporal behaviour corresponding to the nth spatial mode is

$$\mathbf{c}_n G_n'(t) = J\mathbf{c}_n G_n(t) - \lambda_n D\mathbf{c}_n G_n(t) = A_n\mathbf{c}_n G_n(t),$$

say, a system of linear ordinary differential equations with constant coefficients. It has exponential solutions, so look for a solution in the form $G_n(t) = \exp(\sigma_n t)$. (More generally, if there are repeated roots, there may be a solution in the form of a product of a polynomial and an exponential.) The equation becomes, on cancelling through by $\exp(\sigma_n t)$,

$$\sigma_n \mathbf{c}_n = (J - \lambda_n D)\mathbf{c}_n = A_n \mathbf{c}_n, \tag{C.5.52}$$

say, an algebraic eigenvalue problem, i.e. σ_n is an eigenvalue of A_n with eigenvector $\mathbf{c}_n$. These eigenvalues are referred to as the temporal eigenvalues, to distinguish them from the spatial eigenvalues λ_n. The temporal eigenvalues are given by

$$\det(\sigma_n I - A_n) = \det(\sigma_n I - J + \lambda_n D) = 0, \tag{C.5.53}$$

a polynomial of mth degree in σ_n, where I is the identity matrix. Assume for simplicity that each of these polynomials has m distinct roots $\sigma_{n1}, \cdots, \sigma_{nm}$, with corresponding eigenvectors $\mathbf{c}_{n1}, \cdots, \mathbf{c}_{nm}$, and let $G_{ni}(t) = \exp(\sigma_{ni} t)$. We now have a whole set of solutions given by $\mathbf{u}_{ni}(\mathbf{x}, t) = \mathbf{c}_{ni} F_n(\mathbf{x}) G_{ni}(t)$, for $n = 0, 1, \cdots$, and $i = 1, 2, \cdots, m$. The general solution of the system is obtained by taking linear combinations of these,

$$\mathbf{u}(\mathbf{x}, t) = \sum_{n=0}^{\infty} \sum_{i=1}^{m} a_{ni} \mathbf{c}_{ni} F_n(\mathbf{x}) G_{ni}(t). \tag{C.5.54}$$

This also satisfies the initial condition (C.5.51) if

$$\mathbf{u}_0(\mathbf{x}) = \sum_{n=0}^{\infty} \sum_{i=1}^{m} a_{ni} \mathbf{c}_{ni} F_n(\mathbf{x}).$$

Since the F_n are orthonormal, this reduces to

$$\int_{\Omega} \mathbf{u}_0(\mathbf{x}) F_n(\mathbf{x}) d\mathbf{x} = \sum_{i=1}^{m} a_{ni} \mathbf{c}_{ni},$$

for each n, a set of m equations for the m unknowns a_{ni}, $i = 1, \cdots, m$. (NB If the initial conditions involve only one F_n, then so will the solution.) In fact, we do not usually solve initial-value problems. Our main concern is whether any of the σ have positive real part, i.e. whether $u = 0$ is stable or not.

If $\Omega = \mathbf{R}^n$, then any non-negative λ is a spatial eigenvalue, and the temporal eigenvalue problem for σ is

$$\sigma \mathbf{c} = (J - \lambda D)\mathbf{c} = A\mathbf{c}, \tag{C.5.55}$$

say, and σ is a root of the polynomial

$$\det(\sigma I - J + \lambda D) = 0 \tag{C.5.56}$$

for any $\lambda \geq 0$.

C.6 Separating the Spatial Variables from Each Other

In some cases we need not only to separate the time variables from the space variables, but also to separate the space variables themselves. The Laplacian operator with Neumann or Dirichlet boundary conditions is separable on appropriate spatial domains. The boundary conditions have to be specified on lines (or surfaces) where a dependent variable is constant. We shall assume that we have already separated the time and space variables, so that we have an eigenvalue problem in the space variables alone.

Example C.9

In Cartesian coordinates x, y, find the solutions of the eigenvalue problem

$$-\nabla^2 F = -F_{xx} - F_{yy} = \lambda F \text{ in } (0, a) \times (0, b),$$

with boundary conditions

$$F(0, y) = F(a, y) = 0 \text{ for } y \in (0, b),$$
$$\frac{\partial F}{\partial y}(x, 0) = \frac{\partial F}{\partial y}(x, b) = 0 \text{ for } x \in (0, a).$$

We look for separated solutions in the form $F(x, y) = P(x)Q(y)$. The differential equation gives us

$$-\frac{P''(x)}{P(x)} - \frac{Q''(y)}{Q(y)} = \lambda,$$

so that $P''(x)/P(x)$ and $Q''(y)/Q(y)$ must each be constants. The operator is separable. The boundary conditions give us

$$P(0) = P(a) = 0, \quad Q'(0) = Q'(b) = 0.$$

Both the P and the Q problems are familiar, and we immediately deduce that $P_m(x) = \sin \frac{m\pi x}{a}$ for $m \geq 1$, $Q_n(y) = \cos \frac{n\pi y}{b}$ for $n \geq 0$, and the eigenvalues are $\lambda = \frac{m^2\pi^2}{a^2} + \frac{n^2\pi^2}{b^2}$.

Example C.10

This is a problem on the surface of a circular cylinder of radius a, which arises from the tail-pattern analysis of Section 7.5. In cylindrical polar coordinates ϕ, z, solve the eigenvalue problem

$$-\nabla^2 F = -\frac{1}{a^2}F_{\phi\phi} - F_{zz} = \lambda F, \tag{C.6.57}$$

with boundary conditions

$$F_z(\phi, 0) = F_z(\phi, h) = 0, \tag{C.6.58}$$

Neumann (zero-flux) boundary conditions, and

$$F(0, z) = F(2\pi, z), \quad F_\phi(0, z) = F_\phi(2\pi, z), \tag{C.6.59}$$

periodic boundary conditions.

Look for a solution of the form $F(\phi, z) = P(\phi)Q(z)$. Then Equation (C.6.57) becomes, on substituting in and dividing through by $P(\phi)Q(z)$,

$$-\frac{1}{a^2}\frac{P''(\phi)}{P(\phi)} - \frac{Q''(z)}{Q(z)} = \lambda, \tag{C.6.60}$$

so that P''/P and Q''/Q must both be constants. The operator is separable. The boundary conditions give

$$P(0) = P(2\pi), \quad P'(0) = P'(2\pi), \quad Q'(0) = Q'(h) = 0.$$

Both the P and the Q problems are familiar, and know that the solutions are $P_m(\phi) = e^{im\phi}$, $m \geq 0$, $Q_n(z) = \cos\frac{n\pi z}{h}$ for $n \geq 0$, and the eigenvalues λ_{mn} are given by

$$\lambda_{mn} = -\frac{1}{a^2}\frac{P_m''(\phi)}{P_m(\phi)} - \frac{Q_n''(z)}{Q_n(z)} = \frac{m^2}{a^2} + \frac{n^2\pi^2}{h^2},$$

for $m \geq 0$, $n \geq 0$, with eigenfunctions $\cos\frac{m\pi z}{h}e^{in\phi}$.

D
Non-negative Matrices

D.1 Perron–Frobenius Theory

A matrix $M = (m_{ij})$ is positive if all its elements are positive, and non-negative if all its elements are non-negative. Non-negative matrices occur in mathematical biology in several contexts. Population projection matrices, or Leslie matrices, are non-negative; these are described in Chapter 1, and give the stage-dependent birth rates and the transition rates from one stage to another in stage-structured populations. So are contact matrices; these are described in Chapter 3, give the contact rates between members of a structured population, and are used there to analyse the spread of an infectious disease. Perron–Frobenius theory describes the eigenvalues and eigenvectors of such matrices. One of the most important facts about them, used in both contexts above, is the following.

Theorem D.1 (Non-negative Matrices)

Let M be a non-negative matrix. Then there exists one eigenvalue λ_1 that is real and greater than or equal to any of the others in magnitude, $\lambda_1 \geq |\lambda_i|$. This is called the *principal* or *dominant eigenvalue* of M. The right and left eigenvectors $\mathbf{v}_1$ and $\mathbf{w}_1$ corresponding to λ_1 are real and non-negative.

In some cases more can be said about the dominant eigenvalue and its eigenvectors. To present the theory we need to define some terms. We say that there exists an *arc* from i to j if $a_{ji} > 0$; a *path* from i to j is a sequence of arcs starting at i and ending at j; a *loop* is a path from i to itself. A non-negative matrix is either reducible or irreducible; it is *irreducible* if, for each i and j, there exists a path from i to j. In terms of stage-structured populations, a matrix is irreducible if each stage i may contribute at some future time to each other stage j. This is almost always true, unless there are some post-reproductive stages that cannot contribute to any younger stages. An irreducible matrix is either *primitive* or *imprimitive*; it is primitive if the greatest common divisor d of the length of its loops is 1. If $d > 1$, it is called the *index of imprimitivity*. Most population projection matrices are primitive. The only significant exception is in cases like the Pacific salmon, which has a single reproductive stage at two years of age. The population projection matrix with an annual census therefore has index of imprimitivity $d = 2$.

Theorem D.2 (Primitive Matrices)

If M is primitive (and therefore, *a fortiori*, irreducible), then (in addition to the results of Theorem D.1), its dominant eigenvalue λ_1 is

– positive,

– a simple root of the eigenvalue equation $|M - \lambda I| = 0$, and

– *strictly* greater in magnitude than any other eigenvalue.

Moreover, the right and left eigenvectors $\mathbf{v}_1$ and $\mathbf{w}_1$ corresponding to λ_1 are positive. There may be other real eigenvalues besides λ_1, but λ_1 is the only one with non-negative eigenvectors.

Theorem D.3 (Irreducible but Imprimitive Matrices)

If M is irreducible but imprimitive, with index of imprimitivity d, then (in addition to the results of Theorem D.1), its dominant eigenvalue λ_1 is

– positive, and

– a simple root of the eigenvalue equation.

– Although $\lambda_1 \geq |\lambda_i|$ for all i, the spectrum of M contains d eigenvalues equal in magnitude to λ_1, λ_1 itself and $\lambda_1 \exp(2k\pi i/d)$, $k = 1, 2, \cdots, d-1$.

Moreover, the associated right and left eigenvectors $\mathbf{v}_1$ and $\mathbf{w}_1$ are positive.

Hints for Exercises

1.1 This is a particular case of Equation (1.2.4), discussed in example 1.1. There is a transcritical bifurcation point at $(x^*, \lambda) = (0, 1)$; as λ increases past this point there is an interchange of stability from the trivial to the non-trivial steady state $\lambda - 1$. The solution is

$$x_n = \frac{1 - \mu}{\mu + ((1 - \mu)/x_0 - \mu)\mu^n} = \frac{(\lambda - 1)x_0\lambda^n}{x_0\lambda^n + (\lambda - 1) - x_0}.$$

1.2 S is a stable steady state; C is an unstable steady state, and A and B are stable period-2 solutions of the equation.

1.3 a) The linearised equation is $n_{n+1} = h'(N^*)n_n$, and the steady state is linearly stable for $|h'(N^*)| < 1$.

   b) No depensation in the survivorship function for the young, f, or the fertility function for the adults, g.

   c) Fixed points satisfy $a^* = f(y^*)$, $y^* = g(a^*)$, so that they are the intersections of $y = g(a)$ and $y = f^{-1}(a)$. Note that $a_{n+2} = f(y_{n+1}) = f(g(a_n))$, so for stability we must have $|h'(a^*)| < 1$, where h is defined by $h(a) = f(g(a))$. The result follows. Stability depends on whether $y = g(a)$ crosses $y = f^{-1}(a)$ from below to above or from above to below. Stable steady states are S_0, S_2 or an alternation between $(y_2^*, 0)$ and $(0, a_2^*)$.

1.4 a) $N_n/(N_n + S)$ is the probability that a given insect picks a fertile mate, if it picks an insect at random from the population.

   b) If $N^* \neq 0$, $N^* = f(N^*)$ may be solved to give $S = S(N^*) = \frac{R_0 N^*}{1 + aN^*} - N^* = \frac{(R_0 - 1 - aN^*)N^*}{1 + aN^*}$, positive between $N^* = 0$ and $N^* = (R_0 - 1)/a$, where it is zero.

   c) The maximum of S on this curve is at $N^* = (\sqrt{R_0} - 1)/a$, where $S = S_c = (\sqrt{R_0} - 1)^2/a$.

   d) See Figure E.1.

1.5 a) Immediate, on substitution of the given functions.

   b) We obtain $r = \alpha C - \beta$, $K = (\alpha C - \beta)/(\alpha\gamma)$.

   c) If $r < 0$, i.e. $C < \beta/\alpha$, then $N(t) \to 0$ as $t \to \infty$.

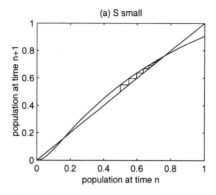
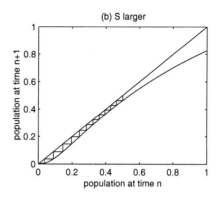

Figure E.1   Cobweb maps for the sterile insect control equation for two values of S. In (a) $S < S_c$, too small to drive the population to extinction, while in (b) $S > S_c$, the saddle-node bifurcation at (N^*, S_c) has taken place, losing the two non-trivial steady states, and the population goes to extinction.

   d) The main advantage of the first approach over the second is its simplicity. The main advantages of the second are the insight gained and its testability, so that, for example, we could predict the effect of a reduction in C on the growth rate and carrying capacity.

1.6 The logistic equation $\dot{N} = rN(1 - N/K)$ is a Bernoulli equation and may be solved by the substitution $M = 1/N$, to obtain $\dot{M} = -\dot{N}/N^2 = -r(M - 1/K)$, which has solution $M(t) = 1/N(t) = 1/K + (1/N_0 - 1/K)e^{-rt}$. The result follows after some algebra, which is made easier if we note that we can take $N_0 = K/2$ without loss of generality.

1.7 The function f is zero at 0, U, and K, negative between 0 and U and positive between U and K. Hence $N(t) \to 0$ if $0 < N_0 < U$, $N(t) \to K$ if $U < N_0 < \infty$, and there is critical depensation, with a minimum viable population size U.

1.8   a) If $f(N) = rN^2$, $f'(N) = 2rN$, an increasing function of N.
    b) By separation of variables, $N(t) = N_0/(1 - rN_0 t)$, and $N(t) \to \infty$ as $t \to 1/(rN_0)$.
    c) The trivial steady state 0 is unstable, and the non-trivial steady state K is stable.
    d) The analysis parallels that of Exercise 1.5.

1.9 Integrating the Bernoulli equation $dL/dt = -(\mu_1 + \mu_2 L)L$ by the substitution $M = 1/L$, or otherwise, from $n + t_1$ to t with initial condition $L(n + t_1) = L_0$, we obtain
$$L(t) = \frac{\mu_1 L_0}{(\mu_1 + \mu_2 L_0)\exp(t - n - t_1) - \mu_2 L_0}.$$
Now, putting $t = n + t_2$ and using the assumptions given, the result follows.

1.10 $Y^* = f(N^*) - N^* = N^*(e^{r(1-N^*/K)} - 1)$, so Y^* is a positive function in $(0, K)$, zero at 0 and K.

1.11   a) The model with fishing is given by $\dot{N} = rN(1 - N/K) - qEN$, which has steady states at 0 and $N^* = K(1 - qE^*/r)$.
    b) $Y^* = qE^* N^* = qE^* K(1 - qE^*/r)$.
    c) For maximum Y^*, $qE^*/r = \frac{1}{2}$, so $Y^*_{\max} = \frac{1}{4}rK$.

1.12   a) Immediate.

b) The yield-effort relationship for critical depensation is as shown in Figure 1.7(b); for non-critical depensation it is similar except that the unstable branch reaches the E^*-axis at a value $E_1 = F(0)/q$ which is non-zero (unless $F(0) = 0$).

1.13 a) The term $m(1 - p)$ represents colonisation of an empty patch from the mainland.

b) The equation is $\dot{p} = f(p)$, where f is a quadratic with $f(0) = m$, $f(1) = -e$, so there is a single steady state in $(0, 1)$, which is stable.

1.14 The non-trivial steady state is $N^* = K$. The equation linearised about this steady state is $\dot{n}(t) = -rn(t - \tau)$, so the characteristic equation of the linearisation is $s = -re^{-s\tau}$, and no instability is possible.

1.15 a) Immediate.

b) Immediate.

c) Substituting into the equation and performing the integration, we obtain the characteristic equation $s = \tau/(1 + \tau s)$, or $\tau s^2 + s - \tau = 0$.

d) The characteristic equation has one positive and one negative root, so the steady state is unstable.

e) The result follows by differentiating the expression for P and integrating by parts.

1.16 The linearisation is again given by Equation (1.7.21). Substituting in $n(t) = n_0 \exp(st)$, and the form given for k, and performing the integration, we obtain $s = 1/(1 + \tau s)^2$. This characteristic equation still has at least one positive root, and the steady state is unstable.

1.17 No stochastic effects, no death, no effects of competition, no predators, discrete time, no effects of ageing on fertility, etc.

1.18 Define $y_n = \sum_{k=0}^{\infty} y_{k,n}$. We shall use the fact that $y_{0,n+1} = y_{1,n} + y_{2,n} + \cdots$. Then

$$y_{n+2} = y_{0,n+2} + y_{1,n+2} + y_{2,n+2} + y_{3,n+2} + \cdots$$
$$= (y_{1,n+1} + y_{2,n+1} + \cdots) + y_{0,n+1} + (y_{1,n+1} + y_{2,n+1} + \cdots)$$
$$= (y_{0,n} + y_{1,n} + \cdots) + (y_{0,n+1} + y_{1,n+1} + \cdots) = y_n + y_{n+1}.$$

Since $y_{0,1} = 1$, $y_1 = 2$, and the resulting sequence is 1, 2, 3, 5, 8, 13, $\cdots$, the Fibonacci sequence shifted by one place (which it must be since no rabbits ever die).

1.19 a) γ is the potential per capita production of offspring, taking into account all factors but over-winter survival and germination. σ is the probability of surviving the winter, α the fraction of one-winter survivors that germinate, β the fraction of two-winter survivors that germinate.

b) Censusing the population at the flowering stage (just before the new seeds are produced), when there are P_n plants and S_n one-year-old seeds, we have

$$\begin{pmatrix} P_{n+1} \\ S_{n+1} \end{pmatrix} = \begin{pmatrix} \gamma\sigma\alpha & \sigma\beta \\ \gamma\sigma(1-\alpha) & 0 \end{pmatrix} \begin{pmatrix} P_n \\ S_n \end{pmatrix}.$$

(Note that this is not in the form of Equation (1.9.26), because there is reproduction and not merely survival between the plant and the seed stage.)

c) The eigenvalue equation is $Q(\lambda) = \lambda^2 - \gamma\sigma\alpha\lambda - \gamma\sigma^2\beta(1-\alpha) = 0$, which has one positive and one negative root, and therefore a solution $\lambda > 1$ if and only if $Q(1) < 0$.

d) $R_0 = \gamma\sigma\alpha + \gamma\sigma^2\beta(1-\alpha)$.

1.20 a) Immediate; γ is the necessary constant of proportionality.

b) The characteristic equation is $(1 - f - \lambda)(-\lambda) - \gamma f = 0$, so the principal eigenvalue is $\lambda_1 = \frac{1}{2}(1 - f + \sqrt{(1 - f)^2 + 4\gamma f})$.

c) Since the problem is linear, $\lambda = 1$ must be a root of the characteristic equation, so $f - \gamma f = 0$, $f = 0$ or $\gamma = 1$.

d) Although C^* is stable, it is not asymptotically stable. Also, the model is not *structurally stable*, meaning that a small change in parameter values can lead to drastic changes in solution behaviour.

1.21 A proof by induction is straightforward.

1.22 The result follows on expanding the determinant by its top row.

1.23 Since $\lambda_1 \mathbf{v}_1 = L\mathbf{v}_1$, we obtain $\lambda_1 v_{1,i} = s_i v_{1,i-1}$ for each i, where the $v_{1,i}$ are the components of $\mathbf{v}_1$, and the result follows.

1.24 a) The context shows that $f_i \geq 0$ for each i, so $\hat{f}$ is the sum of monotone decreasing terms, each of which exhibits the limiting behaviour given.

b) $R_0 = \hat{f}(0)$, while r_1 is the real root of $\hat{f}(r) = 1$. If $R_0 > 1$, $\hat{f}(0) > 1$, so r_1, the real root of $\hat{f}(r)$ is positive, by the properties of $\hat{f}$ in (a). A similar argument holds for $R_0 < 1$, and the results for λ_1 follow immediately.

c) Expanding $\hat{f}$ about zero, we have $1 = \hat{f}(r_1) \approx \hat{f}(0) + r_1 \hat{f}'(0) + h.o.t.$ The result follows, for r_1 small.

1.25 a) $R_0 = \sum_{i=1}^{10} l_i m_i = 5.904$.

b) Using the result of the previous exercise, based on the assumption that r_1 is small, $r_1 = (R_0 - 1)/|\hat{f}'(0)|$, where $|\hat{f}'(0)| = \sum_{i=1}^{10} i l_i m_i$; the result is $r_1 = 0.273$.

c) Use Newton's method or similar.

1.26 a) The result follows from Equation (1.10.29), on multiplying by s^n and summing.

b) Immediate.

c) Immediate.

d) Since $b(s)$ is the ratio of two polynomials $g(s)$ and $1 - f(s)$ with the degree of g less than the degree of $1 - f(s)$, the theory of partial fractions tells us that $b(s) = \sum_{i=1}^{\omega} \frac{B_i}{(1-\lambda_i s)}$, where each $B_i = \lim_{s \to s_i} (1 - \lambda_i s) g(s)/(1 - f(s))$ is a constant. Since $\log(1 - f(s)) = \sum \log(1 - \lambda_i s)$, so $f'(s)/(1 - f(s)) = \sum \lambda_i/(1 - \lambda_i s)$, we may multiply through by $(1 - \lambda_i s) g(s)$ and take limits as $s \to s_i$, to obtain $B_i = \lambda_i g(s_i)/f'(s_i)$, and the result follows on expanding $(1 - \lambda_i s)^{-1}$.

1.27 a) Since $\bar{a} R_0 = -\hat{f}'(0)$, this is just a restatement of Exercise 1.24(c).

b) From part (a), $R_0 \approx 1/(1 - r_1 \bar{a})$, so $\log R_0 \approx r_1 \bar{a}$, and the result follows.

1.28 The first claim follows since the integrand is decreasing for each a, and the claims on limits follow on dividing the range of the integral into $(0, \epsilon)$ and (ϵ, ∞), for ϵ sufficiently small.

1.29 The first claim follows from the definition $R_0 = \int_0^\infty f(a) da$, and the second from the monotonicity and limit properties of the previous exercise.

1.30 a) Multiply Equation (1.10.36) by e^{-st} and integrate, and the result follows from the Laplace transform convolution theorem.

b) The first part is the previous exercise re-stated, and the second from the inequality $|\tilde{f}(s)| \leq \int_0^\infty |f(a)e^{-sa}| da = \int_0^\infty f(a)e^{-(\mathrm{Re}\, s)a} da$.

c) This follows from parts (a) and (b) and the formula for the inverse Laplace transform.

1.31 a) $l(b)/l(a)$ is the probability of surviving from age a to age b, and $\exp(-r(b - a))$ the amount a birth to an individual at age b must be discounted if the individual is now at age a.

b) The reproductive value will increase from birth, reach a maximum some time between first and maximal reproduction, and then decline, becoming zero for individuals of post-reproductive age.

c) Heart disease caused the loss of 250 units of reproductive value, malaria 967, almost four times as much.

1.32 a) We may think of l as the fraction of a cohort surviving to age a.

b) Integrate Equation (1.11.40), with $l = u$ independent of t and $l(0) = 1$.

1.33 a) The rate at which members of a cohort die when the cohort is aged a is $d(a)l(a)$, so the average age of death is $\int_0^\infty a d(a)l(a)da / \int_0^\infty d(a)l(a)da$. The result follows since $\int_0^\infty d(a)l(a)da = 1$ is the probability of eventual death.

b) A similar argument gives the required result, once the number of age a alive now is discounted by e^{-ra} to account for the smaller population in the past.

c) If d is constant, $l(a) = e^{-da}$, life expectancy is $1/d$ and the mean age of those dying simultaneously is $1/(r+d)$. The second expression takes account of the smaller number of older individuals, arising from the fact that the population was smaller when they were born.

2.1 If the effects of host intra-specific competition are negligible during the search period, then we may simply replace R_0 by $R_0(1 + aH_n)^{-b}$ in the H-equation. The solutions are now bounded, and may tend to the steady state (H^*, P^*) or to a closed invariant curve, as we would expect from a Naimark–Sacker bifurcation, in (H, P)-space.

2.2 The searching efficiency is $a_0 P^{-m}$, a decreasing function of P. This stabilises the steady state.

2.3 Assume that hosts always find an available refuge. Then $S_n = \min(\hat{H}, H_n)$ hosts find a refuge, $U_n = H_n - S_n$ do not. The equations become

$$H_{n+1} = R_0(S_n + U_n \exp(-aP_n)), \quad P_{n+1} = cU_n(1 - \exp(-aP_n)).$$

The solutions are now bounded. If there are too many refuges the parasitoid may go extinct, but otherwise the solutions may tend to the steady state (H^*, P^*) or to a closed invariant curve in (H, P)-space.

2.4 a) The eigenvalue equation for the linearised equations is $\lambda^2 + \epsilon u^* \lambda + a u^* v^* = 0$, and the result follows.

b) $\partial/\partial u(Bf) + \partial/\partial v(Bg) = -\epsilon/v < 0$, which does not change sign.

c) It is easy to see that Φ (or, strictly, $\Phi - \Phi(u^*, v^*)$), is positive definite about (u^*, v^*) in the positive quadrant. Now note that $1 - \epsilon u - v = -\epsilon(u - u^*) - (v - v^*)$. It follows that $\dot{\Phi}(u, v) = -a\epsilon(u - u^*)^2 \leq 0$, as required.

d) All solutions in the positive quadrant spiral in to the steady state.

e) Φ is decreasing, bounded below, so $\dot{\Phi} \to 0$, $u \to u^*$. The result follows from the Poincaré–Bendixson theorem, Section B.2.3 of the appendix.

2.5 a) We have to assume that u starts above k; then it never drops below it, so $u - k > 0$. Only these $u - k$ prey are available to the predators.

b) The steady state $(u^*, v^*) = (k + 1, k + 1)$ is stable (node or focus).

2.6 a) Predators switch from handling to searching mode at constant rate β, and from searching to handling on encounter with a prey at rate aU.

b) Immediate.

c) U equation immediate, V equation assumes births proportional to rate of consumption of prey, deaths at constant rate.

2.7 a) On g, u has carrying capacity K. On p, predation reduces population of prey (unless there is none). On q; for the predators, the more prey the better.

b) Equations (2.4.13) satisfy Gause's conditions. If there is a coexistence steady state, the nullclines are as in Figure 2.9; there is no coexistence steady state (in the positive quadrant) if the vertical nullcline $\dot{v} = 0$, or $q(u) = d$, is to the right of the steady state at $(K, 0)$.

c) Trajectories enter $(0,0)$ along the v-axis, leave it along the u-axis; these are the stable and unstable manifolds of the saddle point. $\det J(K,0) = Kg'(K)(-d+q(K)) < 0$ (implying a saddle point) if $q(K) > d$, when a co-existence state exists, $\det J(K,0) > 0$ (implying a stable node) if $q(K) < d$. (Thinking of d as a bifurcation parameter, there is a transcritical bifurcation as d decreases past $d = d_c = q(K)$ when the coexistence state enters the positive quadrant through $(K,0)$ and takes over stability from it.)

d) $\det J(u^*,v^*) = p(u^*)v^*q'(u^*) > 0$, so the condition for instability is $\operatorname{tr} J(u^*,v^*) = u^*g'(u^*) + g(u^*) - v^*p'(u^*) > 0$.

e) The phase plane is similar to that of Figure 2.9(d).

2.8 a) – Prey-predator relationship.

– u can survive at low v populations, v cannot survive at low u populations.

– There is a prey-only steady state.

– For small u, either u has a depensatory growth rate or there is a nonlinear saturating functional response; v is purely compensatory.

– Depensation occurs at the steady state.

b) The phase plane is very similar to that of Figure 2.9(d), except that $g = 0$ has a positive rather than an infinite slope.

c) Boundedness follows by constructing an invariant set. Then use the Poincaré–Bendixson theorem (Section B.2.3 of the appendix).

2.9 a) No relationship between predation term in u equation and any growth term in v equation; problems with singularity at $u = 0$ in v equation.

b) See Figure E.2.

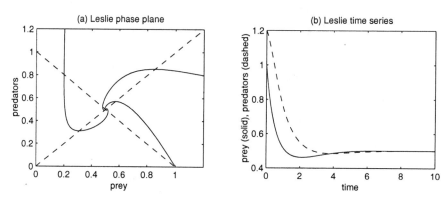

Figure E.2   The phase plane and a typical time series for the Leslie prey-predator equations.

c) All solutions in the positive quadrant tend to the co-existence steady state, which is a stable node or focus.

2.10 The slope of $f = 0$ is $-f_u^*/f_v^*$, etc.

2.11 a) Immediate.

b) $u^{r_2\gamma_2}/v^{r_1\gamma_1} = C\exp(r_1r_2(\gamma_2 - \gamma_1)t)$, where C is a constant of integration. Competitive exclusion follows since u and v are bounded, with u winning if $\gamma_2 > \gamma_1$.

c) u is less affected by the fact that the food is being consumed than v.

   d) n species cannot survive on less than n food resources, modelled in this way.

2.12 a) Take $B(u, v) = (uv)^{-1}$.
   b) $\dot{\Phi}(u, v) = -c((u - u^*)^2 + (a + b)(u - u^*)(v - v^*) + (v - v^*)^2) \leq 0$ since, if the coexistence state is stable, $a < 1$ and $b < 1$, and so $(a + b)^2 \leq 4$.

2.13 The m terms represent the experimental removal. The coexistence results follow directly from the usual stability conditions.

2.14 a)
$$\frac{du}{dt} = u(1 - u + av), \quad \frac{dv}{dt} = cv(1 + bu - v).$$

   b) There is a coexistence steady state $(u^*, v^*) = ((1+a)/(1-ab), (1+b)/(1-ab))$ if and only if $ab < 1$. No solutions in the positive quadrant tend to $(1, 0)$ or $(0, 1)$, since these are saddle points with the axes as stable manifolds. Any set $D = \{(u, v)|0 \leq u \leq Mu^*, 0 \leq v \leq Mv^*\}$ with $M \geq 1$ is invariant, so all solutions are bounded. Dulac's criterion with $B(u, v) = (uv)^{-1}$ shows there are no periodic solutions, so the Poincaré–Bendixson theorem gives the result.

2.15 a) ϕ is the photosynthesis term.
   b) $S_0, (P, H) = (\phi/a, 0)$; $S_1, (P, H) = (P^*, H^*) = (c/(eb), (eb\phi - ac)/(bc))$. The eigenvalues at S_0 are $-a$ and $eb\phi/a - c$, while the Jacobian matrix at S_1 has tr $J^* < 0$, det $J^* > 0$, stable, if $H^* > 0$.
   c) There is a transcritical bifurcation where stability switches from S_0 to S_1 as ϕ increases through $\phi_c = (ac)/(eb)$, and S_1 enters the positive quadrant.
   d) See Figure E.3.

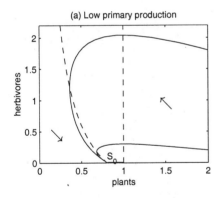
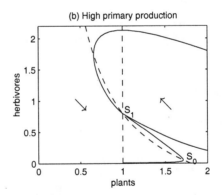

Figure E.3   The phase plane with (a) low and (b) high primary production ϕ. After the bifurcation to S_1, steady state plant density is constant.

2.16 A model for this is given by

$$\frac{dN}{dt} = aP + bZ - cNP, \quad \frac{dP}{dt} = cNP - dPZ - aP,$$

$$\frac{dZ}{dt} = dPZ - bZ - eZC, \quad \frac{dC}{dt} = eZC - fC,$$

with $N + P + Z + C = A$. The steady states are S_0, S_1 and S_2 in the text, with

$C = 0$, and $S_3 = (N^*, P^*, Z^*, C^*)$, where

$$P^* = \frac{e}{d+e}\left(A - \frac{a}{c} + \frac{b}{e} - \frac{(c+d)f}{ce}\right), \quad Z^* = \frac{f}{e},$$

$$C^* = \frac{d}{d+e}\left(A - \frac{a}{c} - \frac{b}{d} - \frac{(c+d)f}{ce}\right).$$

As A increases, we have a sequence of transcritical bifurcations as S_1 takes over stability from S_0, S_2 from S_1, and finally S_3 from S_2.

2.17 The model is

$$\frac{dP}{dt} = \phi - aP - \frac{bPH}{1+kP}, \quad \frac{dH}{dt} = \frac{ebPH}{1+kP} - cH.$$

The steady states are $(\phi/a, 0)$ and (P^*, H^*), where $P^* = c/(c-ek)$, $H^* = (\phi - aP^*)(1+kP^*)/(bP^*)$, in the positive quadrant as long as $c > ek$, $\phi - aP^* > 0$. The Jacobian matrix at the coexistence steady state has $\operatorname{tr} J^* = -a - bH^*/(1+kP^*)^2 < 0$, $\det J^* = eb^2 P^* H^*/(1+kP^*)^3 > 0$, so is always stable.

2.18 a)

$$\frac{dP_1}{d\tau} = (\alpha_1 N - \beta_1)P_1, \quad \frac{dP_2}{d\tau} = (\alpha_2 N - \beta_2)P_2,$$

where $N = A - \gamma_1 P_1 - \gamma_2 P_2$.

b) The result follows on defining $u = P_1/K_1$, $v = P_2/K_2$, $t = (\alpha_1 A - \beta_1)\tau$, $c = (\alpha_2 A - \beta_2)/(\alpha_1 A - \beta_1)$, where $K_1 = (\alpha_1 A - \beta_1)/\gamma_1$, $K_2 = (\alpha_2 A - \beta_2)/\gamma_2$.

c) The slopes are $1/a$ and b.

2.19 a) $S_0 = (0,0)$, $S_1 = (p_1^*, 0) = (1 - e_1/c_1, 0)$, $S_2 = (0, 1 - e_2/c_2)$, $S^* = (p_1^*, 1 - p_1^* - e_2/c_2 - c_1 p_1^*/c_2)$.

b) Immediate; species 2 then cannot survive even in the absence of species 1.

c) See Figure E.4. Species 2 survives whenever S^* is in the positive quadrant,

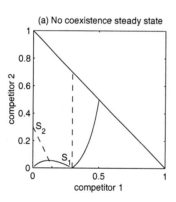
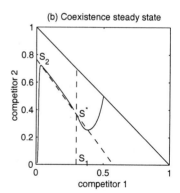

Figure E.4   The phase plane for the metapopulation competition system with and without a coexistence steady state.

i.e. $p_2^* = e_1/c_1 - e_2/c_2 - c_1/c_2(1 - e_1/c_1) > 0$.

d) In particular, it survives if c_2 is sufficiently large.

2.20 Steady states are $(0,0)$, $(1 - e_1/c_1, 0)$, and (p_1^*, p_{13}^*), where $p_1^* = e_{13}/c_3$, $p_{13}^* = (c_1(1 - p_1^*) - e_1)/(c_1 + c_3)$.

Condition for survival is $p_{13}^* > 0$, or $e_{13}/c_3 < 1 - e_1/c_1$.

$\operatorname{tr} J^* = -c_1 p_1^* < 0$, $\det J^* = c_3 p_{13}^*(c_1 + c_3)p_1^* > 0$, so stable.

2.21 The p_{13}-nullcline remains fixed, while the p_1 nullcline moves leftwards. There is a transcritical bifurcation when the coexistence state passes through the prey-only steady state, at $D = D_c = 1 - e_1/c_1 - e_{13}/c_3$, and the predators go extinct.

2.22 We are given coexistence for $D = 0$, $(c_2 - e_2)/(c_2 + c_1) > 1 - e_1/c_1$. It follows that $(c_2(1-D)-c_1)/(c_2+c_1) > 1-D-e_1/c_1$ for all D, and this is the condition for the coexistence state S_2^* to be in the positive quadrant. S^* does not pass through S_1. The only possibilities for bifurcation from the coexistence phase plane is through either S_1 or S_2 passing transcritically through S_0, and the trade-off condition shows that S_1 does so first. If there is such a trade-off, the *better* competitor is always the one that goes extinct first.

3.1 a) Let the required probability be $p(\tau)$. Then $p(\tau + \delta\tau) = p(\tau) - \gamma p(\tau)\delta\tau + O(\delta\tau^2)$. Subtracting $p(\tau)$ from both sides, dividing by $\delta\tau$ and taking the limit as $\delta\tau \to 0$, $\dot{p} = -\gamma p$, and since $p(0) = 1$ the result follows.

   b) By the definition of mean value,

$$\bar{\tau} = \frac{\int_0^\infty \tau \exp(-\gamma\tau)d\tau}{\int_0^\infty \exp(-\gamma\tau)d\tau},$$

   and the result follows.

3.2 a) An individual is infective at time τ if he or she was infected at some time $\tau - \sigma$ and remained infective at least for a time σ. Thus $I(\tau) = \int_0^\infty i(\tau - \sigma)f(\sigma)d\sigma$, and

$$i(\tau) = \beta I S \approx \beta I N \approx \beta N \int_0^\infty i(\tau - \sigma)f(\sigma)d\sigma,$$

   as required.

   b) Trying a solution $i(\tau) = i_0 \exp(r\tau)$, we obtain

$$1 = \beta N \int_0^\infty \exp(-(r + \gamma)\sigma)d\sigma = \frac{\beta N}{r + \gamma},$$

   so $r = \beta N - \gamma$.

   c) Hence $r = \gamma(R_0 - 1)$, familiar as Equation (3.2.3).

3.3 The incidence function is as for the SI disease, but now results in entry to the exposed class. The time in the exposed class is exponentially distributed with mean $1/\delta$, and is followed by entry to the infective class.

3.4 a) At time τ, the I class consists of those who have been infected between $\tau - \tau_I$ and τ.

   b) $I^* = \int_{\tau - \tau_I}^{\tau} \beta I^* S^* d\sigma = \beta I^* S^* \tau_I$, and the result follows.

   c) $R_0 = \beta N \tau_I$.

3.5 At $w = 0$, the derivative of $\exp(-R_0 w) = -R_0$, and the qualitative features follow. u_1 satisfies $R_0 u_1 \exp(-R_0 u_1) = R_0 \exp(-R_0)$, i.e. R_0 and $R_0 u_1$ are two positive values of x where xe^{-x} is equal. Since xe^{-x} increases in $(0, 1)$ and decreases in $(1, \infty)$, the result follows.

3.6 a) If $R_0 w$ is small, then $\exp(-R_0 w) \approx 1 - R_0 w + \frac{1}{2}R_0^2 w^2$. The result follows.

   b) The approximation above holds if $R_0 w_1$ is small, and then $w \to w_1$.

   c) The incidence of death is $p\dot{w}$, where p is the probability of death given infection and w satisfies the logistic equation with $r = R_0 - 1$ and $K = w_1$. This is a sech^2 curve.

3.7 There is a one-parameter family of solution trajectories, given by $R_0(u + v) - \log u = A$, where A is a constant of integration.

3.8 a) Since $\gamma = 1/25$ per year, $R_0 = \beta N/\gamma = 4$.

b) For (i), since $R_0 u_1 < 1$, $w_1 > 1 - 1/R_0$, and $N w_1 > 750$. In fact, numerical calculations show that this is a conservative estimate. For (ii), $R_0' = q R_0 = (0.3)(4) = 1.2$, so $(R_0 - 1)/R_0 = 1/6$ is small, and the approximation of Exercise 3.6 holds; the final size of the epidemic is given approximately by $q N w_1 = 2\frac{1}{6}\frac{5}{6}300 = 83$.

c) The Jacobian at $(1, 0)$ is given by

$$J = \begin{pmatrix} -R_0 v & -R_0 u \\ R_0 v & R_0 u - 1 \end{pmatrix} = \begin{pmatrix} 0 & -R_0 \\ 0 & R_0 - 1 \end{pmatrix},$$

which has eigenvalues $R_0 - 1$ and zero.

3.9 a) $\frac{dS}{d\tau} = \delta R - \beta I S$, $\frac{dI}{d\tau} = \beta I S - \gamma I$, $\frac{dR}{d\tau} = \gamma I - \delta R$.

b) $R_0 = \beta N/\gamma$, since the time spent in I is unchanged.

c) The endemic steady state is

$$(u^*, v^*, w^*) = \left(\frac{1}{R_0}, \frac{1}{1+d}\frac{1}{R_0}, \frac{d}{1+d}\frac{1}{R_0} \right),$$

so the condition is $R_0 > 1$.

3.10 a) Susceptibles enter exposed state on infection, and leave it after an exponentially distributed time of mean $1/\delta$.

b) $R_0 = \beta N/\gamma$.

c) Since we still have $dw/du = -1/(R_0 u)$, the final size is unchanged.

3.11 Parts a) to d) are immediate.

e)

$$f(\sigma) = \int_0^\sigma p \delta e^{-\delta s} e^{-\gamma(\sigma - s)} ds.$$

f)

$$R_0 = \frac{p \delta \beta' N}{\gamma - \delta} \int_0^\infty (e^{-\delta \sigma} - e^{-\gamma \sigma}) d\sigma = \frac{\beta N}{\gamma}.$$

g) R_0 depends on how long is spent in I, not on whether there is a delay in entering it.

3.12 Immediate.

3.13 a) Expresses probability of not dying from disease-unrelated causes before disease age σ.

b) Immediate.

c)

$$R_0 = \beta N \int_0^\infty e^{-(\gamma + d)\sigma} d\sigma = \frac{\beta N}{\gamma + d}.$$

d)

$$f(\sigma) = \int_0^\sigma p \delta e^{-(\delta + d)s} e^{-(\gamma + d)(\sigma - s)} ds, \quad R_0 = \beta N \int_0^\infty f(\sigma) d\sigma.$$

3.14 a) Straightforward.

b) Working with the (N, S, I)-system, the Jacobian at the endemic steady state (N^*, S^*, I^*) is given by

$$J^* = \begin{pmatrix} -d & 0 & -c \\ 0 & -\beta I^* - d & -\beta S^* \\ 0 & \beta I^* & 0 \end{pmatrix},$$

whose eigenvalues are $-d$ and the roots of the quadratic $\lambda^2 + (\beta I^* + d)\lambda + \beta^2 I^* S^* = 0$. Now $(\beta I^* + d)^2 = R_0^2 d^2 \ll 4(R_0 - 1)d(\gamma + c + d) = \beta^2 I^* S^*$, so the roots of this quadratic are $\lambda \approx \pm i\sqrt{\beta^2 I^* S^*}$, as required.

c) $T \approx 2\pi/\sqrt{(365/12)(1/70)13} \to 2\pi/\sqrt{(365/12)(1/70)12} = 2.64 \to 2.75$. The time series from Providence, Rhode Island (Figure 3.6) has 9 peaks in 24 years, a period of 2.67.

3.15 If an endemic steady state exists, $S^* = (\gamma + b)/\beta$, $\beta I^* + b = qbR_0$, so $\beta I^* = b(qR_0 - 1)$.

3.16 a) Immediate.

b) $R_0 = \beta N^*/(\gamma + c + d(N^*))$.

c) Using the (N, S, I)-system, we look at the stability of the disease-free steady state $(N^*, N^*, 0)$. We have

$$J^* = \begin{pmatrix} r'(N^*)N^* & 0 & -c \\ r'(N^*)N^* & -d(N^*) & -\beta N^* \\ 0 & 0 & \beta N^* - \gamma - c - d(N^*) \end{pmatrix},$$

where $r(N) = b(N) - d(N)$, whose eigenvalues are $r'(N^*)N^* < 0$, $-d(N^*) < 0$, and $\beta N^* - \gamma - c - d(N^*)$.

d) $p \geq p_c = 1 - 1/R_0$.

3.17 First note that $\lambda(i)(a, t) = \beta I(t)$, independent of a. Then the integration is straightforward.

3.18 a) Let subscript 1 represent females, 2 males. This is an SIS disease with criss-cross infection; β_{12} is the infectious contact rate for infectious females infecting susceptible males, and β_{21} the contrary. β_{12} may be higher because many infectious females do not know they are infectious; γ_2 may be higher because males know they have the disease and seek treatment; there may also be behavioural differences.

b) $S_1 + I_1 = N_1$, $S_2 + I_2 = N_2$, N_1 and N_2 constant. $R_{01} = \beta_{21}N_2/\gamma_1$, $R_{02} = \beta_{12}N_1/\gamma_2$.

c) The eigenvalue equation for the wholly susceptible steady state is $\lambda^2 + (\gamma_1 + \gamma_2)\lambda - (R_{01}R_{02} - 1)\gamma_1\gamma_2 = 0$, which is unstable if and only if $R_{01}R_{02} > 1$.

3.19 a) Delete the $\gamma_i I_i$ terms from the susceptible equations, and add removed class equations $dR_i/d\tau = \gamma_i I_i$.

b) The eigenvalue equation for the wholly susceptible steady state is simply λ^2 times the eigenvalue equation for the corresponding SIS disease, found in the last exercise, so the condition for an epidemic is again $R_{01}R_{02} > 1$.

c) Separate variables to obtain $dw_1/du_2 = -\gamma_1/(\gamma_2 R_{02} u_2)$, and vice versa, interchanging 1 and 2. Integrating from the disease-free steady state, $u_2 = \exp(-(\gamma_2 R_{02}/\gamma_1)w_1)$, and vice versa. In the limit as $t \to \infty$, $1 - w_2 = \exp(-(\gamma_2 R_{02}/\gamma_1)w_1)$, and vice versa.

3.20 a) W is the *mean* worm burden, so the W equation is equivalent to Equation (3.8.34) divided through by N for humans. I is the *total* number of infected snails, so the I equation is not divided through by N for snails.

b) Always have $(0, 0)$, which is stable; two non-trivial steady states, the smaller one unstable and the larger one stable, exist when $N > N_c = (\delta/c)(d/b + 2\sqrt{d/b})$.

c) As c decreases, the two non-trivial steady states disappear by a saddle-node bifurcation when c passes through $c_c = (\delta/N)(d/b + 2\sqrt{d/b})$, and the level of disease in a population at the endemic stable steady state drops suddenly to zero.

d)

$$\frac{dW}{dt} \approx \frac{bcNW^2}{bW^2 + W + 1} - \delta W.$$

3.21 Plotting $R_0 = (\beta N)/(\gamma_0 c^{-\alpha} + c + d)$ against c, R_0 has an intermediate maximum, and the virulence tends to the value of c that achieves it.

3.22 a) The Jacobian matrix at the endemic steady state $(S^*, I_1^*, 0)$ is given by

$$J = \begin{pmatrix} -\beta_1 I_1^* - d & -\beta_1 S^* & -\beta_2 S^* \\ \beta_1 I_1^* & 0 & 0 \\ 0 & 0 & \beta_2 S^* - \gamma_2 - c_2 - d \end{pmatrix}.$$

Two of its eigenvalues are stable, and the other is given by $\lambda = \beta_2 S^* - \gamma_2 - c_2 - d = (\gamma_2 + c_2 + d)(R_{02}/R_{01} - 1)$, positive if $R_{02} > R_{01}$. There is no coexistence state since $\dot{I}_1 = 0$ and $\dot{I}_2 = 0$ cannot hold simultaneously with neither I_1 nor I_2 zero.

4.1 If, as the information given strongly suggests, the gene is dominant, and Elisabeth Horstmann's daughter was heterozygous for it, then the probability of polydactyly in her children is $\frac{1}{2}$, so the expected number of her children to show the trait is four.

4.2 Let the pure-bred rounded yellow and wrinkled green phenotypes have genotypes $RRYY$ and $WWGG$ respectively. Then all those in the F_1 generation have genotype $RWYG$, and phenotype rounded yellow.

		\multicolumn{4}{c}{Female gametes}
		RY	RG	WY	WG
Male	RY	RY	RY	RY	RY
gametes	RG	RY	RG	RY	RG
	WY	RY	RY	WY	WY
	WG	RY	RG	WY	WG

The F_2 generation is produced by random mating, summarised in the diagram. This shows the phenotype that results from each union of gametic genotypes. Each of these possibilities is equally likely, so that the ratio of $RY : RG : WY : WG$ is $9 : 3 : 3 : 1$.

4.3 Consider a population in Hardy–Weinberg equilibrium; then $x = p^2$, $y = 2pq$, $z = q^2$, so that $y^2 = 4xz$. Conversely, consider a population with $y^2 = 4xz$. Then

$$p^2 = (x + \tfrac{1}{2}y)^2 = x^2 + xy + \tfrac{1}{4}y^2 = x^2 + x(1 - x - z) + xz = x.$$

Similar calculations show that $y = 2pq$, $z = q^2$, so the population is in Hardy–Weinberg equilibrium.

4.4 The phenotypes resulting from the various unions of gametic genotypes are shown in the diagram below.

		\multicolumn{3}{c}{Female gametes}
		A	B	O
Male	A	A	AB	A
gametes	B	AB	B	B
	O	A	B	O

Let the frequencies of the alleles A, B and O be p, q and r, and the frequencies of the blood groups A, AB, B and O be A, AB, etc. If Hardy–Weinberg proportions hold, then $O = r^2$, $B + O = (q + r)^2$, so $r = \sqrt{O} = 62.0\%$, $q = \sqrt{B + O} - \sqrt{O} = 78.0\% - 62.0\% = 16.0\%$, and so $p = 22.0\%$. Then $A = p^2 + 2pr = 32.2\%$, $AB = 7.1\%$, and we are very close to Hardy–Weinberg equilibrium.

4.5 a) The offspring will have genotypes AA, AB, BB in the ratio $1:2:1$, as usual.

b) We have

$$x_{n+1} = x_n + \frac{1}{4}y_n, \quad y_{n+1} = \frac{1}{2}y_n, \quad z_{n+1} = z_n + \frac{1}{4}y_n.$$

Hence

$$p_{n+1} = x_{n+1} + \frac{1}{2}y_{n+1} = x_n + \frac{1}{2}y_n = p_n,$$

so $p_n = p$, $q_n = q$, $y_n = y_0(\frac{1}{2})^n \to 0$ as $n \to \infty$, $x_n = p - y_0(\frac{1}{2})^{n+1} \to p$ as $n \to \infty$, $z_n = q - y_0(\frac{1}{2})^{n+1} \to q$ as $n \to \infty$.

c) Heterozygotes disappear, and the population splits into two non-interbreeding subpopulations. This may be the basis for some speciation events.

4.6 1 in 400.

4.7 Assume random mating and no selection, and other Hardy–Weinberg assumptions. For males, there are two genotypes A and B, with frequencies m and n, say, and

$$m' = p, \quad n' = q,$$

where p and q are female allele frequencies. For females, with the usual notation for genotypes,

$$x' = mp, \quad y' = mq + np, \quad z' = nq.$$

Hence

$$p' = x' + \frac{1}{2}y' = m(p + \frac{1}{2}q) + \frac{1}{2}np,$$

$$p'' = m'(p' + \frac{1}{2}q') + \frac{1}{2}n'p' = p(p' + \frac{1}{2}q') + \frac{1}{2}qp' = \frac{1}{2}p' + \frac{1}{2}p.$$

Solving this,

$$p_n = \frac{2}{3}p_0 + \frac{1}{3}m_0 + \frac{1}{3}(p_0 - m_0)(-\frac{1}{2})^n.$$

As $n \to \infty$, p_n tends to a constant equal to the initial frequency of allele A in the whole population.

4.8 a) Let $w_y < w_z$. Since $\bar{w} > 0$ then Equation (4.3.8) implies that $\delta p < 0$ if $w_x < w_y$, and $\delta p < 0$ as long as $p < p^* = \frac{w_z - w_y}{w_z - 2w_y + w_x}$ if $w_x > w_y$. In either case $\delta p < 0$ for p small, and (p_n) is a positive decreasing sequence, so the steady state $p = 0$ is stable. Biologically, the condition $w_y < w_z$ says that the heterozygote is less fit than the homozygote BB.

b) The steady state $p = 1$ is stable if the heterozygote is less fit than the homozygote AA, $w_y < w_x$.

c) An interior steady state p^* exists if and only if

$$p^* = \frac{w_y - w_z}{(w_y - w_z) + (w_y - w_x)} \in (0, 1),$$

i.e. either

i) $w_y > w_z$ and $w_y > w_x$, the heterozygote is fitter than both homozygotes, or

ii) $w_y < w_z$ and $w_y < w_x$, the heterozygote is less fit than both homozygotes.

It is stable if neither $p = 0$ nor $p = 1$ is, i.e. in case i) above.

4.9 a) From the FHW Equation (4.3.2) with $w_x = 1$, $w_y = 1 + s$, $w_z = 1 - t$,

$$p' = p + pq\frac{-sp + (s+t)(1-p)}{p^2 + 2(1+s)pq + (1-t)q^2} = p + pq\frac{s(1-2p) + t(1-p)}{1 + 2spq - tq^2}.$$

b) There are steady states at $p = 0$, $p = 1$ and $p = p^*$, where

$$s(1 - 2p^*) + t(1 - p^*) = 0, \quad p^* = \frac{s+t}{2s+t}.$$

The cobweb map is as shown in Figure E.5, and it is clear that $p_n \to p^*$ as $n \to \infty$ for any $p_0 \in (0, 1)$.

c) Given $t = 0.8$ (fitness of BB homozygotes 0.2), $p^* = 0.8$, then

$$s = t\frac{1 - p^*}{2p^* - 1} = 0.8\frac{0.2}{0.6} \approx 0.27,$$

so that the fitnesses of AA and AB are in the ratio $1 : 1.27$, and the chance of dying from malaria before maturity is approximately 0.21.

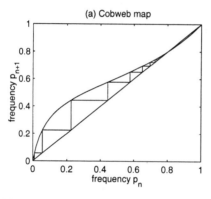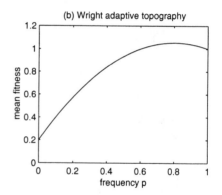

Figure E.5   (a) Cobweb map and (b) Wright's adaptive topography for the sickle cell anaemia gene. Wright's adaptive topography is discussed in Section 4.7 and, for this example, in exercise 4.15.

4.10 Routine. Use the gene ratio $v = q/p$ for the departure from $p = 0$.

4.11 Routine.

4.12 a) From Equation (4.5.15) with $h = 1$, $k = 1$,

$$s = \frac{1}{35}\int_{0.01}^{0.9}\frac{dp}{p(1-p)^2} \approx 0.45.$$

b) Numerical. The weak selection approximation is excellent, despite the relatively large value of s.

4.13 In selection-mutation balance,

$$\delta p = \frac{\alpha_p p}{\bar{w}} - u\frac{w_p p}{\bar{w}} + v\frac{w_q q}{\bar{w}}.$$

a) For a recessive deleterious gene $w_x = 1 - s$, $w_y = w_z = 1$, so that $\bar{w} = (1-s)p^2 + 2pq + q^2 = 1 - sp^2$, $w_p = (1-s)p + q = 1 - sp$, $w_q = 1$, $\alpha_p = w_p - \bar{w} = -spq$, and

$$\delta p = -sp^2 q - up + vq + h.o.t.,$$

assuming $u, v \ll s \ll 1$. At steady state $\delta p = 0$ and $p = p^*$, small, so

$$-sp^{*2} + v \approx 0, \quad p^* \approx \sqrt{v/s}.$$

b) If the deleterious gene is dominant $w_y = 1 - s$, $\bar{w} = 1 - sp$, $w_p = 1 - s$, $w_q = 1 - sq$, $\alpha_p = -sq$, and

$$\delta p = -spq - up + vq + h.o.t.,$$

assuming $u, v \ll s \ll 1$. At steady state $\delta p = 0$ and p is small, so

$$-sp^* + v \approx 0, \quad p^* \approx v/s.$$

c) Here A is recessive, so that (a) holds, and lethal, so that $s = 1$. Since $v = 4 \times 10^{-4}$, $p^* \approx 2 \times 10^{-2}$.

4.14 Differentiating Equation (4.3.5) with respect to p, and noting that $q = 1 - p$, we have

$$\frac{d\bar{w}}{dp} = \frac{\partial \bar{w}}{\partial p} - \frac{\partial \bar{w}}{\partial q} = 2w_x p + 2w_y q - 2w_y p - 2w_z q = 2(w_p - w_q),$$

and Equation (4.7.18) follows. Also,

$$\delta \bar{w} = w_x(2p\delta p + \delta p^2) + 2w_y(q\delta q + p\delta q + \delta p\delta q) + w_z(2q\delta q + \delta q^2)$$

$$= 2(w_p - w_q)\delta p + (w_x - 2w_y + w_z)\delta p^2,$$

since $\delta p + \delta q = 0$. Putting $w_p - w_q = \frac{\bar{w}\delta p}{pq}$, the result (4.7.17) follows.

4.15 From $\bar{w} = 1 + 2spq - tq^2$, we can sketch the Wright adaptive topography, shown in Figure E.5. There is a maximum at p^*, so $p_n \to p^*$ as $n \to \infty$, as expected.

4.16 a) Since p is an interior point, $r = (1+\epsilon)p - \epsilon q \in S_{n-1}$ for ϵ sufficiently small.

b) Since p is an ESS $W(r,p) - W(p,p) \leq 0$, and $W(q,p) - W(p,p) \leq 0$, and since q is an ESS $W(p,q) - W(q,q) \leq 0$. But, since W is linear in its first variable, $W(r,p) - W(p,p) = \epsilon(W(p,p) - W(q,p))$, with $\epsilon > 0$. This is only possible if

$$W(r,p) = W(p,p).$$

It also follows that $W(p,p) = W(q,p)$. Now, since W is linear in its first and second variables, it follows after some algebra that $W(r,r) - W(p,r) = \epsilon^2(W(q,q) - W(p,q))$, so that

$$W(r,r) \geq W(p,r).$$

These two results contradict the alternatives (4.9.23) and (4.9.24) required for p to be an ESS.

4.17 Let p be a mixed ESS, with i in the support of p. We need to prove that $W(e_i, p) = W(p,p)$. We know that for all $q \neq p$,

$$W(q,p) \leq W(p,p).$$

Assume for contradiction that $W(e_i, p) < W(p,p)$. Consider $q = (1+\epsilon)p - \epsilon e_i$. As long as ϵ is sufficiently small, $q \in S_{n-1}$, since i is in the support of p. Now

$$W(q,p) - W(p,p) = \epsilon(W(p,p) - W(e_i,p)) > 0,$$

giving us our contradiction.

4.18   a) If your partner were to play D, you would be better off playing D. If your partner were to play C, you would be better off playing D. Hence you play D; your partner, following the same logic, also plays D. You both obtain pay-off 2, missing out on the benefit of mutual co-operation, with pay-off 3. This is the prisoner's dilemma.

      b) The pay-off matrix if the game is played ten times is given below.

		on encountering this strategy	
		D	TFT
Pay-off to	D	20	22
this strategy	TFT	19	30

        Both D and TFT are ESSs.

      c) Co-operation can persist, but how does it start?

4.19 Let x be the frequency of the hawk strategy. Then (4.10.31) with U as in Section 4.9 gives

$$\dot{x} = \frac{1}{2}x(G - Cx)(1 - x),$$

which is to be solved subject to $x(0) = x_0$, $0 \leq x_0 \leq 1$. Both $x = 0$ and $x = 1$ are steady states, so we need only consider $0 < x_0 < 1$. If $G < C$, then $x(t) \to G/C$ as $t \to \infty$, whereas if $G \geq C$, then $x(t) \to 1$ as $t \to \infty$. The solution agrees with that in Section 4.9, with x^*, the evolutionarily stable *state* of the system, taking the place of p^*, the evolutionarily stable *strategy*.

4.20 The pay-off matrix becomes

$$U = \begin{pmatrix} -\epsilon & 1 & -1 \\ -1 & -\epsilon & 1 \\ 1 & -1 & -\epsilon \end{pmatrix}.$$

  a) Let p be the symmetric strategy $(\frac{1}{3}, \frac{1}{3}, \frac{1}{3})$, and $q = (q_1, q_2, q_3)$ another strategy, $q \in S_2$. Then $W(p,p) = -\frac{1}{3}\epsilon$, $W(q,p) = -\frac{1}{3}\epsilon$, $W(p,q) = -\frac{1}{3}\epsilon$, $W(q,q) = -\epsilon(q_1^2 + q_2^2 + q_3^2)$. It is easy to show that $W(q,q) < W(p,q)$ for $q \neq p$, so (4.9.24) holds and p is an ESS.

  b) The replicator equations are

$$\dot{x} = \left(-\epsilon x + y - z + \epsilon(x^2 + y^2 + z^2)\right)x,$$

with similar equations for $\dot{y}$ and $\dot{z}$. Under these equations,

$$\frac{d}{dt}(xyz) = xyz\left(\frac{\dot{x}}{x} + \frac{\dot{y}}{y} + \frac{\dot{z}}{z}\right) = \epsilon xyz\left(3(x^2 + y^2 + z^2) - 1\right),$$

so that xyz increases with t. It tends to $(\frac{1}{3}, \frac{1}{3}, \frac{1}{3})$, the maximum of xyz on S_2, where the RHS of the equation above is zero, as shown in Figure 4.7. In fact, xyz is a Lyapunov function for the system (see Chapter B of the appendix).

  c) If the penalty is replaced by a reward, then p is no longer an ESS; in fact, it is invadable by any other strategy. The function xyz now decreases with t, and any non-constant solution of the replicator equations with initial conditions in the interior of S_2 approaches a trajectory which visits each apex of the triangle S_2 in turn.

4.21 See the website for one possibility.

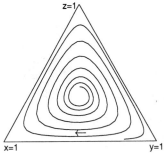

Phase plane, epsilon negative

z=1

x=1

y=1

Figure E.6   Evolutionary dynamics for the lizard (rock-scissors-paper) game with a reward.

5.1

$$\frac{d}{dt}\int_V u dV = \int_V \frac{\partial u}{\partial t} dV = \int_V -\nabla \cdot \mathbf{J} dV = -\int_S \mathbf{J} \cdot \mathbf{n} dS = 0.$$

5.2 The flow is spherically symmetric, so $\mathbf{J} = J(r,t)\mathbf{e_r}$. Conservation of mass for V gives

$$u(r, t+\delta t)4\pi r^2 \delta r = u(r,t)4\pi r^2 \delta r + J(r,t)4\pi r^2 \delta t$$
$$- J(r+\delta r, t)4\pi(r+\delta r)^2 \delta t + f(r,t)4\pi r^2 \delta r \delta t + h.o.t.$$

(The spheres have surface areas $4\pi r^2$ and $4\pi(r+\delta r)^2$ respectively.) Dividing through by $4\pi r^2 \delta r \delta t$ and taking limits as $\delta r \to 0$, $\delta t \to 0$, we obtain

$$\frac{\partial u}{\partial t} = -\frac{1}{r^2}\frac{\partial}{\partial r}\left(r^2 J\right) + f.$$

5.3 The flow is cylindrically symmetric, so $\mathbf{J} = J(R,t)\mathbf{e_R}$. Conservation of mass for V gives

$$u(R, t+\delta t)2\pi R\delta R h = u(R,t)2\pi R\delta R h + J(R,t)2\pi Rh\delta t$$
$$- J(R+\delta R, t)2\pi(R+\delta R)h\delta t + f(R,t)2\pi R\delta R h\delta t + h.o.t.$$

(The cylindrical surfaces have areas $2\pi Rh$ and $2\pi(R+\delta R)h$ respectively.) Dividing through by $2\pi R\delta R h\delta t$ and taking limits as $\delta R \to 0$, $\delta t \to 0$, we obtain

$$\frac{\partial u}{\partial t} = -\frac{1}{R}\frac{\partial}{\partial R}\left(RJ\right) + f.$$

5.4 In two dimensions, some concepts are interpreted slightly differently: for example, u is the amount of substance per unit *area*, and J is the rate at which substance crosses a curve per unit *length* in the direction perpendicular to the flow. Otherwise, the derivation essentially follows from the previous exercise on deleting h.

5.5

$$\frac{dN}{dt}(t) = \int_a^b \frac{\partial u}{\partial t}(x,t)dx = \int_a^b D\frac{\partial^2 u}{\partial x^2}(x,t)dx = \left[D\frac{\partial u}{\partial x}(x,t)\right]_a^b = 0.$$

5.6

$$\frac{dN}{dt}(t) = \int_a^b \frac{\partial u}{\partial t}(x,t)dx$$

$$= \int_a^b \left(-v\frac{\partial u}{\partial x} + D\frac{\partial^2 u}{\partial x^2}\right)(x,t)dx = \left[-vu(x,t) + D\frac{\partial u}{\partial x}(x,t)\right]_a^b.$$

For conservation this must be zero, so $vu - Du_x = 0$ (zero-flux boundary conditions) at $x = a$ and at $x = b$.

5.7 We have $x = z + vs$, $t = s$, so we may think of x and t as functions of z and s. Then, from $U(z, s) = u(x, t)$ by the chain rule,

$$\frac{\partial U}{\partial s} = \frac{\partial u}{\partial x}\frac{\partial x}{\partial s} + \frac{\partial u}{\partial t}\frac{\partial t}{\partial s},$$

or $U_z = vu_x + u_t$. Similarly $U_{zz} = u_{xx}$, and the result follows.

5.8 a) The total flux is given by $\mathbf{J} = J\mathbf{k}$, where $J = -D\frac{du}{dz} - \alpha g u$. The conservation equation is $\frac{dJ}{dz} = -D\frac{d^2 u}{dz^2} - \alpha g\frac{du}{dz} = 0$.

b) The boundary condition at the surface is the zero-flux condition $J = -D\frac{du}{dz} - \alpha g u = 0$; the boundary condition at infinity is u bounded.

c) Integrating, J is constant, and applying the boundary condition, the constant must be zero. Thus $D\frac{du}{dz} + \alpha g u = 0$, which we can integrate to obtain $u = u_0 \exp(-\frac{\alpha g}{D} z)$, where u_0 is the (undetermined) plankton concentration at the surface.

5.9 a) We need to solve the problem

$$0 = -v\frac{du}{dx} + D\frac{d^2 u}{dx^2} \text{ in } (0, L), \quad u(0) = u_0, \quad u(L) = 0.$$

The general solution of the equation is $u = A + B\exp(vx/D)$, and the boundary conditions lead to

$$u(x) = u_0\frac{\exp(vL/D) - \exp(vx/D)}{\exp(vL/D) - 1}.$$

The flux J (equivalent to the diffusion current I) is given by

$$J = vu - D\frac{du}{dx} = vu_0\frac{\exp(vL/D)}{\exp(vL/D) - 1}.$$

b) Since u is a line concentration then the amount of matter N in $(0, L)$ is given by

$$N = \int_0^L u(x)dx = u_0\frac{L\exp(vL/D) - (D/v)(\exp(vL/D) - 1)}{\exp(vL/D) - 1}.$$

c) Hence the transit time τ is given by

$$\tau = \frac{N}{I} = \frac{L}{v} - \frac{D}{v^2}\left(1 - \exp\left(-\frac{vL}{D}\right)\right).$$

d) To find the limits as $v \to 0$, we may either expand the exponentials as Taylor series or use L'Hôpital's rule. Using Taylor series, we have $u \to u_0(1 - x/L)$, $J \to u_0 D/L$, and

$$\tau = \frac{L}{v} - \frac{D}{v^2}\left(1 - (1 - \frac{vL}{D} + \frac{1}{2}\frac{v^2 L^2}{D^2} + \text{h.o.t.})\right) = \frac{1}{2}\frac{L^2}{D} + \text{h.o.t},$$

as required.

5.10 a) Integrating $0 = \frac{1}{r^2}\frac{d}{dr}\left(r^2 D\frac{du}{dr}\right)$, with D constant, we obtain the general solution $u(r) = -\frac{A}{r} + B$. The boundary conditions $u(a) = 0$, $u(b) = u_0$ determine A and B, and

$$u(r) = u_0\frac{b(r-a)}{r(b-a)}, \quad J(r) = -D\frac{du}{dr}(r) = -\frac{Du_0 ab}{r^2(b-a)}.$$

b) The diffusion current, i.e. the rate at which substance flows *out* of the region through the surface $r = a$, whose area is $A = 4\pi a^2$, is given by $I = -AJ(a) = \frac{4\pi Du_0 ab}{b-a}$. Integrating over the spherical shell $a < r < b$, $0 < \theta < \pi$, $0 < \phi < 2\pi$ in spherical polar co-ordinates,

$$N = \int_{\phi=0}^{2\pi}\int_{\theta=0}^{\pi}\int_{r=a}^{b} u(r)r^2\sin\theta\,dr\,d\theta\,d\phi = \frac{4\pi b u_0}{b-a}\left[\frac{r^3}{3} - \frac{ar^2}{2}\right]_a^b.$$

c) Hence the average time it takes a particle to diffuse from a point on $r = b$ to a point on $r = a$ is given by

$$\tau = \frac{1}{D}\left(\frac{b^3 - a^3}{3a} - \frac{b^2 - a^2}{2}\right).$$

d) If $a \ll b$, this is given approximately by

$$\tau \approx \frac{b^3}{3aD}.$$

5.11 a) Integrating $0 = \frac{1}{R}\frac{d}{dR}\left(RD\frac{du}{dR}\right)$, with D constant, we obtain the general solution $u(R) = A\log R + B$. The boundary conditions $u(a) = 0$, $u(b) = u_0$ determine A and B, and

$$u(R) = u_0\frac{\log(R/a)}{\log(b/a)}, \quad J(R) = -D\frac{du}{dR}(R) = \frac{Du_0}{R\log(b/a)}.$$

b) The diffusion current, i.e. the rate at which substance flows *out* of the region through the surface $R = a$, whose area is $L = 2\pi a$, is given by $I = -LJ(a) = \frac{2\pi Du_0}{\log(b/a)}$. Integrating over the annulus $a < R < b$, $0 < \phi < 2\pi$ in plane polar coordinates,

$$N = \int_{\phi=0}^{2\pi}\int_{R=a}^{b} u(R)R\,dR\,d\phi = \frac{2\pi u_0}{\log(b/a)}\left[\frac{R^2}{2}\log(\frac{R}{a}) - \frac{R^2}{4}\right]_a^b.$$

c) Hence the average time it takes a particle to diffuse from a point on $R = b$ to a point on $R = a$ is given by

$$\tau = \frac{N}{I} = \frac{1}{D}\left(\frac{b^2}{2}\log\left(\frac{b}{a}\right) - \frac{1}{4}(b^2 - a^2)\right).$$

d) If $a \ll b$, this is given approximately by

$$\tau \approx \frac{b^2}{2D} \log \left(\frac{b}{a} \right).$$

5.12 Immediate, from Section C.2.1 of the appendix.

5.13 a)

$$N(t) = \int_{R=R_2(t)}^{\infty} \int_{\phi=0}^{2\pi} u(R, t) R \, dR \, d\phi$$

$$= \frac{2\pi M}{4\pi D t} \int_{R_2(t)}^{\infty} R \exp \left(\alpha t - \frac{R^2}{4Dt} \right) dR = M \exp \left(\alpha t - \frac{R_2^2(t)}{4Dt} \right),$$

b) Defining $R_2(t)$ by taking $N(t) = m$, and taking logs,

$$\log \left(\frac{m}{M} \right) = \alpha t - \frac{R_2^2(t)}{4Dt},$$

$$R_2(t) \approx \sqrt{4\alpha D t},$$

for t large, as before.

c) If $\alpha = 0$, $R_1^2(t) \sim t \log t$ but $R_2^2(t) \sim t$.

5.14 a)

$$\frac{\partial u}{\partial t} = \alpha u + D\nabla^2 u.$$

b) We measure t in generations. The RMS dispersal distance for a particle diffusing in two dimensions over one generation is $\sqrt{4D}$, so assumption (iv) gives $\sqrt{4D} \leq 50$. Using the result of Exercise 5.12, the number of trees produced in a generation is e^α, so $e^\alpha \leq 9 \times 10^6$, $\alpha \leq \log(9 \times 10^6) = 16$.

c) The distance R_1 moved in 20000 years is thus $R_1 \leq \sqrt{4\alpha D}\, t = 200t = 200 \times 20000/60$ metres, or about 67 km. Skellam concluded that diffusion was insufficient, and that other dispersal mechanisms such as transport by animals and birds must be important.

d) The estimates sound high for D, outrageously high for α, and low for generation time, but since we are looking for an upper bound to the distance travelled this is not a problem.

5.15 Referring to Figure 5.5, we need to show that T is below the upper boundary $v = mf(u)$ of $D^\cdot$ near $(1, 0)$. But T is below the nullcline $v = (1/c)f(u)$ since it is pointing southeast, and hence is below $v = mf(u)$ if $m > 1/c$, $c > 1/m$. But this holds since $c > 1/m + mK$.

5.16 It is clear that no trajectory may leave D through the lower part of the boundary, where $v = 0$, $u_1 < u < 1$, as s decreases, since v' is negative there. On the upper part of the boundary where $\Phi(u, v) := v - 2(1 - u - w(u)) = 0$, $v' = c(u + w(u) - 1 + v) = \frac{1}{2}cv$, and

$$\Phi'(u, v) = v' + 2u' + 2w'(u)u' = \frac{1}{2}cv - 2\frac{1}{c}uv + 2\frac{1}{R_0 u}\frac{1}{c}uv$$

$$= \left(c^2 - \frac{4}{R_0}(R_0 u - 1) \right) \frac{v}{2c} > \left(c^2 - \frac{4}{R_0}(R_0 - 1) \right) \frac{v}{2c} > 0,$$

since $c^2 > 4(R_0 - 1)/R_0$, and the result follows.

6.1 The equation is
$$\frac{1}{V} = \frac{K_m + S}{V_m S} = \frac{1}{V_m} + \frac{K_m}{V_m}\frac{1}{S},$$
so $1/V_m$ is the intercept and K_m/V_m the intercept of the best straight line fit.

6.2 For the substrate, the inner solution and the common part are both identically equal to 1, so the result follows. For the complex, the common part is $1/(1+K_m)$, and the result follows from $c_{0,\text{unif}}(t) = c_0(t) + C_0(t/\epsilon) - 1/(1 + K_m)$.

6.3 a) Immediate.
   b) The leading order equations are
$$\frac{ds_0}{dt} = \frac{K_d}{K_m}c_0 - s_0, \quad 0 = s_0 - c_0,$$
so $ds_0/dt = (K_d/K_m - 1)s_0 = -Ks_0$, and the result follows.
   c) The inner equations are
$$\frac{1}{\epsilon}\frac{dS}{dT} = \frac{K_d}{K_m}C - S + \epsilon\alpha SC, \quad \frac{dC}{dT} = S - \epsilon\alpha SC - C,$$
so $S_0 = 1$, $dC_0/dT = S_0 - C_0 = 1 - C_0$, and the result follows.
   d) Matching gives $A = 1$, and we obtain $s_{0,\text{unif}}(t) = e^{-Kt}$, $c_{0,\text{unif}}(t) = e^{-Kt} - e^{-t/\epsilon}$.

6.4 a) The term $k_{-2}PE$ should be added to the C equation and subtracted from the E and the P equations.
   b) Immediate, using $E + C = E_0$.
   c) Directly from the P equation, using $E = E_0 - C$ and the expression for C.
   d) Immediate from $dP/d\tau = 0$.

6.5 a) Immediate.
   b) Since
$$Y(S) = \frac{2R_2 + R_1 + T_1 + 2T_2}{2(R_2 + R_1 + R_0 + T_0 + T_1 + T_2)},$$
the result follows from the quasi-steady-state relationships between the dimer and its complexes.
   c) All states of the dimer are R states, which are non-cooperative, and we reduce again to simple Michaelis–Menten kinetics.

6.6 a) Immediate.
   b) The quasi-steady-state hypothesis gives $ES_2^n = K_e X_1$, $S_1 X_1 = K_m X_2$, and the result follows.
   c) Immediate.

6.7 a) The Jacobian is immediate. For stability we require $\text{tr }J^* < 0$, $\det J^* > 0$.
   b) This condition violates $\text{tr }J^* < 0$, which with these conditions on the parameters is the more stringent of the two.
   c) The humps are at $x = \pm 1$.
   d) The existence of the steady state is immediate, and $\text{tr }J^* = 0$ there since $x^* = \gamma$.
   e) We have to show that $\text{tr }J^*$ increases as I increases past I_c. But since $I = -\frac{1}{3}x^{*3} + x^* + (a - x^*)/b$, $dI/dx^* = -x^{*2} + 1 - 1/b = b/c^2 - 1/b > 0$ at the bifurcation point, x^* decreases, and $\text{tr }J^*$ increases.

6.8 a) The drug essentially switches β to zero, so we have
$$\frac{dV}{d\tau} = aY - bV, \quad \frac{dX}{d\tau} = c - dX, \quad \frac{dY}{d\tau} = -fY.$$

b) Then $Y = Y_0 e^{-ft}$, $V = V_0(be^{-ft} - fe^{-bt})/(b - f)$. The behaviour of V follows from the assumption on half-lives, so that $f \ll b$.

6.9 a) W are the uninfectious virus particles, which start to be produced from the infected cells Y after therapy starts. Infectious virus particles are still present, and die as before, but are no longer produced.

b) With $X = X^* = (bf)/(a\beta)$, the equations become a linear system that may be integrated to obtain

$$V = V_0 \exp(-b\tau), \quad Y = Y_0 \frac{fe^{-b\tau} - be^{-f\tau}}{f - b},$$

$$W = V_0 \frac{b}{b - f}\left(\frac{b}{b - f}(e^{-f\tau} - e^{-b\tau}) - f\tau e^{-b\tau}\right)$$

The result follows, using $f \ll b$.

6.10 a) Y_1 is the productive infected class, Y_2 the latent infected class; the probabilities of entering these on infection are q_1 and q_2 respectively. Only the Y_1 cells produce virions, and Y_2 cells leave for Y_1 at a per capita rate δ.

b) Y_2 cells produce Y_1 cells at a rate δ for a time $1/(\delta + f_2)$. Hence, adding contributions from Y_1 and Y_2 cells, $R_0 = \frac{\beta c}{db}\left(q_1 + q_2 \frac{\delta}{\delta + f_2}\right)\frac{a}{f_1}$.

7.1 a) Define $x = \pi\tilde{x}/L$, $t = \pi^2\tilde{t}/L^2$, and u by $u(x,t) = \tilde{u}(\tilde{x},\tilde{t})$, We obtain the required equation with $\gamma = L/\pi$.

b) Immediate from the Taylor expansion of f, since $f(u^*) = 0$.

c) By Fourier analysis, the solution must contain a term in $\sin nx$ for $n = 1$ only. Because the equation is linear with constant coefficients, we try $v(x,t) = \sin x \exp(\sigma t)$, to obtain

$$v(x,t) = \sin x \exp\left((\gamma^2 f'(u^*) - D)t\right).$$

d) From the expression for v, with $\gamma = L/\pi$, it may be seen that increasing D tends to stabilise the solution (making it more likely to decay to zero), whereas increasing L tends to de-stabilise it.

7.2 Only the ϕ derivatives in ∇^2 are non-zero. Hence the spatial eigenvalue problem is given by

$$-\nabla^2 F = -\frac{1}{a^2}\frac{d^2 F}{d\phi^2} = \lambda F,$$

with periodic boundary conditions $F(0) = F(2\pi)$, $F'(0) = F'(2\pi)$. From Section C.4 of the appendix, the solutions of this are $F(\phi) = \exp(in\phi)$, for any non-negative integer n, with corresponding eigenvalue $\lambda_n = n^2/a^2$. The result follows from the general theory.

7.3 a) Since $\operatorname{tr} J = 0$, $\det J = 4$, then the eigenvalues of $\mathbf{u}_t = J\mathbf{u}$ are $\pm 2i$, so any solution is of the form $\mathbf{u}(t) = \operatorname{Re}\{\mathbf{A}\exp(2it)\}$, and the trivial steady state is stable.

b) The spatial eigenvalue problem is $-F''(x) = \lambda F(x)$ on $(0, \pi)$ with boundary conditions $F'(0) = F'(\pi) = 0$, one of the standard problems in Section C.4 of the appendix, which has eigenvalues n^2, n a non-negative integer, and corresponding eigenfunctions (spatial modes) $\cos nx$. Quoting Equation (7.4.33), the mode n is unstable for any n with $\underline{\lambda} < \lambda_n < \overline{\lambda}$, i.e. $\underline{\lambda} < n^2 < \overline{\lambda}$, where $\underline{\lambda}$ and $\overline{\lambda}$ are positive real roots of $a_2(\lambda) = 0$, and

$$a_2(\lambda) = D_1 D_2 \lambda^2 - (D_2 f_u^* + D_1 g_v^*)\lambda + \det J^* = 9\lambda^2 - 24\lambda + 4.$$

But this has two positive real roots with $0 < \underline{\lambda} < 1$, $2 < \overline{\lambda} < 3$, and the only unstable mode is $n = 1$.

7.4 a) It is simple algebra to show that $A = g_v^{*2} > 0$, $B = 2f_u^* g_u^* - 4(f_u^* g_v^* - f_v^* g_u^*) < 0$, and $C = f_u^{*2} > 0$, so that

$$B^2 - 4AC = 4f_u^{*2} g_v^{*2} - 16 f_u^* g_v^* f_v^* g_u^* + 16 f_v^{*2} g_u^{*2} - 4 f_u^{*2} g_v^{*2}$$
$$= -16 f_v^* g_u^* \det J^* > 0.$$

b) Hence the quadratic $AD_1^2 + BD_1 D_2 + CD_2^2 = 0$ has two real positive roots (for D_2/D_1), and the result follows.

c) Immediate.

d) Immediate from the fact that the Turing bifurcation curve is the envelope of the hyperbolae $a_2(\lambda) = 0$, (which in turn is immediate from its construction by eliminating λ between $a_2(\lambda) = 0$ and $a_2'(\lambda) = 0$).

e) Only the second mode is unstable, so we would expect this to grow exponentially until nonlinear terms became important. We would expect the final solution to be close to a multiple of the second mode.

7.5 a) If a_2 is a perfect square, with repeated root λ_c, then $a_2(\lambda) = D_1 D_2 (\lambda - \lambda_c)^2$, so $D_1 D_2 \lambda_c^2 = \alpha^2 \det J^*$, from Equation (7.4.22), and the result follows.

b) Immediate.

7.6 a) The positive spatially uniform steady state is given by $(u^*, v^*) = (1/b, 1/b^2)$. The Jacobian matrix here is given by

$$J^* = \begin{pmatrix} 2u^*/v^* - b & -u^{*2}/v^{*2} \\ 2u^* & -1 \end{pmatrix} = \begin{pmatrix} b & -b^2 \\ 2/b & -1 \end{pmatrix}.$$

b) For asymptotic stability of (u^*, v^*) to spatially uniform perturbations, we require $\operatorname{tr} J^* = b - 1 < 0$, $\det J^* = b > 0$, i.e. $0 < b < 1$.

c) For spatially non-uniform perturbations, we look at the roots of $a_2(\lambda)$, which is given by
$$a_2(\lambda) = d\lambda^2 + (1 - bd)\lambda + b.$$
The curve of marginal stability is the part of $(bd - 1)^2 = 4bd$ where $bd > 1$. Solving the quadratic $(bd - 1)^2 = 4bd$, $bd = 3 \pm 2\sqrt{2}$, and applying the condition $bd > 1$, the plus sign must be taken.

d) See Figure E.7.

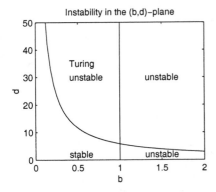

Instability in the (b,d)–plane

Turing unstable
unstable
stable
unstable

Figure E.7 The region of Turing instability in (b, d)-space. Right of the line $b = 1$, (u^*, v^*) is unstable to spatially uniform perturbations, but in the region marked "Turing unstable" it is only unstable to spatially nonuniform perturbations.

e) From Equation (7.4.27) $\lambda_c = \sqrt{b/d} = \sqrt{bd/d^2} = (1 + \sqrt{2})/d$, or from Equation (7.4.35), $\lambda_c = \frac{1}{2}(b - 1/d) = \frac{1}{2}(bd - 1)/d = (1 + \sqrt{2})/d$.

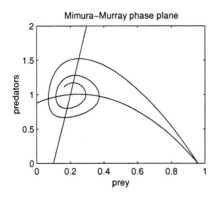

Mimura–Murray phase plane

Figure E.8   A possible phase plane for the Mimura–Murray prey-predator model. The steady state here is stable to spatially uniform perturbations, but can become unstable to spatially non-uniform perturbations and exhibit ecological patchiness.

7.7 a) See Figure E.8.
     b) The Jacobian at the spatially uniform steady state (u^*, v^*) is given by

$$J^* = \left(\begin{array}{cc} u^* f'(u^*) & -u^* \\ v^* & -v^* g'(v^*) \end{array} \right),$$

which is a cross-activator-inhibitor system as long as $f'(u^*) > 0$, so that the steady state is to the left of the hump on the nullcline $v = f(u)$.

7.8 The spatially uniform steady state is given by

$$(u^*, v^*) = \left(\frac{a+1}{b}, (\frac{a+1}{b})^c \right).$$

The Jacobian here is given by

$$J^* = \left(\begin{array}{cc} -b + cu^{*c-1}/v^* & -u^{*c}/v^{*2} \\ cu^{*c-1} & -1 \end{array} \right),$$

which gives a pure activator-inhibitor system near (u^*, v^*) as long as $cu^{*c-1}/v^* > b$, i.e. $c > bu^* = a + 1$.

7.9 a) $(u^*, v^*) = (a + b, b/(a + b)^2)$.
     b) The Jacobian matrix here is given by

$$J^* = \left(\begin{array}{cc} -1 + 2u^* v^* & u^{*2} \\ -2u^* v^* & -u^{*2} \end{array} \right),$$

which gives a cross-activator-inhibitor system near (u^*, v^*) as long as $2u^* v^* > 1$, $2b/(a + b) > 1$, $b > a$.

     c)
$$r_1 = \sqrt{\frac{2D_1(b + a)}{b - a}}, \quad r_2 = \sqrt{\frac{2D_2}{(b + a)^2}}.$$

     d)
$$l_c \approx 2\pi \sqrt{\frac{2D_1(b + a)}{b - a}}.$$

7.10 a) The nth mode is unstable if $a_2(\lambda_n) = a_2(n^2) < 0$, where

$$a_2(\lambda) = 10\lambda^2 - 14\alpha\lambda + 4\alpha^2 = 2(5\lambda - 2\alpha)(\lambda - \alpha).$$

For instability, $\frac{2}{5}\alpha < n^2 < \alpha$.

b) The bifurcation diagram is as in Figure 7.6(b), with successive bifurcation points 1, 5/2, 4, 9, 10, 16 (not shown) and 25 (mode 5 bifurcation not shown).

7.11 a) Immediate.

b) Must have $\lambda_{m,n} = \lambda_c$, and the result follows.

7.12 a) The steady state is given by $(n^*, c^*) = (\sqrt{\gamma}, \theta_0\sqrt{\gamma}/\beta)$, as long as

$$\theta_0\sqrt{\gamma}/\beta > c_0.$$

If this condition does not hold, the concentration c decays to zero.

b) We have $Q(\sigma) = \sigma^2 + a_1(\lambda)\sigma + a_2(\lambda) = 0$, where $a_1(\lambda) = (D_n + D_c)\lambda + 2\gamma + \beta$,

$$a_2(\lambda) = D_n D_c \lambda^2 + (\beta D_n + 2\gamma D_c - \theta_0\chi\sqrt{\gamma})\lambda + 2\gamma\beta.$$

A sketch of $a_2(\lambda) = 0$ in the (λ, χ)-plane is again as in Figure 7.9, and the result follows, as long as inequality in (a) also holds.

c) The critical wave-length $l_c = 2\pi/\sqrt{\lambda_c}$, where $\lambda_c = \sqrt{2\gamma\beta/(D_n D_c)}$.

7.13 a) Here $a_1(\lambda) = (D_n + D_c)\lambda + \beta$,

$$a_2(\lambda) = D_n D_c \lambda^2 + (\beta D_n - \alpha\chi n^*)\lambda.$$

Since a_1 is always positive, and $a_2 = 0$ always has a root $\lambda = 0$, potential instability occurs if the other root of $a_2 = 0$ is positive, $\alpha\chi n^* > \beta D_n$.

b) As χ increases beyond $\chi_c = \beta D_n/(\alpha n^*)$, long wave-length patterns are formed; their wave-length decreases as χ increases further.

c) In this case $\lambda = n^2\pi^2/L^2$, and

$$a_2(\lambda_n) = \frac{n^2\pi^2}{L^2}\left(D_n D_c \frac{n^2\pi^2}{L^2} + \beta D_n - \alpha\chi n^*\right).$$

For $\chi < \chi_c$, there are no bifurcations, and the spatially uniform steady state remains stable. For $\chi > \chi_c$, the first mode to become unstable is mode 1, at

$$L = L_1 = \pi\sqrt{\frac{D_n D_c}{\alpha\chi n^* - \beta D_n}}.$$

Other modes become unstable in turn as L increases further, but this remains the most unstable one.

8.1 If $u = \log(N/K)$,

$$\frac{du}{dt} = \frac{1}{N}\frac{dN}{dt} = -bu,$$

so that $u = u_0 e^{-bt} = \log(N_0/K)e^{-bt} = -Ae^{-bt}$, and the result follows.

8.2 a) Immediate, on defining $K = (\alpha/\beta)^{1/\nu}$.

b) With $u = (N/K)^{-\nu}$,

$$\frac{du}{dt} = -\beta\nu(1 - u),$$

so $u = 1 - (1 - u_0)e^{-\beta\nu t}$,

$$\frac{1}{u} = \frac{1/u_0}{(1/u_0)(1 - e^{-\beta\nu t}) - e^{-\beta\nu t}},$$

and the result follows.

c) With $u = \log(N/K)$,

$$\frac{du}{dt} = \frac{1}{N}\frac{dN}{dt} = \beta(e^{\nu u} - 1).$$

Separating variables,

$$\beta t = \int \frac{du}{e^{\nu u} - 1} = \int \frac{e^{-\nu u}du}{e^{-\nu u} - 1} = -\frac{1}{\nu}\log\left(\frac{e^{-\nu u} - 1}{e^{-\nu u_0} - 1}\right),$$

and the result follows.

8.3　a) Immediate, from

$$\frac{1}{R}\frac{d}{dR}\left(R\frac{dc}{dR}\right) = \frac{k}{D}.$$

b) If there is no necrotic core, the boundary conditions $\frac{dc}{dR}(0) = 0$, $c(r_2) = c_2$ give $c(R) = -\frac{1}{4}\frac{k}{D}(R_2^2 - R^2) + c_2$, which is valid as long as $c(0) \geq c_1$, $R_2^2 \leq R_c^2 = 4(c_2 - c_1)\frac{D}{k}$.

c) For $R_2 > R_c$, we have

$$c_1 = \frac{1}{4}\frac{k}{D}R_1^2 + A\log R_1 + B,$$

$$c_2 = \frac{1}{4}\frac{k}{D}R_2^2 + A\log R_2 + B, \quad 0 = \frac{1}{2}\frac{k}{D}R_1 + \frac{A}{R_1}.$$

d) Subtracting the first of these from the second and substituting for A from the third,

$$c_2 - c_1 = \frac{1}{4}\frac{k}{D}R_2^2\left(1 - \frac{R_1^2}{R_2^2} - 2\frac{R_1^2}{R_2^2}\log\frac{R_2}{R_1}\right).$$

As $R_2 \to \infty$, the quantity in parentheses must tend to zero, so $R_2/R_1 \to 1$. But then, if $\delta R = R_2 - R_1$,

$$c_2 - c_1 = \frac{k}{4D}\left(R_2^2 - (R_2 - \delta R)^2 + 2(R_2 - \delta R)^2\log(1 - \frac{\delta R}{R_2})\right)$$

$$\approx \frac{k}{4D}\delta R^2,$$

and so $\delta R^2 \to h^2 = \frac{4D}{k}(c_2 - c_1)$ as $R_2 \to \infty$.

8.4 The model for the necrotic layer $0 < r < r_1$ is unchanged. In the quiescent layer $r_1 < r < r_2$, $0 = -k_1 + D\nabla^2 c$, in the proliferative layer $r_2 < r < r_3$, $0 = -k_2 + D\nabla^2 c$, with $k_2 > k_1$. Continuity of concentration and flux could be applied at $r = r_1$ and $r = r_2$.

8.5 While $r_2 < r_c$, there is no necrotic core, and $v = \frac{1}{3}Pr$. Thus $\frac{dr_2}{dt} = \frac{1}{3}Pr_2$, giving exponential growth.

8.6 The model for the nutrient concentration is as above. For the velocity field, $\nabla \cdot \mathbf{v} = 0$ in the quiescent layer $r_1 < r < r_2$, with continuity of the velocity field at r_1 and r_2.

8.7　a) We have two relations between r_1^* and r_2^*. From Equation (8.4.20),

$$\frac{r_1^*}{r_2^*} = \left(\frac{P}{P+L}\right)^{1/3}.$$

Now, substituting into Equation (8.3.14),

$$c_2 - c_1 = \frac{k}{6D}\left(1 + 2\left(\frac{P}{P+L}\right)^{1/3}\right)\left(1 - \left(\frac{P}{P+L}\right)^{1/3}\right)^2 r_2^{*2},$$

which gives r_2^*.

b) In the limit as $L/P \to 0$, this becomes

$$c_2 - c_1 \approx \frac{3k}{6D}\left(1 - \left(1 + \frac{L}{P}\right)^{-1/3}\right)^2 r_2^{*2} \approx \frac{k}{2D}\frac{L^2}{9P^2}r_2^{*2},$$

and the result follows.

8.8 See Figure E.9.

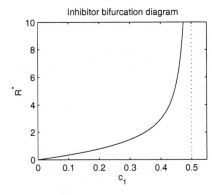

Inhibitor bifurcation diagram

Figure E.9 Using the inequality (8.5.29), the bifurcation curve is given by $c_1 = (\lambda/\mu)(1 - 1/f(\alpha R^*))$. The curve has positive slope at the origin, and tends asymptotically to the line $c_1 = \lambda\beta/(\mu(1+\beta))$. Here we have taken $\alpha = \beta = \lambda = \mu = 1$.

8.9 a) The inhibitor is produced at constant rate λ within the tumour, decays at specific rate μ everywhere, and diffuses with diffusion coefficient D everywhere.

b) Continuity of concentration and flux.

c) The concentration inside the tumour is given by $c(x) = \lambda/\mu + A\cosh(\sqrt{\mu/D}x)$, using the symmetry condition $c'(0) = 0$. Outside the tumour, it is given by $c(x) = B\exp(-\sqrt{\mu/D}|x|)$, since c is bounded as $|x| \to \infty$. The continuity conditions now give two equations for the constants of integration A and B, which may be solved to give $A = -(\lambda/\mu)\exp(-\sqrt{\mu/D}L)$, $B = (\lambda/\mu)\sinh(\sqrt{\mu/D}L)$.

d) The concentration of inhibitor in the tumour is least at the surface, so the tumour stops growing at $L = L^*$ when the concentration at the surface increases to c_1, where $c_1 = (\lambda/\mu)\sinh(\sqrt{\mu/D}L^*)\exp(-\sqrt{\mu/D}L^*)$. This is easily solved to give $2\sqrt{\mu/D}L^* = \log(\lambda/(\lambda - 2\mu c_1))$, and the bifurcation diagram is similar to that in Figure E.9 above with asymptote $\frac{1}{2}\lambda/\mu$.

8.10 a) Immediate.
      b) See Figure E.10.

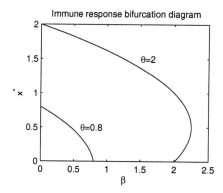

Figure E.10 Bifurcation diagram for tumour immune response. The equation gives a parabola with nose pointing in the positive β-direction, whose slope $\frac{dx^*}{d\beta}$ is negative or positive at $x^* = 0$ according to whether θ is less than or greater than 1. There is a transcritical bifurcation in each case, and also a saddle-node bifurcation (at the nose of the parabola) if $\theta > 1$.

   c) It is clear from the figure that for $\theta < 1$, x^* rises continuously from 0 as β drops below θ, but for $\theta > 1$ it jumps from 0 to a positive value; the production of tumour cells is suddenly turned on.

Index